The Art of Reading Buildings

THE ART OF READING BUILDINGS

JOHN MITTENDORF & DAVE DODSON

Fire Engineering

> **Disclaimer:** The recommendations, advice, descriptions, and the methods in this book are presented solely for educational purposes. The authors and publisher assume no liability whatsoever for any loss or damage that results from the use of any of the material in this book. Use of the material in this book is solely at the risk of the user.

Copyright © 2015 by
PennWell Corporation
1421 South Sheridan Road
Tulsa, Oklahoma 74112-6600 USA

800.752.9764
+1.918.831.9421
sales@pennwell.com
www.FireEngineeringBooks.com
www.pennwellbooks.com
www.pennwell.com

Marketing Manager: Sarah De Vos
National Account Manager: Cindy J. Huse

Director: Mary McGee
Managing Editor: Marla Patterson
Production Manager: Sheila Brock
Production Editor: Tony Quinn
Cover Designer: Karla Womack
Illustrations by Paul Bunch
Photographs by the authors (unless otherwise noted)

Library of Congress Cataloging-in-Publication Data

Mittendorf, John, 1940-
  The art of reading buildings / John Mittendorf and Dave Dodson.
     pages cm
  Includes bibliographical references and index.
  ISBN 978-1-59370-342-4
  1. Fire extinction. 2. Building layout. 3. Fire risk assessment. I. Dodson, David W. II. Title.
  TH9310.5.M58 2014
  628.9'2--dc23

2014024469

All rights reserved. No part of this book may be reproduced, stored in a retrieval system, or transcribed in any form or by any means, electronic or mechanical, including photocopying and recording, without the prior written permission of the publisher.

Printed in the United States of America

3 4 5   19 18 17 16

This book and its primary focus is dedicated to all past, present, and future fire service personnel who willingly risk their lives in firefighting suppression operations within structures that are under demolition by fire and gravity in order to maximize the safety of any trapped occupants and minimize the loss to property. The unselfish devotion of the American fire service to the people they serve is one of the hallmarks of the United States of America.

Additionally, every firefighter owes an immense amount of gratitude to two well-known and respected authors from the American fire service who have worked to chronicle the increasing dangers of building construction. The late Francis Brannigan devoted countless hours to his monumental work and numerous editions of *Building Construction for the Fire Service* (first published in 1971), which focused attention on the multitude of dangers associated with building construction when applied to structural firefighting operations. Retired Deputy Chief Vincent Dunn, Fire Department of the City of New York (FDNY), also devoted numerous hours and publications to building construction, specifically his text *Collapse of Burning Buildings: A Guide to Fireground Safety*.

The authors hope that *The Art of Reading Buildings* becomes an adjunct that will respectfully continue the traditions of Brannigan and Dunn.

CONTENTS

Acknowledgments xi

FESHE Objectives Correlation xiii

SECTION 1: BUILDING YOUR FOUNDATION

Chapter 1: Introduction 1
    Objectives 1
    Why This Book? Why Now? 1
    Overhead Issues and Concerns 2
    Maximizing this Book 8
    Chapter Review Exercise 10

Chapter 2: Essential Building Concepts 11
    Objectives 11
    Why Should You Read This Chapter? 11
    Loads 12
    Building Material Characteristics 14
    Specific Building Materials 15
    Chapter Review Exercise 26

Chapter 3: Anatomy of a Building—A Map 29
    Objectives 29
    Communication Skill-Building for Buildings 29
    Structural Elements 30
    Structural Assemblies 40
    Structural Hierarchy 42
    Chapter Review Exercise 43

Chapter 4: Classifying Buildings—NFPA 220 System 49
    Objectives 49
    It's Politically Incorrect to Profile Anything 49
    Classifying Buildings 50
    NFPA 220 Overview: The Five Classic Building Types 52
    Chapter Review Exercise 83

Chapter 5: Classifying Buildings—Hybrid, Era, Use, Type, and Size Considerations 85
    Objectives 85
    Thinking Beyond the Five Types 85
    Hybrid Buildings. 86
    Classifying Buildings by Era, Use, Type, and Size 88
    Chapter Review Exercise 109

Chapter 6: Alternative and Evolving Construction Trends 117
    Objectives 117
    Performance Design Has Nothing to Do with Firefighters 117
    Alternative Building Methods 119
    Evolving Building Methods/Materials 129
    Chapter Review Exercise 136

SECTION 2: BUILDING COMPONENTS AND FIREFIGHTERS—PRACTICAL LESSONS

Chapter 7: Foundations, Floors, Ceilings, and Walls 139
    Objectives 139
    The Box That Surrounds You 139
    Foundations 140
    Floors 146
    Ceilings 149
    Walls 152
    Chapter Review Exercise 164

Chapter 8: Reading Roofs 165
    Objectives 165
    The Importance of Reading Roofs 165
    Eight Most Common Roof Styles 167
    Unique Roof Construction Considerations 186
    Roof Coverings 193
    Roof Appendages 195
    Chapter Review Exercise 200

Chapter 9: Building Features and Concerns 203
    Objectives 203
    The 800-Pound Gorilla in the Room 203
    Windows 204
    Doors 214
    Utility Systems 217
    Alternative Energy Systems 223
    Overhead Hazards. 227
    Renovations and Remodels 234
    Light Wells, Skylights, and Atriums 240
    Miscellaneous Hazards 243
    Chapter Review Exercise 250

Chapter 10: Reading Buildings: How to Size Up a Building 257
   Objectives 257
   Time to Put It All Together. 257
   Building Size-up 258
   The Six Tactical Challenges for Buildings. 269
   Perspectives On Building Triage and Predicting Collapse. 275
   Chapter Review Exercise 280

SECTION 3: RAPID STREET-READ GUIDES

Chapter 11: Using the Rapid Street-Read Guides 289
   52 Buildings 289
   Design Features. 290
   Using the Rapid Street-Read Guides 290
   Rapid Street-Read Guide Index 292
Rapid Street-Read Guides 295
   Single-Family Dwelling (SFD) 295
      SFD 1: Colonial and Georgian 295
      SFD 2: Victorian/Queen Anne and Cape Cod 297
      SFD 3: Craftsman and American Four Square. 299
      SFD 4: Prairie Style 301
      SFD 5: Split Level. 303
      SFD 6: Modern Lightweight. 305
      SFD 7: McMansion 307
      SFD 8: Manufactured (Mobile) Home 309
   Multifamily Dwelling (MFD). 311
      MFD 9: Brownstone 311
      MFD 10: Tenement 313
      MFD 11: Row Frame 315
      MFD 12: Railroad Flat 317
      MFD 13: Center Hallway Structure 319
      MFD 14: Garden Apartment. 321
      MFD 15: Project Housing—Low Density 323
      MFD 16: Project Housing—High Density 325
      MFD 17: Legacy Townhome/Condo/Apartment. 327
      MFD 18: Lightweight Townhome/Condo/Apartment. 329
   Main Street Commercial (COM) 331
      COM 19: Pre-WWI Ordinary 331
      COM 20: Pre-WWII Ordinary (Taxpayer) 333
      COM 21: Industrial/Legacy Strip-Style 335
      COM 22: Modern Strip-Style 337
      COM 23: Fast Food 339
      COM 24: Mega-Box 341
      COM 25: Big-Box Store 343

Manufacturing/Warehouse (MANF) 345
    MANF 26: Block/Masonry 345
    MANF 27: Steel 347
    MANF 28: Concrete Tilt-Up 349
    MANF 29: Wood 351
    MANF 30: Converted Mill. 353
    MANF 31: Public Storage—Single Story 355
    MANF 32: Public Storage—Multistory 357
Office Building/Hotel (OFF). 359
    OFF 33: Pre-WWII—Low Rise 359
    OFF 34: Post-WWII—Low Rise 361
    OFF 35: 21st Century 363
    OFF 36: High Rise—1st Generation. 365
    OFF 37: High Rise—2nd Generation 367
    OFF 38: High Rise—3rd Generation 369
Institutional Building (INST). 371
    INST 39: School 371
    INST 40: Hospital. 373
    INST 41: Detention (Jail) Facility. 375
    INST 42: Attended Care Facility 377
Public Assembly (PUB) 379
    PUB 43: Restaurant. 379
    PUB 44: Stadium/Arena 381
    PUB 45: Auditorium/Theatre 383
    PUB 46: Meeting Hall 385
    PUB 47: Church 387
Miscellaneous Building/Structure (MISC). 389
    MISC 48: Pole Barn 389
    MISC 49: Kit Building 391
    MISC 50: Silo 393
    MISC 51: Historical Building—Dwelling 395
    MISC 52: Historical Building—Commercial. 397

Glossary. 399
Acronyms. 409
Index 411
About the Authors 429

ACKNOWLEDGMENTS

Thank you to the various organizations and individuals for their willingness to provide resources and information for this first edition of *The Art of Reading Buildings*. An endeavor such as this can only be accomplished with the help of many individuals who share ideas, provide feedback, and lend technical assistance. The number of individuals who have influenced the content of this book exceeds the space we have here, although we'd like to specifically thank the following for sharing their expertise and time:

The City of Bend Fire Department, OR; Firefighter Dion Evans, Compton Fire Department, CA; Battalion Chief Jim Forquer and Deputy Chief John Nohr, Portland Fire & Rescue; Battalion Chief Joe Castro, Los Angeles Fire Department; Jerry Knapp, Rockland County Hazardous Materials Team, NY; Training Officer Jason Hoevelmann and Fire Marshal Steve Gettenmeier, Florrisant Valley Fire Protection District, MO; Lieutenant Christopher Flatley, Fire Department of the City of New York; Firefighter Ric Jorge and the crew of Engine 33, Palm Beach County Fire-Rescue, FL; Chief Mark J. McLees and Battalion Chief Todd Milton, Syracuse Fire Department, NY; Deputy Chief Kenneth Morgan, Clark County, NV; Captain William Gustin, Metro-Dade Fire Department, FL; Captain Rick A. Haas Jr., Cosumnes CSD Fire Department, CA; Battalion Chief John A. Alston, Jr., Jersey City, Department of Fire, NJ; and Chief Dan Petersen, Jackson County Fire District #3, OR.

Special thanks goes to John J. Lewis and the other photographers who have provided photographic support. Likewise, thanks to Paul Bunch for his talents in creating our graphics.

Individually, John Mittendorf would like to thank his wife, Janice, for her patience and support during his numerous hours of contemplation, typing, and taking numerous photographs. Additionally, a special thank you for the opportunity to work with Dave Dodson—and still remain friends after this project.

Likewise, Dave Dodson would like to thank Chief Bobby Halton, editor, *Fire Engineering* magazine; Deputy Chief Phil Jose and Captain Steve Bernocco, Seattle Fire Department; Battalion Chief (ret.) Katherine Ridenhour, Aurora Fire Department, CO; Training Chief Mike West, South Metro Fire Authority, CO;

and Division Chief Brian Kazmierzak, Penn Township, IN. A special thanks to his wife, LaRae, and to Becky Stafford for their tolerance, patience, and unconditional support. Working with John was an amazing experience!

We cannot pass up the opportunity to acknowledge the various instructors who have spent a great deal of time preparing and delivering training programs related to the importance of building construction on a national level, and that as a result have significantly increased the safety and operational effectiveness of the fire service. Instructors such as Chief John Norman (*Fire Officer's Handbook of Tactics*), Chief John F. "Skip" Coleman (*Managing Major Fires*), Chief Anthony L. Avillo (*Fireground Strategies*), Chief Michael A. Terpak (*Fireground Size-Up*), and a host of others have willingly devoted their time, talents, and expertise to help the fire service.

The gang at PennWell are simply amazing. Mary McGee, Marla Patterson, and Mark Haugh deserve high fives for their encouragement, persistence, and direction. We'd also like to show gratitude for the PennWell review and production teams.

There are probably other individuals and departments that we have forgotten, but our collective old age and fading memories will have to take precedence.

Finally, and most importantly, we'd like to thank Captain Mike Gagliano of the Seattle Fire Department for his inspiration and efforts that helped bring together the wandering thoughts of your authors.

FESHE OBJECTIVES CORRELATION

This text is written to meet the course outcomes for the Building Construction for Fire Protection curriculum established in the Fire and Emergency Services Higher Education (FESHE) Model Curriculum. Outcomes are met in the following chapters:

Objectives	Chapter(s)	Comments
1. Identify various classifications of building construction.	4, 5, and 6 plus rapid street-read guides (RSRGs)	
2. Understand theoretical concepts of how fire impacts major types of building construction.	1–11 plus RSRGs	All chapters contribute. RSRGs highlight potential fire impacts of 52 common building types.

Outcomes

1. Describe building construction as it relates to firefighter safety, buildings codes, fire prevention, code inspection, firefighting strategy, and tactics.	1–11 plus RSRGs	All chapters contribute. RSRGs focus on firefighter safety and strategy and tactics. Fire prevention and code inspections are outside the scope of the book.
2. Classify major types of building construction in accordance with a local/model building code.	4, 5, and 6 plus RSRGs	This is a local agency outcome. The book can help local agencies classify buildings in agreement with national and international codes.
3. Analyze the hazards and tactical considerations associated with the various types of building construction.	1–11 plus RSRGs	All chapters contribute. RSRGs focus on hazards and tactical considerations.
4. Explain the different loads and stresses that are placed on a building and their interrelationships.	2, 3, 7, 8, 9, and 10	
5. Identify the function of each principle structural component in typical building design.	3, 7, 8, 9, and 10	
6. Differentiate between fire resistance, flame spread, and describe the testing procedures used to establish ratings for each.	4	NFPA ratings and definitions are explained. Testing procedures are outside the scope of this text.
7. Classify occupancy designations of the building code.	5, 10, and RSRGs	Occupancy types are defined in chapter 5 (NFPA and IBC). Tactics/hazards associated with occupancy types are expanded in chapter 10 and the RSRGs.
8. Identify the indicators of potential structural failure as they relate to firefighter safety.	2–11 plus RSRGs	Structural failure indicators are included throughout entire book and are highlighted in chapter 10.
9. Identify the role of GIS as it relates to building construction.	RSRGs	The role of GIS is driven by local agencies and outside the scope of the book. However, the RSRGs provide an excellent GIS template for buildings.

SECTION 1
BUILDING YOUR FOUNDATION

INTRODUCTION

OBJECTIVES

- List the three influencing factors that allow firefighters to perform interior fire operations.
- Identify the three new building trends that are reducing firefighter safety.
- Outline the commitment necessary to help achieve safe fireground operations.
- Describe the challenges that exist for conducting relevant building size-ups.

WHY THIS BOOK? WHY NOW?

Using questions as an introductory title probably violates some rule somewhere—but we did it anyway! We could have easily started this book by citing numerous firefighter injury and death statistics, line-of-duty death (LODD) investigative reports, and case studies designed to sell you on the need to read this book. Rather, we want you to know why we wrote another building construction book when several great ones already exist. The answer: Society has changed the fire and society has changed the building. Basic curriculums utilized to teach building construction haven't kept pace with the societal changes. Further, we feel there has been a missing element in construction curriculums: rapid recognition of building factors that impact first-due decision making. Don't misinterpret what we are saying. We absolutely appreciate the classic fire service books dedicated to building construction. What we do believe, however, is that there is a middle ground between how buildings are built (Brannigan) and how they fall down (Dunn). That middle ground is rapid recognition and tactical relevance.

For the most part, firefighters are dialing into the changes in fire behavior: higher heat release rates, higher temperatures, compressed time to flashover, and the explosive nature of ventilation-limited fires. We can thank the works of

the National Institute of Standards and Technology (NIST) and the Underwriters Laboratory (UL) for their tireless work quantifying modern fire behavior. The second societal influence seems to be a bit more elusive—the changed building. Sure, we've heard our fire service leaders talk about the new lightweight buildings and the challenges they present to firefighters, yet we still find too many firefighters trying to apply a 1970s tactic to a newly constructed building. We don't fault the firefighters—we fault the painfully slow process of updating the training standards, curriculums, and practices that have guided our fireground decision making. Jason Hoevelmann, training officer for the Florrisant Valley Fire Protection District in Missouri, summed up the issue perfectly:

> *Since well before my very first attendance of an organized firefighter certification class, instructors have stood in the front of classrooms lecturing about the critical relationship of building construction and fire behavior. Instructors would pound the pulpit and state the same common phrase that we have all since heard over and over again, "the building is the enemy!" With this essential point made, students are then held in a trance while discussing the five different types of building construction.*
>
> *Building construction and fire behavior lays the groundwork for fire service knowledge. But, what has been missing in a great number of these classes, is the tactical relevancy of the subject for firefighters and fire officers. Too many times these classes and texts have one common theme—"the building is bad"—as a generalization. We can, and must, do better at making building construction mean something. We must make it tangible for firefighters and officers to use every time they respond to and operate at a building fire.*

The ability to "read" a building has never been more critical. As Jason points out, we need to take building construction knowledge and make it a practical and tangible part of building fire responses. This book is designed to do just that. In fact, our overall approach to the book was to combine the best of previous fire service building construction books into a single desk reference that also addresses modern construction, rapid-recognition assistance, and practical relevance.

Before we jump into the journey, we'd like to present a few overhead issues and concerns that are impacting structure fires today (vs. yesteryear). We'll follow that with a brief explanation of how you can maximize the information contained within this book (our methodology).

OVERHEAD ISSUES AND CONCERNS

Through incident responses and building familiarization activities, firefighters are discovering new and innovative building materials and methods that collectively reduce the amount of time that firefighters have to safely mitigate interior fires. Unfortunately, new construction methods are not usually designed to assist fire suppression operations. Considering the cost of labor, equipment, and building materials, it is not economically feasible today to construct a structure the same as those from the 1800s and early 1900s. In many cases, heavy timbers have been replaced by 2×4s of varying lengths of finger-jointed and glued sections of wood in lightweight truss configurations, and petrochemical-based compounds have replaced many conventional building materials, regardless of building type or size.

As a current example, "engineered wood" products are beginning to replace concrete and steel in many structural applications. As modern architects reduce the amount of mass and change the chemical composition of building materials, the fire service is losing three valuable factors that allow us to perform interior fire operations: (1) *safety*, (2) sufficient *time* to conduct suppression operations,

and (3) the ability to conduct a *relevant building size-up*. Let's briefly look at these three factors.

Safety

National Fire Protection Association (NFPA) technical reports indicate that firefighter injuries and deaths related to structural incidents are a result of three primary factors: (1) it falls on you, (2) you fall into it, and (3) flashover. The ability to read a building and translate that information into fireground operations that are safe and relevant is the key to reducing and/or eliminating the chance of the building falling on you, you falling into it, or you being caught by flashover.

Although some older methods of building construction are still being utilized, many construction methods are being replaced by new and more cost effective techniques. This new approach to construction is typically accomplished by maximizing engineering principles (math) and minimizing material use (cost). As mentioned previously, new construction methods are not usually designed to assist fire suppression operations and/or enhance fireground safety. The loss of material mass and the engineered shape of essential structural elements is a double whammy for firefighters. The reality is that new construction methods have increased the chances that a building will fall on you or you will fall into it. Further, newer construction methods and materials are designed to better insulate the building against weather elements and reduce heating and cooling costs. Those insulation efforts can actually speed the occurrence of flashover. New building trends have thus created a triple whammy for firefighters or, to borrow a baseball cliché, three strikes and you're out.

- Strike 1: Lower material mass
- Strike 2: Engineered shapes for structural elements
- Strike 3: Better insulation

These three strikes have the impact of reducing firefighter safety. So how do we address safety with the new construction trends? There is a wealth of information that is readily available when you take the time to look at your *fireground office* when driving through your district (either returning from responses or doing area familiarization), conducting fire prevention inspections, or just spending a few hours on the weekend to see what's new. There is also a tremendous wealth of resource information available in books and online to help you understand buildings.

Safety is not something that just happens automatically at a fireground and/or something that always arrives unannounced when necessary. It is a commitment you make to be familiar with your surroundings, operate in a predictable and standard way, and train at every opportunity. Remember that training is the most important nonemergency function that we do, and typically pays immense dividends in the form of *safe fireground operations*.

Time

When the word "time" is mentioned, some of the initial factors that come to mind are how advanced is the fire, how long did it take someone to become aware of the fire and report it, what was the dispatch time, and what was the length of time for response of resources. While these factors are certainly applicable to fireground suppression operations, we must also add in our setup (evolution) time and the time it takes to make a visible impact on the fire. This fireground clock starts when the first-due apparatus sets the parking brake and we go to work. We address the fireground clock in more detail in chapter 10, but suffice it to say that the prearrival and fireground clocks are ticking when a building is on fire. Time is a constant—you can't buy it or sell it—and you must be aware of its passage. Each tick of the clock has the potential of weakening the gravity-defying nature of the building. Eventually, the building will succumb to gravity and begin to collapse in partial or general fashion. As we know, fires are dynamic—they are not static. Destruction is taking place (often exponentially) until we stop it. Unfortunately, each building has its own set

of factors that govern the amount of time it can withstand the fire (and our operations). With that said, we must also say that the time we have to make an impact on a fire (and save the building) has been reduced for *all* buildings. We've made the point that our safety has been impacted by new building trends: engineered shapes, lower mass materials, and better insulation. Yet the higher rate of heat release in today's fires accelerates building degradation in older buildings as well (blame the abundance of synthetic contents). Reading buildings and tracking the passage of time has never been more important.

Relevant size-up

Webster's Dictionary defines *relevant* as "pertinent, applicable, important, pertaining to," and the fire service often defines *size-up* as "a mental evaluation of incident factors that determines the course of action necessary to achieve a desired goal." From a simplistic viewpoint, if we combine the terms relevant and size-up into one definition, it can be summarized as "the ability to quickly determine an applicable impression of an incident (particularly buildings) that lays the foundation for a safe abatement strategy that continues until the conclusion of the incident." When applied to *reading a building*, it is easy to see that if buildings did not collapse and/or flashover did not occur, firefighter injury and death statistics would dramatically decrease. However, building collapse that can consist of walls, floors, roofs, and/or a complete building has been a consistent and prominent cause of injury and death to firefighters that are involved in structural suppression operations. The ability to quickly read a building to develop an effective foundation for safe suppression operations does not come overnight. It takes diligent study and time to become familiar with the relevant attributes of building construction, particularly when applied to the area of your responsibility.

Fundamentally, the fire service is fighting fires in buildings the same ways and using the same basic principles that were used over 50 years ago. Hose lines are still advanced into involved buildings, searches are still conducted in hazardous conditions with little or no visibility, and ventilation operations are conducted to improve the internal environment. But are modern buildings the same as the buildings that were constructed just a few short years ago? They are *not even close*, and the gap between construction methodologies of older and newer buildings is continuing to widen due to evolving technology—to the detriment of the fire service and firefighters, whether paid or volunteer. Face it, in a simpler time, we could teach firefighters the strengths and weaknesses of five general types of buildings and outline a dozen or so visual clues that could help with size-up. That is no longer the case. The palette of buildings we face has grown exponentially (figuratively and literally). Therefore, fireground personnel who can quickly recognize and evaluate the relevant strengths and hazards of differing buildings will increase their own safety and efficiency. A working knowledge of building construction provides not only the necessary expertise to conduct a quick and accurate size-up of a building, but as we have previously mentioned, also the foundation for safe, effective, and timely fireground operations.

Although it is relatively easy to challenge personnel to develop a working knowledge of building construction, it is becoming more difficult to achieve that goal due to evolving technology and the vast differences in methodologies and materials that have been used in building construction since the founding of this country. As an example, fire departments whose districts are predominantly comprised of older construction should be able to anticipate what should be encountered, such as the beneficial strengths of full-dimensional lumber, conventional construction and materials, the identifiable characteristics of older construction/buildings, and the common hazards of balloon frame construction, knob-and-tube wiring, multiple layers of roofing materials, and so on.

Interestingly, this same viewpoint also applies to departments whose areas of responsibility are principally comprised of buildings of newer

construction. In this case, these departments should not be surprised by a lack of the beneficial strengths of conventional materials/construction and the rapid and catastrophic collapse of varying types of lightweight metal and wood construction in minimal time frames. The substantial challenge that most fire departments must recognize is the diversification of old and new construction that can be found in most areas of this country, particularly when buildings have been renovated to include new construction or remodeled to appear different from their original configuration.

An example of this dilemma is illustrated in figure 1–1. This commercial building appears to be a modern type of building that could easily be considered as constructed with current methods and materials. However, after stepping inside the building, a far different picture emerges. It quickly becomes apparent the building was constructed in the early 1900s due to the presence of unreinforced masonry walls and a gable roof constructed from solid wood truss structural members (fig. 1–2).

Fig. 1–2. The interior of the building in figure 1–1 is significantly different from the exterior.

Fig. 1–1. This building appears to be recently constructed.

Another example is the common understanding that a large, well-involved fire within any building can lead to collapse. Although this is a true assessment, it is also becoming obvious that common-appearing buildings that incorporate varying degrees of modern construction techniques and materials can be responsible for creating an unexpected hazardous environment and/or building collapse potential from seemingly small fires. The following example is a recent incident that occurred in Clark County, Nevada, and is reprinted with the permission of Clark County Deputy Chief Kenneth Morgan:

> We have all read about or heard of the dangers of engineered wood joists—or I-joists—when exposed to fire, but they don't need to be on fire to degrade. Developed in 1969, I-joists have become a construction staple because of their economic value and ability to shoulder large loads over long spans. Originally constructed of a plywood web and solid flanges (chords), the flanges were replaced with laminated wood in 1977. The switch to orientated strand board (OSB) webs occurred in 1990. Just how much does it take to damage these engineering marvels? When exposed to heat, they degrade rapidly, so the importance of knowing the

structures in your first-due response area becomes alarmingly evident.

An occupancy fire in Clark County, Nevada, in the early morning hours of March 5, 2009, was a reminder to the Clark County (NV) Fire Department of how even a small fire can place crews in imminent danger. This occupancy, constructed in 1992, was in an area of the county known for light commercial businesses that borders the city of Las Vegas and on the southern edge of a larger complex of similar structures. It was of typical block wall construction with a flat roof and was split into two occupancies. At 0422 hours, a caller advised the fire alarm office that there was an orange glow coming from under the garage door with gray smoke showing from a small, single-story commercial structure at 3111 S. Valley View Boulevard. A fire attack engine was assigned, and a second engine arrived and forced entry through the roll-up doors. In all, eight crews were called to the scene.

Once crews gained access, they advanced a 1¾-inch line, advised command of poor visibility and low heat, and requested ventilation to assist them in locating the fire. A truck company advanced to the roof to vent; the officer from the company used a thermal imaging camera (TIC) to attempt to locate a heat signature, but it indicated no hot areas on the roof other than at the base of a heating, ventilation, and air-conditioning (HVAC) duct. The company close to an area around the duct decided to open the roof, thinking that it may follow the HVAC ducting. There was no indication that the roof was compromised. Shortly after the fire attack crew made entry, the fire was found and easily extinguished. The smoke level began to dissipate, revealing a small debris pile consisting of some boxed textile material.

When the smoke cleared enough to see the interior roof, a horrific sight appeared. The roofing system was constructed of engineered I-joists, and many of the joists were completely compromised. Although the fire was confined to the debris pile of origin, the I-joists were severely degraded. In figure 1–3, the height of the flame can be noted on the wall. In figure 1–4, the height of the heat pattern can be noted along with the degradation of five adjacent I-joists over the fire, caused primarily by convected heat. Not only were the joists damaged, but the connections to the wall were also compromised; heat had attacked and degraded the roofing system. There was no notable alligatoring on the flanges, yet the webs in over half of the joist span were completely gone.

Fig. 1–3. This was a small rubbish fire with the height of the flame visible on the wall. (Photo by Kenneth Morgan.)

Fig. 1–4. The height of the heat pattern is visible in addition to a degradation of adjacent roof I-joists. (Photo by Kenneth Morgan.)

This incident presented the indications of a small fire, but it held a secret that had the potential to kill firefighters. The crews learned and reinforced several valuable lessons that morning, which included the following:

- No fire is routine! Do not become complacent.

- The value of knowing the first-due area cannot be overstated. If this building had been identified as having an engineered roofing system, alternate, safer tactics could have been used, and no personnel would have been placed on the roof.

- It does not take much fire to make an I-joist fail; heat alone can do this. There is some indication of fire in the truss around the wall area, but the majority of the damage is strictly heat related.

- Consider identifying roofing systems and placarding the buildings so fire crews don't have to guess what they are walking on.

- Thermal imaging cameras did not help locate an area of concentrated heat. This roof was composite foam over plywood, which hid any heat signature. The lack of heat signature and no indication of compromise in the foam suggested a small fire and no involvement of the roofing system. This can give you a false sense of security. If the roof sheathing had been OSB instead of plywood, the roofing system might have completely failed.

Although this was a relatively small article that appeared in the August 2010 issue of *Fire Engineering* magazine, it illustrates an expanding predicament that every firefighter engaged in structural firefighting operations can easily encounter. The increasingly common situation is less fireground time and rapid collapse of floor *and* roof systems as a result of heat and also due to the dynamic changes in building construction that are constantly being pioneered by the building industry in an effort to reduce costs and improve profits. If history can be an indicator of future events (and it often is), it should be noted that all fireground personnel—whether a firefighter on an initial attack nozzle or an incident commander at a command post—should be familiar with this article and incorporate its concerns into their fireground decision-making process. Additionally, the preceding examples that consisted of the remodeled building and the small debris fire in a light commercial area in Las Vegas graphically illustrate the importance of being able to conduct a relevant building size-up that results in a workable awareness and knowledge of what is about to be encountered in suppression operations.

It is our belief that (1) the type of building construction that is being exposed to fire and/or a noteworthy amount of heat and (2) the approximate amount of time of exposure should be incorporated into an initial and continuing size-up provided to an incident commander at every structure fire (when appropriate). As an example, assume you force entry into an older-appearing single-story strip mall (mini-mall) of masonry construction and encounter high heat and moderate visibility. Prior to advancing into the structure with an initial attack line, you quickly use a pike pole to pull the ceiling above of your position and discover that fire has extended into the attic space above your position, and the roof structural members are comprised of 2×4 in. glued, parallel chord trusses. Does this

fictional scenario immediately raise a concern that compared to older conventional 2 × 6 in. construction (or larger), you have a minimal amount of time before a structural collapse of modern lightweight roof assemblies? Additionally, should this information also be quickly be relayed to the incident commander (who is probably looking at the exterior and thinking this building is an older Type III building of conventional construction)? Obviously, the answer to both questions is a resounding yes!

Quick summary

- Firefighters are losing the three valuable factors that allow us to perform interior fire operations: safety, sufficient time for operations, and the ability to conduct a relevant size-up.
- Firefighter safety has been diminished in newer construction as result of lower material mass, engineered shapes for structural elements, and increased insulation.
- A commitment to being familiar with your surroundings, operating in a predictable/standard way, and constant training is essential to improving fireground safety.
- The passage of time is constant and fires are dynamic—the two combine to degrade a building and eventually cause its collapse.
- The amount of time that a building can resist a fire is governed by many factors, but has been reduced in almost all buildings due to the higher heat release rate of burning synthetic contents.
- Conducting a relevant building size-up has become challenging due to evolving technology and the vast differences, methodologies, and materials that have been used in building construction since the founding of this country.
- Classic building fire attack tactics (interior attack supported by rooftop ventilation) can no longer be a default. Prefire planning, building familiarization, and developed skills in reading a building are essential to fireground decision making.

MAXIMIZING THIS BOOK

The primary intent of this book is to provide a reference text that helps you understand and predict how the buildings in your response district behave when they are burning. If you can do that, then you will make better fireground decisions. Obviously, we haven't seen all the buildings in your district—you have to do that. We can, however, provide a detailed look at common types of building construction methods/materials and their potential risks to fireground personnel from a practical perspective. That perspective is gleaned from decades of personally fighting fires in buildings, reviewing case studies and investigative reports, researching evolving technologies, talking to construction industry professionals, and making countless building visits (under construction, under demolition, while in use, and after fires).

Organizing all that information is a challenge, but we think we hit on a logical, step-building approach that will help you. The three sections in this book build on each other. We also throw in some features that can help you apply the information in each section.

Section 1: Building Your Foundation

The first step in understanding any building is to capture the language and concepts that are used to explain how (and why) a building is built the way that it is. Knowing the relationships of materials and loads helps you see (know) how weight is carried and transferred—and how they react in a fire. Next, you discover the basic anatomy of any building—which gives you the X-ray vision necessary to see behind what is typically covered up.

As buildings evolved, it was discovered that there were many ways to make the basic anatomy function. Occupancy needs (use and size), fire and earthquake resistance, resource (material) economies, world wars, and inventiveness have

all factored into the evolution. Given all of that, we know of many types of buildings—although efforts to classify a building can help us quickly understand its strengths and weaknesses from a fire spread and collapse perspective. Section 1 discusses the various ways to classify a building and is full of issues, concerns, and tips that can help you make better decisions with each type of building.

Section 2: Building Components and Firefighters—Practical Lessons

If you're like most firefighters, you'll want to skip right by section 1 and jump into practical stuff. We get that—but encourage you to resist the urge because section 2 is written using the language established and defined in section 1. We all know that the fire service has developed a unique language of slang and that there are regional or geographic nuances (East coast, West coast, no coast); section 2 will make more sense having invested in section 1.

In section 2, we get down to the nitty-gritty components that fill most buildings and discuss the challenges they present whether you are stretching a line, pulling a ceiling, cutting a roof, or securing utilities. Taking that further, we end the section with several perspectives on how to rapidly assess (size up) a building. In doing so, we present some building triage ideas, risk (rescue) profiling clues, universal size-up considerations, and collapse prediction algorithms.

Section 3: Rapid Street-Read Guides

The last section of the book contains a handy reference tool that can help you index the buildings in your jurisdiction. The 52 Rapid Street-Read Guides that are assembled include pictures, construction characteristics, and tactical issues that should be considered for each. To help categorize, we group the guides based on the *street language* we use to initially describe buildings we respond to, such as single-family dwelling and office building. We encourage you to use the street guides to not only help you rapidly recognize buildings, but to develop your own guides for preplans and training activities using the suggested format. With your input, we are sure to expand the rapid street-read guides for future editions.

Other features

In addition to the typical learning objectives, diagrams, photos, and glossary you find in most books, we've added some features that we hope you find useful:

- Case studies
- Feature perspectives from other fire service leaders
- Quick summaries to help you capture key points (as you study for your next promotion!)
- A master acronyms list
- Review activities
- Resources for further study (books, websites, etc.)

We hope that this introduction underscores the importance of fireground personnel being able to understand the basic concepts of building construction, and being able to apply its relationship to formulating safe and effective strategic and tactical decisions on the fireground. We further hope that you find something in this book that is very practical and relevant to your efforts to make it safer for you and your fellow firefighters.

CHAPTER REVIEW EXERCISE

Answer the following:

1. What are the three influencing factors that allow firefighters to perform interior fire operations?

2. New construction methods have impacted firefighter safety in a negative way. What are the three trends contributing to that impact?

3. How can safe fireground operations be improved by individual firefighters?

4. What modern influence is reducing building integrity "time" for all buildings?

5. What does the phrase "relevant building size-up" mean?

6. List several factors that challenge the ability to conduct a relevant building size-up.

ESSENTIAL BUILDING CONCEPTS 2

OBJECTIVES

- Define loads and imposition of loads as they relate to buildings.
- List three types of forces created from imposed loads.
- Explain the effects of fire on common building materials.
- List the factors that determine the suitability of a building material for a given application.
- Describe the relationship of surface-to-mass ratio and fire degradation for building materials.
- Differentiate the terms native wood, traditional wood products, and engineered wood products.

WHY SHOULD YOU READ THIS CHAPTER?

The fire officers who read this chapter are better prepared to not only "read" a building, but better prepared to communicate building problems, hazards, and collapse potential. Likewise, fire officers need to learn a certain language that is used in the building construction/engineering field. That language helps the fire officers teach new firefighters, prepare more complete prefire plans, and interact with building representatives and engineers during incidents.

The intent here is not necessarily to make you a building engineer—far from it. We do, however, define basic terms and concepts that come from the building construction world and help provide a bridge to the 2:00 a.m. fire suppression world. Additionally, this chapter helps address an evolving issue that is related to younger generations of firefighters. The fire service is experiencing an evolutionary change in the types of individuals who serve as firefighters. In the 1960s and 1970s it was very common for firemen (as they were called) to possess a significant understanding of building construction trade knowledge. They were

carpenters, plumbers, electricians, roofers, and masons as their primary career (volunteers) or as part-timers on their off-duty days (career). Those who didn't possess that experience/knowledge often learned building construction by helping other firefighters with home projects, trading favors, firehouse do-it-yourself projects, and such. The bottom line—most fire training programs didn't have to emphasize building construction because it was a common knowledge base. Unfortunately, that is not true today.

While skilled building tradespeople still exist in the fire service, their numbers are dwindling quickly. Many of today's new recruits have little building trade background. Terms such as ledger, torsion, lintel, sill, ductile, and rafter tie plate have little meaning to them, and they are unable to translate these terms into determining strengths and weaknesses when reading a building. Thankfully, these new recruits are well versed in gigabytes, phone apps, extreme sports, and multitasking—which show they have a tremendous capacity to learn and absorb knowledge.

This chapter outlines some essential terms and concepts as it relates to loads, forces, and materials that are used to assemble the building that may ultimately fail/collapse during fire suppression operations.

LOADS

To be sound, a building must be designed, engineered, or otherwise assembled to resist a load. **Loads** are static and dynamic weights that come from the building itself and anything that is placed within or acts upon a building. Gravity is responsible for creating most loads—as any natural or human-created thing that has weight is being pulled to the planet's surface by gravity. Likewise, gravity is trying to flatten buildings—whether the building is on fire or not. Other loads come from atmospheric conditions such as wind, rain, and snow. To help classify all these loads, the building industry starts with simple definitions, then adds some specificity for certain applications and situations.

Dead loads. The weight of the building itself and anything *permanently* attached to the building.

Live loads. Any load applied to a building other than dead loads. Live loads are typically transient, moving, impacting, or static (like furniture).

Live and dead loads can be further classified as to their application nature. For example:

Concentrated load. A load that is applied within a small area or at one point.

Distributed load. A load spread over a large surface area or over multiple points.

Impact load. A moving or sudden load applied to a building in a focused or short time interval. For example, wind, large crowds, and fire stream water are all impact loads.

Repeated load. Loads that are transient or intermittently applied (like people on an escalator).

Static load. A constant load that rarely moves.

Suspended load. A load that is hanging from something above.

Wind/snow load. Atmospheric loads that stress a building.

The fire service uses the term **fire load**, which is the potential amount of heat energy (measured in British thermal units—BTUs) that may be released when a material is burning. The term fire load is not a building engineering term—it's purely a fire service one.

Load imposition

A firefighter team making a panel cut on a pitched wood roof for heat ventilation (while working from a roof ladder) can be classified as a live, impact, *and* distributed load (the firefighters are the live load, the movement of the firefighters are the impact load, and the ladder beams are

distributing the firefighters' weight as a distributed load). That live, impact, and distributed load is *imposed* on the roof and its supporting elements. The imposition of loads refers to the contact orientation of the load to the material(s). Loads can be imposed three ways (fig. 2–1):

Axial load. Load is imposed through the center of the material.

Eccentric load. Load is imposed off-center, causing a material to want to bend.

Torsion load. Load is imposed in such a way that causes a material to twist.

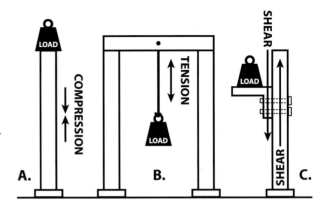

Fig. 2–2. There are three forces created when loads are imposed on materials: (A) compression, (B) tension, and (C) shear.

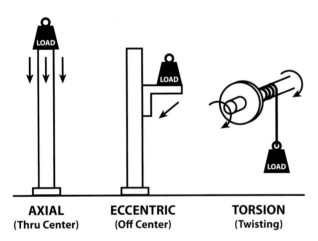

Fig. 2–1. Loads can be imposed three ways: axially, eccentrically, and torsionally.

Forces resisting loads

Imposition of loads on a given material causes stress within the receiving material. These stresses are called forces. Forces help resist the load. There are three primary forces created within materials (fig. 2–2):

Compression. A stress that causes a material to flatten or crush.

Tension. A stress that causes a material to pull apart or stretch.

Shear. A stress that causes a material to tear or slide apart.

Obviously, the material chosen to receive the imposed load must have characteristics that resist the forces that are created within it. Further, the material must transfer that load and eventually deliver the load to earth in compression.

Note: There are very few building/structure applications that deliver a load to earth in any force other than compression. Guide (guy) wire anchors that are drilled deep into the earth for a suspension bridge or tower antennas are noted exceptions (they are delivered in tension and/or shear). Next, we discuss the relationship between loads and material characteristics.

Quick summary

- Loads are static and dynamic weights that exist within or on a building.
- There are basically two loads that exist within every building: live and dead loads. Live and dead loads can be further classified as impact, suspended, distributed, and concentrated.
- Fire load is primarily a fire service concept.
- Loads are imposed three ways: axially, eccentrically, and torsionally.
- Imposed loads create forces within materials: compression, tension, and shear.

BUILDING MATERIAL CHARACTERISTICS

The previously mentioned firefighters making the ventilation panel cut on the pitched roof are a live, impact, and distributed load being imposed in an eccentric manner that is creating a compressive force on the roofing materials that the ladder beams are resting upon. If you can picture this in your mind's eye, then you are well on your way to reading buildings. This whole load/imposition/force equation relies on certain material characteristics to prevent failure.

When discussing the characteristics of materials, we have to understand that there are many factors that determine the suitability of the material for a given application. These factors include the following:

- Type of material (wood, steel, concrete, etc.)
- Shape of the material (round, square, rectangular, etc.)
- Orientation or plane of the material (vertical, horizontal, etc.)
- Mass of the material (surface-to-mass ratio, density, depth, etc.)
- Material surface (rough, slippery, hard, or adhesion/connection ability)

Additionally, the fire service looks at how these materials react during a fire and how their ability to resist a load changes during fire conditions. Of importance to firefighters is the concept of *surface-to-mass ratio*. The more mass a material has relative to its exposed surface area, the more it is resistant to heat. We learned in fire behavior class the concept of heat flow: Heat flows from hot to cold (heat seeks cold). Materials exposed to the radiant heat of other burning materials serve as a heat sink. The amount of heat the material can absorb before it starts breaking down is directly proportional to its surface-to-mass ratio. In essence, mass is heat resistance, and heat resistance is time.

From a firefighter's point of view, the more mass a material has in a given surface area, the more time (or heat) is required before the material starts to degrade. When a material does degrade it may also deform—thus changing its engineered shape and the way loads are resisted. A change in a material's shape causes a change in load imposition—potentially resulting in rapid failure (fig. 2–3).

Fig. 2–3. A change in load imposition can result in rapid failure.

The building engineering community also considers mass, but from a different perspective. Economics, sustainable-resource concerns, and technological advances have challenged the building community to maximize strength with reduced material mass. Basically, most new buildings are high-strength/low-mass buildings. If mass is time during a structure fire, then time has been lost to fight fires in new buildings with modern lightweight construction methods and materials. Granted, the building may be stronger, but that assumes that building materials are not being heated, have not changed shape, and are loaded within their engineered design limits.

The engineering community also classifies materials as being *brittle* or *ductile* based on their reaction to imposed loads and resistive forces.

Brittle. A material that will fracture or fail as it is deformed or stressed.

Ductile. A material that will bend, deflect, or stretch as a load is applied—yet retain some strength.

Simply stated, a brittle material breaks before it bends and a ductile material bends before it breaks. Wood, plastics, and most metals are ductile, whereas masonry, tile, and cast iron are brittle. With an understanding of building characteristics, we can now discuss individual materials.

> ## Quick summary
>
> - Factors that can help materials to work or fail can be defined as type, shape, orientation, mass, and surface.
> - The surface-to-mass ratio of a material is especially important to firefighters. Mass is heat resistance. Loss of mass (high surface-to-mass ratio) means loss of time during fires.
> - Most new buildings are comprised of high-strength/low-mass materials such as lightweight wood construction.
> - Materials can also be classified as either brittle (breaks before bending) or ductile (bends before it breaks).

SPECIFIC BUILDING MATERIALS

In the past, four basic materials were used to erect buildings: wood, steel, concrete, and masonry. Today, advanced material technologies have created composites of the aforementioned materials as well as new plastics, graphite, wood derivatives, and exotic metals. Let's look at the four basic building materials (wood, steel, concrete, and masonry) as well as some of the new composites, particularly wood composites (often referred to as *engineered wood*). In this section, we take a closer look at wood as it is the most predominant building material of the past, present, and likely the future. Engineered wood products are not only progressively replacing native wood, but also concrete and steel that have previously been used as structural members in building construction.

Native wood (cut or sawn lumber)

Native woods (or sawn lumber)—whole lumber pieces cut from a tree—are not created equal. There are hard woods, soft woods, tight grained, knotty, old-growth, and new-growth woods. Each has interesting characteristics and the true craftsman woodworker knows how to maximize the strength and application of each. However, it would be easy to assume that wood taken from the trees of today is roughly the same as wood that was taken from trees yesterday. Unfortunately, that is not the case as it is relatively easy to overlook the fact that the wood currently being used in modern building construction is significantly different from the wood that was used in older construction. From a logging perspective, old-growth trees that were common a hundred years ago are just a memory, with new-growth trees (or second-growth trees) and plantation trees normally replacing the older trees.

Interestingly, it is common for today's timber industry to harvest trees similar to corn or wheat. As an example, pine and spruce trees can often be cut 25 years after they are planted. This is why modern lumber trucks routinely carry numerous smaller logs (fig. 2–4) instead of yesterday's logging trucks that routinely carried several large logs. This has resulted in wood that is not only different, but wood that also burns significantly hotter and faster. Old-growth trees produce a wood that is denser and has a reduced level of pitch (which burns like a petroleum-based product). Additionally, in the past it was not uncommon for wood—particularly wood that was to be used for structural members—to be cut from the heart of a tree (which has maximum density and minimal pitch). Conversely, new-growth trees are less dense and have a higher concentration of pitch. This has resulted in wood that is lighter in weight (which can reduce strength), is capable of burning more rapidly and with a higher British thermal unit (Btu) heat

level, and in cheaper types of wood being used for structural members.

Fig. 2–4. New-growth logs are typically smaller in diameter and length and less dense than old-growth logs.

As a result, it should not be surprising that around 1986 the lumber industry changed its rating system from Utility (utl), Standard (std), Construction grade (cons), and Select (sel), to #3, #2, and #1 (best). Additionally, Douglas fir was the standard for exterior bearing walls (the good stuff), with white fir or hemlock used for interior walls (the cheaper stuff). Today, the building construction industry routinely uses the aforementioned cheaper grade of lumber for exterior load-bearing walls. To compound the problem, today's sawn wood is known as *nominal dimension* lumber as opposed to yesteryear's *full-dimensional* lumber. Previously, a 2 × 4 was really 2 in. by 4 in. A nominal dimension 2 × 4 is 1½ in. by 3½ in. When the preceding factors are combined, it should not be a surprise that modern structures built with this type of lumber (that is used to replace conventional construction and also used in various lightweight configurations) are collapsing faster and burning hotter than the structure fires of yesterday.

Despite this fact, while from an engineering point of view wood has marginal resistance to forces compared to its weight, it does the job and is the most used building material. We also know that wood burns and when it does, mass is lost. The more mass a section of wood has, the more material must burn away before its strength is lost.

In the past, firefighters could gauge when a solid hardwood beam would collapse because it started to sag a bit before it failed. Today, lightweight wood construction can catastrophically fail in a short period of time (compared to conventional lumber) and without prior warning signs such as sagging.

Traditional wood products

Used here, the phrase *traditional wood products* refers to the century-old development and improvement of manufactured wood products for a specific application that cut lumber cannot fill. Namely, we are talking about heavy timber, glue-laminated beams/columns and sheathing. Within this context, *engineered wood* refers to modern, technologically advanced wood products and is covered separately after this section.

Glued laminated heavy timber. Originally, the common availability of large trees meant that large pieces of sawn lumber were available for use in sizeable buildings that required large supporting and/or structural members (this is readily observable in large older buildings). As large trees began to become more rare, the original glued laminated beam or column (*glulam* for short) was born from the fact that trees only grow so wide and so thick. It was quickly discovered that a heavy timber beam could be created by using a number of smaller cut lumber pieces (fig. 2–5).

Fig. 2–5. Glulams are a popular modern replacement for solid lumber.

These pieces were originally strapped or bolted together before suitable glues were developed. As described here, glulams are heavy timbers and can absorb lots of heat prior to failure. They also burn forever! However, under heavy (or lengthy) fire conditions, glulams can fail and often cause a failure of large sections of a building (fig. 2–6). The glues are deeply impregnated and protected by the shear mass of the wood pieces used. However, remember that the glues or adhesives can, depending on their chemical composition, emit toxic gases when burned or exposed to heat.

Fig. 2–6. When used as primary structural members, the failure of glulams can cause failure of large sections of a building.

Sheathing. We can think of sheathing as merely a cover for something. While floor and roof sheathing needs to have some strength, wall sheathing requires very little. This led to the development of traditional wood sheathing products to maximize the waste parts of a tree left when lumber was cut. All sheathing products can be considered high surface-to-mass. Traditional wood sheathing includes plywood, particle board, and decorative (paneling).

- **Plywood.** Often called the original engineered wood product, plywood is made from layering sheet veneers of wood such that grain directions alternate 90° with each layer. (This is similar to the engineered wood products that are listed in the following section.) These layers are glued to each other as they layer together. There are various grades of strengths and applications of plywood based on its grain density, thickness, and gluing/coating process. When exposed to fire, the plywood layers start to char and burn away, layer by layer. When exposed to serious heat (as opposed to flame), the layers dry out and begin to curl. The destruction is usually easy to detect. Obviously, the thicker the plywood sheet, the longer it can withstand heat and flame, but similar to other wood products that use adhesive compounds for bonding strength, a toxic smoke can be emitted under fire conditions. Generally speaking, plywood has been replaced with a true engineered wood product—oriented strand board (OSB)—which is covered in the following section.

- **Particle board.** Wood sheathing made from a coarse sawdust and glue is known as particle board (PB). Particle board appears very smooth and consistent and has no wood grain. The sheathing is relatively heavy due to the compaction of the sawdust and glue during manufacturing. Even with this density, PB sheathing is actually quite weak with low resistance to trauma (it cracks and crumbles easily). In fact, its own weight can cause a large sheet to crack just from flexing. Because of its fragility, PB used as a floor or roof cover must be well supported with closely-spaced joists (beams) and include a durable surface covering. Fire and heat will easily destroy particle board. PB breaks down so easily in heat conditions that it is the sheathing of choice for the pyrolyizing (off-gassing) fuel source in flashover simulators. The smoke produced from the degradation of PB is full of wood particles and sticky aerosols (hydrocarbon-based glues) that are quite flammable and toxic.

- **Decorative sheathing.** Thin wood paneling used to finish interior walls or the outside of cabinets are classic examples of decorative sheathing. These products are not intended to resist loads and are merely decorative. This sheathing can range in thickness from ⅛ in. to ⅜ in.—meaning a high surface-to-mass ratio. Because of the rapid flame spread characteristic of decorative wood paneling, most are

not allowed by code for interior wall finishing. Additionally, if adhesives are used (which is likely), they will emit toxic gases under fire conditions.

> ## Quick summary
> - Old-growth trees were widely used for large timber structural members in older buildings. These members can resist the effects of fire for longer time frames than newer construction materials due to mass.
> - Newer lumber is typically harvested from new-growth trees, which results in a softer wood with a higher pitch content.
> - Older conventional roofs can often sag before collapsing, while newer lightweight wood trusses do not prior to catastrophic collapses.
> - Traditional wood products refer to older sawn lumber products like glulams and plywood sheathing.
> - Plywood, particle board, and decorative sheathing use adhesives that may emit flammable and toxic gases when exposed to heat or fire conditions.

Engineered wood products

While no official definition exists, the term **engineered wood** is used by the fire service to describe a host of wood products that use modern methods to transform wood chips/slivers, veneers, shavings, and even recycled wood products into components that replace sawn lumber, sheathing, and other composite structural materials. The wood used to make engineered wood products (EWP) is typically derived from new-growth forests and rapid-growth tree farms, although in some cases it is possible to manufacture similar engineered cellulosic products from other lignin-containing materials such as hemp stalks, wheat straw, and other vegetable fibers. In either case, the wood used for EWP is loose grained, has lots of pitch, and is amazingly lightweight compared to natural forested woods.

Once harvested, the wood is milled into veneers, wood chips/slivers, or shavings (shredded wood fiber or pulp). The milled product is then processed into forms, emulsified in binding agents (glues/adhesives), then autoclaved (application of heat and pressure) to set the binding agent. The glues that bind engineered wood products require only heat to break down and are also toxic, combustible, and will contribute to burning.

Engineered wood products are currently being used as a replacement for solid sawn wood materials (cut lumber) in common applications due to their advertised advantages of higher strength; greater stability over longer spans; resistance to shrinking, crowning, twisting, and warping; ease of manipulation; and efficiency in using more portions of a tree. Following are common examples of engineered wood products that rely on adhesives for bonding strength.

Oriented strand board (OSB). Known mostly by its acronym, OSB is sheathing that is formed with wood shavings and a urea-formaldehyde adhesive. The wood chips are oriented such that the grain directions are randomly oriented and layered. An adhesive locks these layers in place such that multidirectional and uniform strength is achieved. OSB is used extensively in new construction as a structural sheathing to form roof and floor assemblies (when glued to trusses) and as the web portion of a wooden I-beam (fig. 2–7). OSB is subject to degradation by direct sunlight (UV rays), moisture, and heat. The heat of fire or smoke can cause rapid destruction of OSB. Likewise, direct flame contact will cause OSB to ignite and burn rapidly and emit toxic gases from adhesives.

The strength and economy of OSB has led to a proliferation of its use in residential buildings. As an example, a typical 1,800-sq-ft two-story new home would use a concrete foundation, I-beams for floor joists (that use OSB for the webbing), OSB sheathing for the flooring, OSB I-beams for second floor joists, OSB stair treads and kickers/risers,

exterior OSB sheathing, and OSB roof sheathing attached to trusses that may also include OSB stiffeners. As a result, a major portion of this home would be comprised of wood chips that are bonded with an adhesive.

Fig. 2–8. Laminated veneer lumber is often used in place of cut lumber.

Fig. 2–7. OSB is used for sheathing and for the web of engineered wooden I-beams.

Laminated veneer lumber (LVL). To form LVL, thin sheet veneers of native wood are stacked with grains aligned and then glued with a phenolic resin. LVL is used in place of cut lumber for beams (fig. 2–8). LVL is also used to form the chords (or flanges) that are glued to the OSB web of engineered wooden I-beams. LVL is designed to have the load imposed axially and perpendicular to the grain. While the mass of LVL is typically higher than OSB, it is still degraded by the heat of a fire or smoke.

Because LVL is formed with native wood veneers, the individual sheets hold together until the wood burns. The glue that binds each layer tends to cause delamination of the veneer sheet by sheet when heated. LVL is commonly used as a replacement for conventional sawn lumber and timber for beams, joists, rafters, columns, studs, and rim boards.

Laminated strand lumber (LSL). LSL is a structural composite lumber manufactured from flaked and chipped strands of native wood blended with an adhesive. Mostly, LSL uses strands oriented in a parallel fashion (also known as parallel strand lumber—PSL). PSL is similar to LVL in its use. The primary difference is that LVL uses sheet veneers of native wood whereas PSL uses flaked wood strands. The glue may be phenolic resin, urea-formaldehyde, or phenol formaldehyde. All of these glues are derived from crude oil. Fire behavior wisdom suggests that LSL/PSL will fail before LVL. LSL/PSL can be used as beams, headers, studs, and rim boards.

Cross-laminated timber (CLT). CLT is an engineered wood product using several layers (three to seven or more) boards that are layered crosswise (typically rotated 90°) and glued. CLT is used as a structural element for columns (much like a glulam is used for beams). CLT uses actual timber boards cut from smaller trees to form a panel and the crosswise layers. In many ways, CLT is a structurally sound form of plywood but thicker. CLT can also be used for long spans and structural assemblies that are used for roofs, walls, and floors.

This product is gaining widespread acceptance due to its improved acoustics over sawn lumber and its reduced carbon emission footprint. For example, for every 1 ton of wood, it takes 5 times more energy to produce 1 ton of concrete, 24 times

more energy for 1 ton of steel, and 126 times more energy for 1 ton of aluminum. CLT also advertises a greater resistance to fire. This claim is based on the premise that due to the solid nature of the material, it will char at a slow and predictable rate (similar to mill construction type members). The char on wood forms a crust that slows the burning rate and helps shield the wood from further degradation.

Glued laminated timber (GLT). GLT is comprised of multiple layers of dimensional timber bonded together with moisture-resistant adhesives. GLT is a more modern form of the traditional glulam heavy timber covered above. GLT can be used as horizontal beams and vertical columns, and can also be produced in curved shapes, which makes this product very attractive to interior designers who want visible structural members that are more decorative than straight members.

Finger-jointed lumber (FJL). FLJ has become a common method to produce long lengths of wood members from multiple short pieces of native wood lumber. When joining these short pieces, the joining ends are mitered in an interlocking fingers configuration and pressed together with an adhesive as a bonding agent. Using the FJL process, wood manufacturers can create a long, straight, and solid wood joist or stud from a bunch of scrap mill ends. FJL can also be used to join one section of lumber to another (such as a 90° angle, as shown in figure 2–9) with an adhesive that is used as a bonding agent.

Fig. 2–9. Finger-jointed lumber is commonly used to join one section of lumber to another.

The proliferation of engineered wood products over solid wood is based on multiple advantages that include greater strength and stiffness, pound for pound strength that is greater than steel, more efficient use of wood (use of smaller pieces, wood with defects, wood chips, etc.), and conformance to emerging "green" considerations. From a fire service perspective, however, there are huge disadvantages associated with the use of EWP.

- Some products may burn faster than solid lumber; they have a high surface-to-mass ratio.
- Some adhesives are toxic, and some resins can release formaldehyde (urea-formaldehyde resins). Currently, there are four basic types of adhesives that are used in engineered wood products: urea-formaldehyde (most common), phenol-formaldehyde, melamine-formaldehyde, and methylene diphenyl diisocyanate (expensive). So, it is easy to see that three out of the four adhesives are formaldehyde based.
- As adhesives pyrolize, the resulting gases can become flammable.

Quick summary

- Engineered wood products are those made from wood chips/slivers, veneers, shavings, and recycled wood that have been bonded using various adhesive methods.
- The wood used for EWP is often harvested from rapid-growth tree farms. These trees are then milled (shredded into a pulp). The pulp is very lightweight, loose grained, and contains lots of pitch.
- OSB is the most prolific of the EWPs. It is used extensively in modern residential construction for beams, structural sheathing, and stair assemblies. OSB is very susceptible to heat degradation and burning in fire conditions—leading to rapid failure.
- Although engineered lumber can offer many advantages over sawn lumber, its primary disadvantage to the fire service is threefold.

> It has already begun to change the way buildings are being constructed when steel and concrete are replaced by wood products (see chapter 6), some applications can burn faster and with more intensity than sawn lumber (OSB as an example), and many of the adhesives that are used as a bonding agent will emit toxic gases such as formaldehyde.

Steel

Steel has been a staple building material for commercial buildings for almost two centuries. It is used extensively for columns and beams (the true bones of a building), especially in applications where strength, long spans, or tall walls are needed. The classic I-beam and H-column are most associated with steel. More recently, lightweight steel C-channel is being used to replace wood studs in occupancies that have noncombustible or low dead load requirements.

Steel is made from iron ore, carbon, and an alloy agent (metallic solid solution). During manufacturing, iron ore is crushed and made molten using a blast furnace and smelted with coke (a carbon source that is a derivative of coal). Alloying agents (like chromium, nickel, etc.) are added to help achieve strength and ductility. The molten solution can then be formed into pieces by casting, hot rolling, or cold rolling. Casting is just that—the molten steel is poured into a desired mold to form the finished product. Hot-rolled steel is the result of molten steel shaped at temperatures above the crystallization stage, which allows thinner sheets or shapes. Hot-rolled steel is often called extruded steel. I-beams and H-columns are typically hot-rolled extruded. Cold-rolled steel is shaped as it cools (below crystallization temperatures, forming stronger steel). Some call cold-rolled steel *cut* or *rolled steel*. Nuts, bolts, cables, rebar, and wires are examples of cold-rolled steel. Lightweight C-channel studs are also cold-rolled steel.

Steel has excellent resistance to compression, tension, and shear forces. Its strength-to-mass ratio is excellent. Additionally, steel has factory versatility; that is, it's relatively easy to fabricate different shapes, sizes, and strengths during production. For this reason, steel is a popular choice for large commercial structures. From a fire service viewpoint, steel has two weaknesses: it is engineered for very specific applications (thickness, length, shape, and strength) and it softens and elongates when heated.

In a fire, steel acts as a collector of heat—it conducts heat readily. Steel loses strength as temperatures increase; the specific range of temperatures at which it loses strength depends on how the steel was manufactured. As a general rule, cold-drawn steel like cables, bolts, rebar, and lightweight fasteners loses 55% of its strength at 800°F. Hot-rolled structural steel used for beams and columns loses 50% of its strength at 1,100°F. Structural steel also elongates or expands as temperatures rise. At 1,000°F, a 100 ft beam can elongate 10 in. In cases where a steel beam is affixed at both ends and heated, the steel can't elongate, thus it will twist, sag, or buckle as it tries to expand. This deformation can cause an immediate and often general collapse of floors and roofs.

Cast iron

Cast iron is a material usually formed from pig iron, which is a high carbon content iron. The iron is heated until it liquefies and is then poured into molds to solidify into desired shapes. Because of the high carbon content and lack of alloys, cast iron is brittle. Cast iron has good compressive strength qualities and acceptable shear strength if the cast iron was formed with significant mass. In the 1800s and early 1900s, cast iron was used in structural applications such as columns and door/window frames. Many historical buildings still have gorgeous, ornate storefront cast iron columns (fig. 2–10).

Fig. 2–10. Cast iron columns were used on the storefronts of many older commercial buildings.

Unfortunately, cast iron is brittle and history has shown that it can crack from aging, eccentric and/or torsional loading, and trauma. Cast iron has great resistance to slow heating and cooling. In a hostile fire environment, cast iron can initially resist heat fairly well but tends to fracture (or crumble) more easily when an eccentric load is applied (sagging of a floor or being struck with a powerful fire stream). For this reason, cast iron is no longer used for structural applications.

Research shows that there is still debate on whether the application of cold water to fire-heated cast iron causes explosive fracturing[1]. The debate centers on two arguments:

- The rapid cooling causes sudden contraction that implodes the brittle material.
- The physical force of the fire stream is an impact load that rapidly separates the brittle material.

Regardless, most agree that cast iron is a brittle material that can fail more easily when heated by fire.

Other metals

Aluminum. Aluminum is a natural element that exists in many minerals and ores. In fact, aluminum is the most abundant metal that exists on earth. Unfortunately, the reactivity of aluminum requires massive amounts of energy and refinement to produce the beer cans we are familiar with. As a building material, aluminum is considered a soft metal that is high strength-to-weight, highly ductile, noncorrosive with air/water, and nonmagnetic. Because of production costs, aluminum is rarely used for the main structure of a building but is used extensively for trim, brackets, finishes, sheeting, and special applications where lightweight, noncorrosive materials are needed. During fires, the low mass and ductile nature of aluminum causes rapid failure.

Titanium. Like aluminum, titanium is an abundant metal found in many minerals and is lightweight, low density, noncorrosive, and nonmagnetic. Titanium alloys are known for a high strength-to-weight ratio and tremendous resistance to heat. Most firefighters consider "lightweight" as a recipe for rapid failure. Titanium is an exception to the rule. For most buildings, titanium is too expensive to be used as a building material although its light weight and high strength makes it ideal for innovative architectural designs (soaring beams, twisting columns, etc.). As material technologies progress, more variants of titanium will be found in building materials.

Quick summary

- Steel has been a staple of the building industry for many years, and is a structural material that is used for both lightweight studs (interior and exterior) and lightweight roof structural members as well as primary structural members in the form of I-beams and H-beams.

- Steel is made from iron ore, carbon, and an alloy agent that provides increased strength and ductility. Steel products are either hot-rolled (extruded), cold-rolled (cut steel), or cast.

- Steel has excellent properties such as versatility and resistance to tension, compression, and shear forces.

- Steel will act as a collector of heat in a fire and lose much of its strength above 800°F, which can cause it to elongate, twist, and ultimately fail.

- Cast iron has been used in the building industry since the 1800s; however, as opposed to steel, it is a brittle material.
- Cast iron can fracture when heated in a fire and then exposed to water from suppression operations.
- Aluminum and titanium are abundant minerals that have excellent strength-to-weight ratios. Due to manufacturing costs, neither has been used for the structural bones of a building although that is starting to change. Aluminum fails quickly during fires whereas titanium shows significant resistance to heat.

Concrete

Concrete is a mixture of Portland cement, sand and aggregate (usually gravel), and water that cures into a solid mass. The curing process creates a chemical reaction that bonds the mixture to achieve strength. During the mixing process, gravel can be added as a volume and strength expander. The final strength of concrete depends on the ratio of these materials—especially the ratio of water to Portland cement. *Low-slump* concrete is stronger and has a lower water-to-cement ratio, while *high-slump* concrete is wetter and flows easier.

Cured concrete has excellent compressive strength but poor tensile and shear strength. Pure concrete is considered a brittle material. For this reason, steel is often added to concrete as reinforcement when the concrete is being used in a way that will subject it to those forces (like a floor or roof). When steel is added to concrete, the finished material is considered a composite—brittle with some ductile properties. Steel can be added to concrete in many ways during its casting:

Reinforced concrete. Concrete that is poured over steel rebar, which becomes part of the cured concrete mass.

Pre- and post-tensioned concrete. Concrete that has steel cables placed through the plane of the material and then tensioned, compressing the concrete to give it the required strength. Cables can be pre-tensioned (at a factory) or post-tensioned (at the job site).

Precast concrete. Slabs of reinforced concrete that are poured at a factory and then shipped to a job site (fig. 2–11). Precast concrete can be used for walls, floors, or roofs. Common applications of precast concrete are the venerable tilt-up slabs that are used for walls and twin-T slabs used for floors and roofs.

Fig. 2–11. Precast reinforced concrete slabs are being transported to a job site.

Monolithic buildings are concrete buildings built on location using a steel rebar frame and wood or composite material forms to shape the concrete. Concrete is then pumped into the forms encasing the steel—creating a reinforced concrete building. Monolithic is derived from the Latin word for single

stone. Monolithic buildings are typically built one floor at a time. The columns are built ahead of the floors and utilize a slip form, which moves slowly upward as each level is poured (fig. 2–12). Floors are then anchored into cured columns. The floors are built upon a scaffold-like platform (called falsework). Once a floor cures, the falsework is removed and rebuilt on the next level.

Fig. 2–12. A wooden slip form is used as a concrete mold for the construction of a monolithic building.

Unlike steel, concrete is a heat sink and tends to slowly absorb and retain heat rather than conduct it. This heat is not easily reduced. All concrete contains some moisture and continues to absorb and wick moisture (humidity) as it ages. When heated, this moisture content expands, causing the concrete to crack or spall. **Spalling** refers to a pocket of concrete that has crumbled into fine particles through the exposure to heat. Spalling can reduce the critical mass of the concrete—the mass used for strength. Steel rebar that becomes exposed to a fire after spalling can easily conduct heat within the concrete mass, causing catastrophic spalling and failure of the structural element. Concrete can also stay hot long after the fire is out, causing additional thermal stress to firefighters performing overhaul. As an example, these conditions were present in the Central Library fire in Los Angeles, as described in chapter 4.

Masonry

Masonry is a common term that refers to brittle materials like brick, tile, concrete block, and stone. The classic concrete masonry unit (CMU)—some call it a cinder block—is the most common material used for building a masonry wall (fig. 2–13). Masonry is used to form load-bearing walls because of its compressive strength, but it can also be used to build a veneer wall (one that bears only its own weight). Individual masonry units are held together using mortar. **Mortar** is a workable paste made from a mixture of sand, cement or lime, and water. Once cured, the mortar serves as a binding agent for masonry blocks. These mixes have little to no tensile or shear strength; they rely on compressive forces to give the masonry strength. It is important for firefighters to know that a masonry wall actually gets stronger as axial loads are applied and compressive forces increase. Obviously, there is an absolute maximum weight that can be applied before the brittle material fails.

Fig. 2–13. Concrete masonry units are a popular building material for walls.

Individually, brick, CMU, and stone have excellent fire-resistive qualities. Oftentimes masonry walls will still be standing after a fire has ravaged the interior of the building. Even though the wall is still standing, the loss of the roof means the wall no longer has the compressive forces needed for strength—the wall is unstable and can collapse quickly when an eccentric load (like wind) is applied. The masonry wall also has an Achilles' heel: the mortar used to bond the individual units. Mortar is subject to spalling, age deterioration, and washout. During a fire, masonry blocks (or bricks) can absorb more heat than the mortar used to bond them, creating different heat stresses that can crack the binding mortar. Whether from age, water, or fire, the loss of bond causes a masonry wall to become very unstable.

Composites

New material technologies are posing interesting challenges for the firefighting community. The term *composite* can be used for many things but in this case refers to a combination of the previously mentioned basic materials, as well as various plastics, glues, exotic metals, and assembly methods. One thing is certain, these materials are designed to offer maximum strength with minimal material mass—a dangerous proposal in the structural firefighting environment. Given that, several composites are commonly used for building materials.

Plastic. Simply stated, a plastic is synthetic or semisynthetic material that is made of moldable polymers (a molecule with many connected atoms). Most plastics are derived from petroleum. The hydrocarbon chemical chains found in crude oil can be altered with other chemicals to form many different products. While there are thousands of chemically-named plastics used in everyday life, most plastics can be divided into thermoplastics and thermosetting plastics. Thermoplastics can be heated and reshaped without losing the inherent composition found in the plastic. Thermosetting plastics use heat to harden or set the plastic. Reheating a thermosetting plastic will change the composition and will likely result in breakdown. Most plastics are considered ductile.

The building industry is the second leading consumer of plastics in the U.S. (packing and shipping is first). Since the mid-1960s, plastics have been increasingly used for just about everything except the structure itself. That is now changing. Plastics are now being used to reinforce wood and concrete. Several all-plastic buildings have been constructed to demonstrate the potential of plastics to replace wood, steel, and concrete. Clearly, the trend is for more plastics to be included as building materials. From a firefighter's perspective, this trend can have negative impacts on building stability during fires as most plastics melt at relatively low temperatures and emit very explosive gases that add tremendous heat-release rates when the hydrocarbons burn.

Carbon-fiber reinforced polymer (CFRP). CFRPs are composite materials that include a reinforcing material (the carbon fibers) that is bound together with a polymer (like epoxy). Carbon fiber is amazingly strong and can be woven or shaped in many forms. To put things in perspective, the carbon content of steel helps gives the steel its strength. Manufacturers have figured out a way to take that essential strength element, crystallize the bonds, and form a fiber. The fiber is a fraction of the thickness of human hair, pliable, and heat, corrosion, and rot resistant. Because of cost, CFRPs are not prevalent as a building construction material, although engineers are finding application of CFRPs for reinforcing concrete and steel. As the cost comes down, more applications will be found.

Under fire conditions, CFRPs offer initial heat resistance until the polymer degrades. Once the actual carbon fibers are exposed to flame, they separate and release microscopic carbon particles that can burn. The particulates from CFRP smoke are especially destructive to microelectronic circuit boards as they can form a conductive path between components.

The expanding use of composites for many types of building materials will certainly change the basic perceptions of fireground operations that have been taken for granted for many years. This is a primary reason why firefighters *must* keep abreast with technology and resultant changes in building construction, particularly in your area of responsibility.

To finish this section, it is important to know that technological advances in material science are ongoing and new combinations of materials are finding their way into buildings. In chapter 6, we further explore some of the evolving building methods and the material composites that are being used as well as some firefighter tactical considerations we face in them.

Quick summary

- Concrete has been a primary building material for hundreds of years and is a combination of, sand and aggregate, water, and Portland cement.
- Concrete is known for its high compressive strength (hence its use in foundations), but has poor tensile and shear strength.
- Steel is added to concrete as reinforcement for applications that require tensile and shear qualities. Reinforced concrete can be formed as monolithic, pre- or post-tensioned, or precast.
- Concrete will absorb and also radiate stored heat, and can spall when exposed to heat from fire.
- Masonry often refers to brick, tile, concrete block, and stone.
- Similar to concrete, masonry products have good compressive strength, and are generally resistant to heat from fire.
- In particular, CMUs (also known as cinder blocks) have become very popular due to their strength, resistance to fire, and minimal ongoing maintenance. These factors are a benefit to the fire service.
- Standard materials that are commonly used in building construction are constantly evolving in concert with advances in technology, and as a result, the term composites is becoming more familiar to building construction methodologies and materials. Two notable examples are plastics (being added to engineered wood products) and carbon-fiber materials.

CHAPTER REVIEW EXERCISE

Answer the following:

1. What is a "load"?
2. How are loads imposed on materials?
3. What are the differences between compression, tension, and shear forces?
4. What influences the suitability of a material for a given building application?
5. How are brittle and ductile materials different?
6. Describe the relationship of surface-to-mass ratio and fire degradation on building materials.
7. Match the wood products in the left column to the associated wood type in the right column:

Plywood	Engineered wood product
OSB	
LVL	
Sawn Douglas fir	Traditional wood product
Glulam	
Heart wood	Native lumber

8. At what temperature does hot-rolled (extruded) steel begin to fail?
9. What is meant by "spalling" and how is it caused?
10. During a fire, two conditions will cause a masonry wall to become unstable. List the two conditions.

RESOURCES FOR FURTHER STUDY

- APA—The Engineered Wood Association, "A Guide to Engineered Wood Products," www.apawood.org.
- APA—The Engineered Wood Association, "A Glossary of Engineered Wood Terms," www.apawood.org.
- Brannigan, Francis L., *Building Construction for the Fire Service*, 3rd ed., Quincy, MA: National Fire Protection Association, 1992. Authors' note: Brannigan's third edition is no longer in print. A revised fourth edition was coauthored with Glenn Corbett following Professor Brannigan's death and is available through Jones and Bartlett Publishing.
- Canadian Wood Council, "Wood Products," www.cwc.ca/Products/EWP/.
- CertiWood Technical Centre, www.certiwood.com.
- Engineered Wood Products Association, www.ewpa.com.
- "Modern Construction Considerations for Company Operations," DVD Training Program and Instructor Guide, ISFSI, Pleasant View, TN, 2010. Available through www.ISFSI.org.
- "Report on Structural Stability of Engineered Lumber in Fire Conditions," Project Number 07CA42520, Underwriters Laboratories Inc., Northbrook, IL, 2008. Available at www.ul.com.
- Zoltek Commercial Carbon Fiber, "Carbon Fiber," www.zoltek.com/carbonfiber.

NOTE

1. The authors searched dozens of building fire collapse reports where cast iron columns were cited as a contributing factor. Several concluded that the introduction of cold water streams caused the collapse. Most, however, did not make that conclusion and cited poor casting, gravity connections, and sagging as the collapse cause. Noted fire service building construction author Francis Brannigan calls the cooling cause "questionable."

ANATOMY OF A BUILDING—A MAP 3

OBJECTIVES

- Define foundations, columns, beams, and connections.
- Identify the parts of a truss.
- List the three types of structural connections.
- Define structural assembly.
- Define the difference between panel, partition, and curtain walls.
- Describe the structural hierarchy.

COMMUNICATION SKILL-BUILDING FOR BUILDINGS

When launching an aggressive interior fire attack on a building, you and your crew are relying on building integrity—that is, the building's resistance and reaction to heat, fire, smoke, firefighting impacts, and gravity. We know that a building can only stand so much fire and heat assault before gravity takes over and the building starts to fail. Before you can predict building failure, it is important to understand basic building anatomy and the appropriate terms and concepts that can help you communicate collapse concerns. In the previous chapter we made the case for understanding common building construction/engineering terms and materials. In this chapter, we strive to help you understand the components of a typical building. While this sounds easy enough, the importance of this understanding and the use of appropriate terms and labels cannot be over emphasized. For example, confusing the term column with beam, or truss with rafter, can have detrimental consequences when communicating with an incident commander or safety officer. Likewise, understanding the interplay of structural components can help you make better decisions regarding safe or unsafe positions from which to operate or complete your tactical assignment.

Lastly, the critical communication information found in this chapter can help you understand the language used in the following chapters of this book where we talk about different construction types and the considerations related to fighting fire in, on, or around buildings.

STRUCTURAL ELEMENTS

Buildings contain many elements, components, features, finishes, and systems that come together to form an enclosure for a given purpose. Many of the aforementioned things are not necessary for the building to stand with integrity. From a firefighting perspective, attention must be given to the underpinnings of a building that must be monitored when evaluating the building's integrity to stand—the *structural* elements. Further defined, **structural elements** are those essential underpinnings of a building that allow it to stand erect and resist imposed loads and gravity (fig. 3–1). Structural elements work together to deliver all loads to earth. In the simplest form, there are four structural elements of any building:

- Foundations
- Columns
- Beams
- Connections

Fig. 3–1. Structural elements work together to deliver a building's loads to the earth.

At first glance, you may argue that walls, floors, and roofs are missing from the structural elements list. In some cases, walls, floors, and roofs are structural elements, but not always. In many buildings, you can remove the walls, floors, and roofs and the building will still stand. This is where the phrase "structural elements" differs from building coverings, finishes, and features. If you remove or damage a structural element, the building will begin to fail. Granted, this may be semantics, but once again, the use of appropriate language will help you better communicate issues during your firefight.

Understanding whether or not a building component is a structural element begins with the definition of foundations, columns, beams, and connections and the interplay each must have. In this chapter we define structural elements (and how they affect each other), then define other building components. In later sections of this book, we discuss the firefighting concerns associated with them.

Foundations

The **foundation** of a building can be defined as the building's anchor to earth and base for all elements built above that anchor. This anchor must have properties that allow it to deliver all imposed building loads and deliver them to earth in compression. Further, foundations must be designed in a way that will help keep a building level, and resist the chance of sinking, twisting, or leaning. The foundation of a building can be formed using footers, foundation walls, slabs, and/or pilings.

Footers: Footers (or **footings**) are weight-distributing pads that serve as the bottom of foundations. Footers are typically the lowest/deepest part of any building and directly contact earth. Footers can serve as a perimeter base for slabs and foundation walls or as solo pads to support columns (fig. 3–2).

Foundation walls: These are walls installed below grade to serve as structural support for other structural elements and also to hold back soil and other materials. Foundation walls also typically serve

as the perimeter basement walls of a building. Most foundation walls are poured-in-place, steel rebar reinforced concrete, although masonry block, precast panels, or heavy timbers and planking can be used. Often, foundation walls incorporate footers.

Fig. 3–2. Footers, solo pads, and foundation walls

Slabs: Slabs (when used as a foundation) are flat horizontal elements that simply rest on the ground. Some call this a slab-on-grade foundation. Most slab foundations still incorporate footers.

Note: Foundation walls and slabs are covered from another viewpoint in chapter 7.

Pilings: Pilings are vertical posts that are driven down into the earth to serve as the foundation or foundation anchor of buildings.

Columns

A **column** is defined as any structural element that is loaded axially, along its length, in compression. In most buildings, columns deliver the weight of beams and other columns to the foundation. Columns can take on the form of a wall or a post (fig. 3–3). When a wall is used as a column, it is often referred to as a load-bearing wall or, more appropriately, a wall column. In earlier times, the term *pillar* was used synonymously with column. More recently, a **pillar** is defined as a freestanding vertical post, monument, or architectural feature.

Fig. 3–3. A column delivers the weight of beams and other columns to a foundation.

The load that a column can carry is dependent on many factors such as the material used, the length (height if vertical), and its cross-sectional shape. The cross-sectional shape of a column is important because columns carry compressive loads axially through their length. The best shape for a column is one where the compressive load is shared equally (and spread further) from the center axis of the material being used. Rectangles, squares, and cylinders (like a pipe) are preferred ways of shaping columns (fig. 3–4).

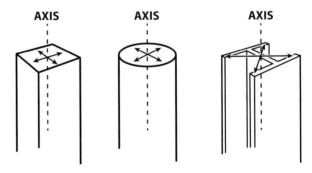

Fig. 3–4. Rectangles, squares, and cylinders are preferred ways of shaping columns.

Hot-rolled steel is often used for hollow columns that are either square or cylindrical, or a cross-sectional shape like the letter "H" (hence the term H-column). The use of an "I" cross-sectional shape is not ideal for a column—it becomes prone to buckling because the load is not shared equally

around the axial center. Hollow columns incorporate a cap of some type to help connect the column to beams and distribute the load evenly around the perimeter shape. Concrete masonry blocks (or CMUs) used to build a wall column are typically rectangular, suggesting that they don't disperse weight equally around their center axis. Closer examination of these CMUs shows that they are formed with multiple squares (two, three, or four cubes) that are joined together to achieve uniform distribution.

Columns are considered a critical element of any building—they hold up floors and roofs. For that reason, columns must be designed and built with resistance to lateral forces that could knock them over. The material used and the application of compressive load through the length of the column help to resist lateral loads.

Buttress and pilaster: A **buttress** is an exterior wall bracing feature used to assist with lateral forces created where roof beams or trusses rest on a wall. Buttress are structural in nature and can take on numerous shapes (a diagonally ascending stack of stone or brick is most common, as in fig. 3–5). Some textbooks use the term *pilaster* to also describe a buttress. Historically, a **pilaster** is a decorative column that protrudes in relief from a wall to give the appearance of a separate post column. Over time, the fire service began using the term pilaster to describe any interior or exterior thickening of wall used to add lateral support for roof beams and trusses.

For our purposes, we accept that both pilasters and buttresses can be structural and that they can be differentiated by shape. A pilaster appears as an interior or exterior vertical stack that thickens a wall column, whereas a buttress is a separate, diagonally-stacked brick, stone, or concrete wall that protrudes perpendicularly from the wall column supporting the roof.

While most columns in a building are vertical in attitude, it's important to note that columns can also be diagonal or horizontal. The guiding principle is that a column is compressively loaded through its length. Figure 3–6 shows horizontal columns that are being used to keep two brick wall columns from falling into the alleyway. Horizontal columns are called **struts** and diagonal columns are called **rakers**. A typical application for a raker is a post driven diagonally into the ground to help hold shoring for earthen excavations and trenches. Exterior load-bearing walls under construction are often wind-braced with rakers until a floor or roof load is applied.

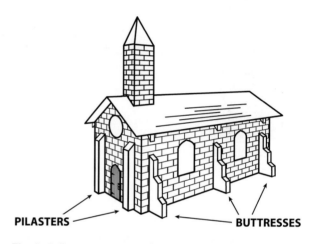

Fig. 3–5. Buttresses are used to counteract lateral forces in a vertical wall and pilasters are used to strengthen a wall.

Fig. 3–6. Horizontal columns that are compressively loaded through their length

Quick summary

- Buildings are comprised of many structural components that combine to form an enclosure for a given purpose.
- The primary structural elements of a building are foundations, columns, beams, and connections. These elements work together to transfer all loads to the earth.
- A foundation is a building's anchor to the earth.
- Columns are a primary structural element that delivers the weight of beams and columns to a foundation.
- Buttresses and pilasters are used to enhance the stability of an exterior wall. Buttresses assist with lateral forces and pilasters help to strengthen a wall.

Beams

Beams are used to create a covered space—usually between columns. Roofs, floors, and most loads placed in buildings are picked up by beams and delivered to columns that deliver a load to the foundation. By definition, **beams** are structural elements that deliver loads perpendicularly to their imposed load and in doing so, create opposing forces within the element. Any load placed on a beam (and the weight of the beam itself) causes the beam to deflect. That is, the top of the beam is subjected to a compressive force while the bottom of the beam is subjected to tension (fig. 3–7). In between the compressive and tensile load is a neutral plane. The neutral plane creates an area where there are no stresses. This is the area where a small hole can be punched through the beam to accommodate piping, electrical wires, or for other purposes.

The distance (or depth) between the top of the beam and the bottom of the beam dictates the amount of load the beam can carry or the distance the beam can span. The amount of load that a beam can carry for a given span is actually proportional to the square of its depth. If you double the top-to-bottom depth of the beam, you can carry four times the load. If you triple the depth, you can carry nine times the load. For example, a simple 2 × 4 in. wood beam might carry 50 lb. If you attach a second 2 × 4 *next* to it—you merely double the weight it can carry to 100 lb. If you replace the side-by-side attached beams with one that is 2 × 8 in. (increase the depth), it can carry 200 lb (fig. 3–8)! The length that a beam can span is directly proportional to its depth. If you double the top-to-bottom beam depth, you can double the span. These depth relationships are simplistically sound although the engineering community uses complex formulas that take into account shapes, materials, spacing, and safety factors when designing beams.

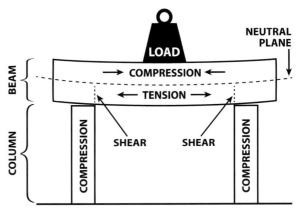

Fig. 3–7. Beams transfer load using opposing compressive and tensional forces.

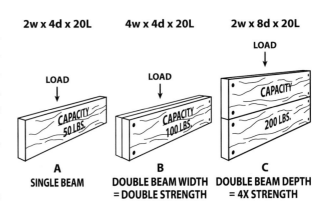

Fig. 3–8. Beam strength is proportional to its depth.

Beams are typically labeled by their application and by their designed shape/material arrangement. We'll discuss shape/material a bit later in this chapter. Beams labeled by application include the following:

Simple beam: A beam supported by columns at the two points near its ends.

Continuous beam: A beam supported by three or more columns.

Cantilever beam: A beam supported at only one end. (Or a beam that extends well past a support in such a way that the unsupported overhang places the top of the beam in tension and the bottom in compression.)

Lintel: A beam that spans an opening in a load-bearing wall, such as over a garage door opening (often called a "header"). Lintels can also be commonly found over windows and doors in unreinforced masonry construction and in newer CMU construction (fig. 3–9).

Fig. 3–9. Lintels are used to span an opening in a load-bearing wall.

Girder: A beam that carries other beams.

Ledger: A beam attached to a wall column that serves as a shelf (ledge) for other beams or building features.

Joist: A wood or steel beam used to create a floor or roof assembly that supports sheathing or decking. Joists span between primary supporting members such as foundations, load-bearing walls, or structural beams.

Rafter: A sloped wood joist that supports roofing coverings between a ridge beam and wall plate on peaked and hipped roofs.

Ridge beam: The uppermost beam of a pitched roof. Rafters attach to the ridge beam.

Purlin: A beam placed horizontally and perpendicularly to trusses or beams to help support roof sheathing or to hang ceilings.

Suspended beam: A beam that has one or both ends supported from above by a cable or rod (sometimes called a hung beam).

Most beam applications are such that the beam is laid in a horizontal attitude. Beams can also be vertical. A retaining wall for landscaping or grading is essentially a vertical cantilevered beam. The same can be said for communication antennas. A highway billboard has to be built like a beam to withstand wind—another example of a vertical beam.

Beams are further classified by their material arrangement and shape. A solid wood or reinforced concrete beam is simply that—a solid beam. A steel beam that is cross-sectionally shaped as a rectangle is also considered a solid beam, although technically it is hollow inside. Beams can also be created or built with various shapes and pieces. The term "I-beam" reflects the shape of the beam viewed from either end. The top of the "I" is known as the *top chord* (or flange); the bottom of the "I" is the *bottom chord* (or flange). The piece used to span the distance between the chords is known as the *web* (sometimes called the *stem*). The two most popular I-beams are the popular engineered wooden I-beams (see fig. 8–12) and extruded steel I-beams. Both of these beams have a solid or closed web, although small holes can be punched through the neutral plane of the web to accommodate utilities.

Truss: A truss is an engineered structural element that uses groups of rigid triangles to distribute and transfer loads. The triangles create an open web space. Trusses are used in lieu of solid beams in many buildings. Although the initial perception of this definition may seem slightly complex, the basic

design of a truss is rather simplistic and has been used in this country since the early 1800s. The basic concept of the truss is not new, but the materials that are currently being used in its construction *are* new as heavy timbers have been replaced with lightweight wood and/or metal structural members. This has resulted in a significant change within the building industry and has dramatically changed how many buildings burn and fail when exposed to heat and/or fire. Because truss construction has, in many instances, become the norm for many structural applications, let's take a quick look at the concept of a truss.

A truss is nothing more than a structure that consists of one or more triangles formed by straight members whose ends are connected at joints that are referred to as nodes. The triangular configuration is the key to the success of a truss as the structural stability of the triangular design/shape is the simplest geometric figure that will not change shape when subjected to a load if the lengths of the sides of a truss are fixed to retain its shape. Remember, the strength of a beam is greatly enhanced by depth—the distance between the top and bottom chords. Interestingly, it is the depth of a truss (or the distance between the upper and lower chords) that also results in an efficient structural design. However, because the perimeter of a triangle is an inherently strong shape, it also results in a hollow center. As a comparison, a solid beam of equal strength would result in a substantial increase in weight, size, and cost—hence the popularity of modern lightweight truss construction.

The simplest form of a truss is a *planar truss*, which is a single triangle and is comprised of a bottom chord and two top chords (fig. 3–10A). In this form, the two top chords are in tension and the bottom chord is in compression. Triangles (or other hollow shapes) are labeled as *open web*. Common modifications that are used to increase load capacities and design configurations of a planar truss configuration are to add diagonal and vertical members within the triangular shape. These members are referred to as *truss web members*. A simple example of a truss with a single web member is called a *king post truss*, which consists of two angled supports that intersect a common vertical support (fig. 3–10B) that is joined to the bottom chord.

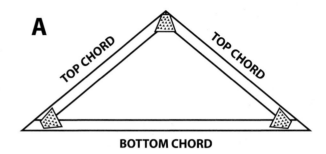

Fig. 3–10A. A single triangle is an example of a simple planar truss.

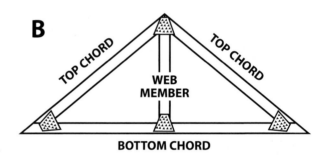

Fig. 3–10B. When a single web member is added to a planar truss, it becomes a king post truss.

Trusses can be built in many configurations using a multitude of materials and shapes—and with that another set of terms:

- **Triangular truss:** The most common type of truss used to form a peaked roof.
- **Parallel chord truss:** A parallel chord truss is one in which the top and bottom chords run in the same plane. Parallel chord trusses can be engineered where just the top chord is attached to the support (column/other beam) or traditionally resting on the bottom chord. The top chord of a parallel truss is loaded in compression and the bottom cord in tension. Parallel chord trusses can be manufactured using wood, metal, or a combination of wood and metal. The term **bar truss** (or **bar joist**) refers to steel

parallel chord truss assembled with angle iron for the chords and cold-drawn round billet for the web. The pieces are tack-welded together to form the truss unit (fig. 3–11).

Fig. 3–11. Metal bar joist trusses

- **Arched truss:** An arched truss is one where the top chord is arched and the bottom chord is straight (horizontal). Arched trusses are often called bowstring trusses, which could be confusing when understanding how arched trusses fail. Arched trusses can transfer loads to columns in multiple ways, which help us better label the type of arched truss you may be dealing with at a structure fire. Arched trusses can be labeled as rigid arch or bowstring.

- **Rigid arch truss:** Also known as a *rib arch truss*, a rigid arch truss has a curved, self-supporting top chord (not tied by the bottom chord) and horizontal bottom chord along with web members that are all rigidly connected. The load of the truss is delivered axially downward through the bottom chord and onto support walls or columns. These trusses can be constructed with steel or heavy timber—the latter creating a very tough roof with predictable fire reaction.

- **Bowstring truss:** A tied truss with an arched upper chord and a horizontal tension bottom chord that connects the ends of the arched cord, creating compression in the top chord. Diagonal web members are added to help transfer loads. The top chord of the truss abuts to the support wall or column. A true bowstring truss typically requires buttresses or pilasters for masonry walls to help accept the lateral forces that may be developed as the live loads are gained or lost on the roof.

Other types of arched roofs exist—namely tied-arch and lamella—but they are not true trusses. These are covered in chapter 8.

Fire service texts don't always agree on these descriptions and labels for arched trusses; they are presented here in a way that helps the fire officers determine how an arched roof will fail and the potential for roof and wall collapse. We include a section titled "Historical Perspective on Arched Trusses" to help dissect the terminology enigma. Additionally, figure 3–12 shows how each of the arched trusses are supported and the forces are created.

Quick summary

- Beams are structural elements that deliver a load perpendicularly to its imposed load.

- The distance between the top and bottom of a beam dictates the amount of load it can carry.

- Beams can be classified by use, shape, arrangement, and/or materials used.

- Lintels are used to span an opening in a load-bearing wall and can be made from wood, steel, concrete, or stone.

- A truss is a type of beam that uses geometric shapes to form an open web, and is nothing more than a structural element that consists of one or more triangles.

- The hollow shape of a triangle is referred to as open web.

- Geometrically configured trusses have largely replaced sawn lumber when used as structural members.

- Two common arched trusses that can be encountered at structure fires are the rigid arch truss and the bowstring truss.

Chapter 3 — ANATOMY OF A BUILDING—A MAP

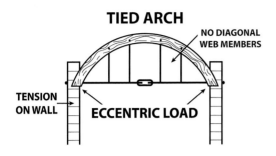

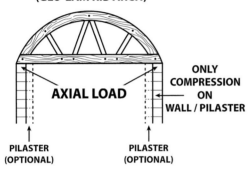

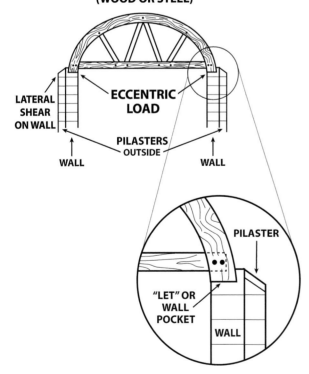

Fig. 3–12. Differences between a rigid arch truss, bowstring truss, and a tied arch

HISTORICAL PERSPECTIVE ON ARCHED TRUSSES

It is common for firefighters to classify an arch-shaped roof as a bowstring truss roof. The reality is that an arched roof can be created using many different construction methods. Four notable methods are: bowstring truss, rigid arch truss, lamella, and tied arch. Each is unique and can behave differently during building fires, as described in chapter 8.

So where did the terminology go awry? A quick historical perspective may answer the question.

Bridge building in America took on a fevered pace in the late 1700s and early 1800s. Wood was abundant. Using the inherent design strength of arches and trusses, all manner of wooden bridges were built. Specific bridge engineering was one of discovery—some worked, some didn't. Most discovered that wood had serious drawbacks—mostly due to weather and stress/strain issues.

Innovation continued. Of note is the Pratt truss, which uses a trapezoidal shape (unequal parallel chords) and diagonal tension members between each panel (or section) of the truss. The timber bridge truss roof construction, which is presented in chapter 8, was derived from Pratt's design.

Another innovator, Squire Whipple, improved on Pratt's design using wrought iron as diagonal tension members that cross multiple panels. Whipple, known as the father of modern bridge construction, decided that wood was not well suited for bridge construction and set about designing and constructing a bridge from cast and wrought iron. In doing so, he created a tied arch truss that used a cast iron arched top chord that was splayed to receive tensioning

rods from a wrought iron bottom chord, which kept the top chord in compression. He reinforced the design with diagonal tension bars that helped keep tension on the bottom chord. Each side of the top chord attached to a support abutment—typically stone or concrete with a cast iron pad. The innovation was quickly labeled a bowstring truss bridge after the classic archer's bow. Thus, the bowstring truss was born and innovators, wary of Whipple's 1841 patent, began to improve the design.[1] Bowstring was all the rage for anything structural—they looked good, could span distances, and were quite durable.

Applied to buildings, it is important to note that arched roofs of the day were made of cut timber (fastened with iron pins and plates) using a rigid (non-tied) truss approach. In many ways, these timber arches were more like Pratt's bridge truss in that they relied on a stout bottom chord to sit on walls and receive the stresses of the top chord and web members. It was quickly discovered that the wooden bottom chord fatigued from the constant strain, leading to cracks and splits in the wood (especially after significant snowstorms). A quick remedy was to reinforce the bottom cord with an iron tie-rod and turnbuckle (much cheaper than replacing the roof) and thus making the rigid truss a true bowstring. The remedy is still being used today to fix older rigid arch and bowstring trusses.[2]

Learning the lesson, builders began assembling bowstring trusses using various combinations of wood and iron (and later steel). The use of bowstring truss roofs was quite prevalent well into the 1950s—especially in warehouse or repair facilities that required open span spaces. Still, rigid arch trusses and tied arches were being used—especially the wooden glulam rigid arch truss that was designed to produce a desired architectural feature that was functional. Modern arched trusses can be found in two configurations: a bowstring truss where the top chord abuts to the wall with the bottom chord tied to the top chord, and a rigid arch truss where the bottom chord sits upon the wall.

Connections

Obviously, structural elements must be connected to one another in such a way to effectively transfer loads. Arguably, structural connections can be considered one of the most important structural elements (discussed in the section titled Structural Hierarchy). One thing is sure—they are small, low-mass points that must transfer a great deal of force. As we know, low-mass means poor resistance to heat and therefore earlier failure during a fire. For this reason, structural connections are often considered one of the weak links in construction.

As can be imagined, connections are subjected to numerous forces that can include compression, tension, and shear. For this reason, steel and other metals are the most common materials used. Generally speaking, there are three types of connections: pinned, rigid, and gravity.

Pinned connection: Pinned connections are those that use a screw, nail, nut and bolt, rivet, or similar device to pass through the elements being connected. Pinned connections concentrate transferred loads to a single point.

Rigid connection: Materials that are bonded together are considered to be rigidly connected. Examples include bead welds in steel, glues, poured concrete over steel, and the like. Rigid connections tend to spread transferred loads over a greater area. A spot weld is more of a single point tack and should be considered a pinned connection.

Gravity connection: In a gravity connection, materials rely on the gravitational weight of the upper element to hold them together. Beams that are set into a wall pocket on a brick or masonry wall are perhaps the most common gravity connection (see the feature block titled "Fire Cut Beams"). Wood

beams/rafters/joists that sit in a brick or stone wall pocket used to be "fire cut" to help them release from the wall once they started to sag after fire exposure.

Fire Cut Beams

Gravity connections were quite prevalent in the pre-WWI building. Masons would build walls of brick or cut stone and include pockets for the insertion of wooden floor timbers. It was discovered during fires that the tight fitting gravity connection of the wood and masonry had an undesirable consequence. As the wood floor beam started to sag during a fire, it exerted upward force on the wall pocket. Because the wall was designed to only transfer a downward compressive load, wall failure ensued. To correct the problem, builders discovered that a simple diagonal coffer cut on the top side of the wood beam (where it inserted into the wall pocket) would help the wood beam release from the wall if it sagged (fig. 3–13). This idea was expanded to include floor joists and roof rafters.

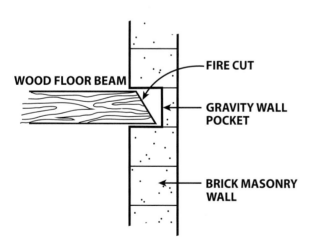

Fig. 3–13. A wall pocket and fire cut beam

From the builders' perspective, this was smart: the floor/roof would release and the wall would be saved. Following the fire, a builder could come in and insert new beams into the preserved wall—saving lots of time and money. From a firefighter's perspective, this is a self-releasing floor and/or roof!

Once a building is finished, it is not evident if fire cut beams exist. From a historical perspective, it is also difficult to say if fire cut beams are included for any given building. However, in many cases, exterior tie plates have been installed to tie the floor beam/joists and/or roof rafters to an exterior wall to attempt to strengthen these buildings, and they are normally visible on the exterior (fig. 3–14). As a rule, however, firefighters should treat late 1800s and early 1900s load-bearing brick and stone buildings like they have self-releasing floors. Any visual or audible evidence of floor sagging should trigger a switch to exterior-only operations.

Fig. 3–14. Visible tie plates indicate the location of floor beams/joists and roof rafters that are likely fire cut.

Pinned and rigid connections are considered restrained connections—moving forces of one element are transferred to the other. A gravity connection is considered unrestrained. A gusset plate or joist hanger uses small metal teeth to puncture the wood grain of lumber to form a friction connection. Technically, these are pinned connections, although that may be dubious. Because the teeth don't anchor completely through wood, they may easily pull out of the wood when minimal moving forces are applied. For that reason, firefighters should consider these unrestrained connections.

STRUCTURAL ASSEMBLIES

A **structural assembly** can be defined as an engineered collection of interconnected building components that form a cohesive structural unit such as a roof or floor. In more simple times, buildings were constructed with individual parts that could stand alone to serve a purpose. In some ways, older buildings were over-built and had redundancy in structural elements and therefore more strength. If one rafter failed, no big deal—the rafters on each side were still sound. Perhaps the first structural assembly was the wooden stick-framed wall. Frontiersmen discovered that a series of small dimension wall studs attached to a top and bottom sill formed a decent wall and used much less wood than a log cabin or heavy timber pole barn.

As the world became more industrialized, architects, engineers, building supply businesses, and financiers realized that there are many ways to build a building and use less material—and build for less cost or for more profit. Engineers met the challenge of retaining desired strength by designing an interdependence system. Building materials and technologies evolved with this realization and thus gave birth to the concept of *assembly-built* or *performance-designed* construction. The basics of performance-designed construction is that individual pieces of an assembly are codependent on other pieces to form a sound unit. For example, a single truss must be attached to sheathing that is attached to another truss and another (and so on) to form an assembly that meets the strength needs of a roof or floor. In the case of roofs, the assembly-built evolution allowed house builders to increase spacing between trusses provided the trusses were all joined by OSB sheathing through an engineered staple and/or glue schedule. As an example, the home under construction in figure 3–15 has been rough framed and then sheathed with ½ in. OSB (which will stabilize the framing). The roof trusses were then placed on top of the walls and are in the process of being tied together with ½ in. OSB sheathing. When finished, the entire building will be stabilized by ½ in. OSB sheathing.

Fig. 3–15. An assembly-built structure and roof

Structural assemblies are now commonplace in new construction. In the context of this chapter, structural elements are defined as individual components. As you can see, structural assemblies are indeed structural elements.

Sheathing

Sheathing describes all manner of materials used to cover or encase walls, ceilings, and roofs of framed structures. It is the first layer of covering for studs, joists, trusses, or rafters. In assembly-built buildings, sheathing is a structural element that helps lock the framed walls and roofs together. Prior to the assembly-built building, sheathing was not considered a structural element although it did add rigidity to a building. Some use the word "decking" or "sheeting" in place of sheathing. Technically, decking is the durable cover used to accept loads on floors, and sheeting is typically 4 × 8 ft sheets of various materials (gypsum, OSB, plywood, etc.). For continuity, we use sheathing to describe materials used to encase framed walls, floors, and roofs. Sheathing is usually classified by the materials used to create the platform.

Wood. Wood sheathing can take on many forms. Plywood and OSB sheets are probably the most used forms of wood sheathing. These sheets are glued, nailed, or stapled to the joists below. When wood is used as floor sheathing, it is often referred to as the *subfloor*. Subflooring is then covered with some form of durable covering. Tongue and grove wood sheathing uses milled lumber planks

that connect together by the tongues and groves on either edge of the wood planks.

Sheathing is also laid perpendicularly or diagonally across joists to assemble the platform (fig. 3–16). In figure 3–16, notice the 1 × 6 in. wood roof sheathing has been installed at a 45° angle to the 1 × 8 in. rough-cut rafters and the presence of scissor bracing between each rafter (which is used to prevent twisting of the rafters). Diagonal sheathing is stronger than straight sheathing as each piece of sheathing covers more rafters. Now, compare the construction in figure 3–16 to modern lightweight truss construction and you can see why modern construction can collapse in less time than older conventional construction! The attractive finished wood flooring found in many buildings is not a structural element. Finished wood floors act as a durable cover over some form of sheathing that rests on the joists below.

Fig. 3–16. Diagonal roof sheathing

Metal. Thin panels of corrugated metal placed across joists form a platform known as metal sheathing. Metal floor sheathing then serves as a pan for a poured concrete, cementitious, or epoxy-resin floor finish. Metal roof sheathing can be covered in various ways to achieve a watertight finish. For example, built-up tar and gravel has been used for decades. More recently, metal deck roofs are covered by expanded polystyrene sheets (for insulation) that are sealed with an impervious membrane.

Composites. Advances in material technologies have led to a plethora of new materials used for floor and roof sheathing. Cementitious sheathing includes panels made from proprietary blends of concrete dust and a binding agent. They are lightweight, quite strong, and noncombustible. Gypsum-based panels used for sheathing are actually blends of gypsum with mesh, fibers, or beads that give the gypsum sheathing strength.

Walls

Previously we mentioned that walls built to hold beams and floor or roof assemblies are called load-bearing or wall columns—a structural element. Walls that are non-load-bearing typically serve to enclose a space and may not be true structural elements. Here we define various wall types that are not necessarily structural elements.

Panel wall. A single-story exterior wall used to enclose a space.

Curtain wall. An exterior wall used to enclose multiple stories.

Partition wall. A wall used to divide areas or rooms into smaller areas or to separate one portion of an area from another and usually not load-bearing.

Party wall. A wall shared by two buildings or two occupancies within the same building. If the party wall carries beams or structural assemblies, it is a structural element.

Shear wall. A reinforcement wall that adds building stiffness to help resist the impact load of wind.

Veneer wall. A decorative-only wall added to help improve the building's appearance.

STRUCTURAL HIERARCHY

Together, structural elements defy gravity and make a building sound. During fires, earthquakes, and other destructive events, the building elements are being challenged to resist attack. When certain elements fail, other elements may also be affected. This cause and effect sets the stage for the **structural hierarchy**, a concept that defines the progressive order in which building loads are delivered to earth. In a sound building, the hierarchy states that decking provides a platform for beams to receive loads (people, contents, snow, etc.). The beam delivers the loads to girders. Girders deliver the loads to columns (posts or load-bearing walls), and columns deliver loads to the foundation. When being attacked, the structural elements will fail in consecutive degrees of severity that follow the hierarchy. Following that logic, a failure of the foundation will cause the most severe destruction of the building. Fortunately, foundations are very well protected (buried below grade and contacting earth). That leaves the columns as perhaps the most important part of the hierarchy. Granted, a failure of sheathing or beam can be severe—especially if you're on it or below it when it falls!

The structural hierarchy of a building can be affected by the way structural elements are constructed. In the next chapter, we look at some very specific construction methods. Here, we discuss some general building methods used to create the hierarchy.

Framed. A building constructed on site one piece at a time—also known as stick-built. The building is enclosed by simple siding attached right to the framing.

Monolithic. A poured-in-place concrete and steel building that forms a "single stone."

Post and beam. The same as a skeletal frame building.

Skeletal frame. A building constructed with a series of post columns and beams (no load-bearing walls). The building is enclosed by panel or curtain exterior walls.

Tilt-up. A structure built using prefabricated, load-bearing wall sections (typically reinforced concrete) that are tilted upright, then pinned together.

Wall-bearing. A building where beams or roof/floor assemblies rest on the load-bearing walls (as opposed to posts).

These general types of hierarchy are just that—general. Firefighters need to classify buildings in a much more definitive way. The next chapter does just that.

Quick summary

- Structural connections are considered one of the weak links in construction.
- Connections can be categorized as pinned, rigid, and gravity.
- Fire cut beams are designed to pull out of a wall during a collapse without forcing the corresponding wall outward.
- A structural assembly is an assembly of interconnected building components that form a cohesive structural unit.
- The concept of assembly-built relies on individual pieces that are codependent on other pieces to form a sound unit.
- Structural hierarchy defines the progressive order that building loads are delivered to earth.

CHAPTER REVIEW EXERCISE

Name the component

In the following photos, provide the correct term for each of the numbered elements.

Photo exercise 3–1

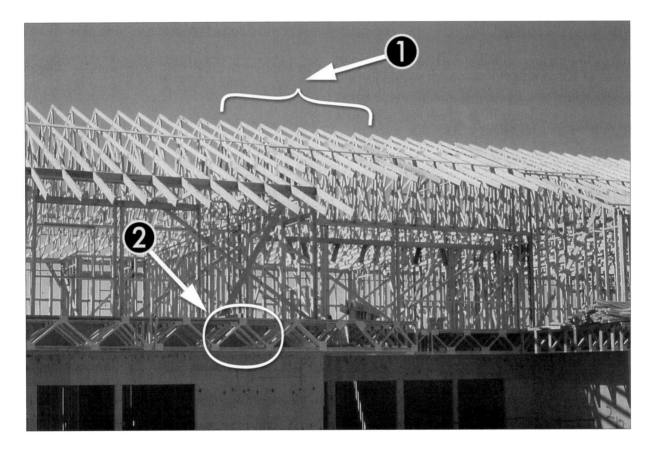

Element 1 =

Element 2 =

Photo exercise 3–2

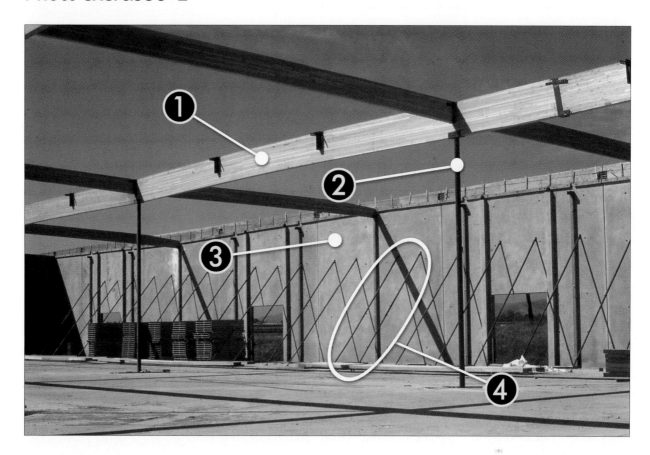

Element 1 =

Element 2 =

Element 3 =

Element 4 =

Photo exercise 3–3

Element 1 =

Element 2 =

Element 3 =

Photo exercise 3–4

Element 1 =

Element 2 =

Element 3 =

RESOURCES FOR FURTHER STUDY

- "Introduction to Structural Design," University of Virginia School of Architecture.
- "Classical Truss Theory," University of Virginia School of Architecture, http://urban.arch.virginia.edu.
- Hibbeler, Russell Charles, *Engineering Mechanics: Statics*, 13th ed., Upper Saddle River, NJ: Prentice Hall, 2012.
- Maginnis, Bernard Owen, *Roof Framing Made Easy*, New York: The Industrial Publication Company, 1903.
- Ricker, Nathan Clifford, *A Treatise on Design and Construction of Roofs*, New York: J. Wiley & Sons, 1912.
- Structural Building Components Association (SBCA) website, www.sbcindustry.com.
- Thallon, Rob, *Graphic Guide to Frame Construction*, 3rd ed., Newtown, CT: Taunton Press, 2008.

NOTES

1. Guise, David, *Abstracts and Chronology of American Truss Bridge Patents, 1817-1900*, Society for Industrial Archeology, 2009, www.sia-web.org.

2. Gilham, Paul C., Chief Engineer, "Bowstring Trusses 'Fail' to Meet Current Code Requirements," Western Wood Structures, Inc., www.westernwoodstructures.com.

CLASSIFYING BUILDINGS— NFPA 220 SYSTEM

OBJECTIVES

- Summarize the various methods of classifying building construction and discuss the traps in such classifications.
- List and describe the structural features of the five classic construction types found in NFPA 220.

IT'S POLITICALLY INCORRECT TO PROFILE ANYTHING

The previous two chapters were designed to give you an understanding of common building engineering principles, the basic parts of a building, and the language used to identify and communicate the interplay of those principles and parts. Now, we present the various construction methods used to assemble—and therefore classify—a building. The history of construction methods is one of evolution influenced by myriad factors, including the following:

- Fires and fire deaths
- Other natural disasters
- Deterioration and aging
- Economics and profits
- Politics and codes
- Conservation
- Population increases
- Occupancy needs
- Architectural innovation
- Material technologies
- Engineering advances

With so many influences, it's easy to imagine that there are multitudes of construction methods that have led to many building *types*. For the fire service, this presents quite a challenge for classifying—or profiling—buildings. This chapter helps show how building construction methods are classified and dwells

The Art of Reading Buildings

a bit on the traps associated with profiling building types. Before we dive in, let's reinforce why this chapter is so important.

As firefighters we *must* consider the construction type so we can make judgments regarding strengths and weakness of the building that can help or hinder our suppression efforts. As you know, one of the initial steps in the mitigation of structural fires is understanding a particular building and/or being able to conduct a size-up in a rapid, logical manner that is applicable to a safe and timely conclusion to the incident. To achieve this familiarization and/or rapid size-up, firefighters must invest in a three-step process:

1. Pre-incident study
2. Prefire familiarization of actual buildings
3. On-scene pre- and post-incident experience

This chapter (and subsequent ones) provides the technical foundation for understanding building construction methods and features. That is the "pre-incident study" in your three-step process. Additionally, you must gain practical experience from actually walking through and evaluating buildings (building familiarization and prefire planning). Clearly, there is nothing to replace the personal knowledge gained from a previous building walk-through when involved in an attack or search operation inside a building at 2 a.m. (fig. 4–1)! Finally, you must blend your own experience at actual incidents to gain a working knowledge of how buildings are put together, how building construction affects fire and the extension of its by-products, and how buildings come apart during fire conditions.

Obviously, the on-scene incident experience can be an unpleasant one if sufficient expertise has not been accumulated prior to actual fireground operations. In this chapter we present some shared experiences dealing with various construction types (from personal and historical archives) to help those with minimal incident experience. Using the three-step process above, you'll improve your ability to read a building in a rapid/logical manner and improve your ability to apply that information for appropriate strategic and tactical decisions that are fundamental to safe and effective suppression efforts.

Fig. 4–1. There is nothing to replace walking through a building and evaluating its applicable characteristics.

CLASSIFYING BUILDINGS

As we stated above, there have been many influences that have led to myriad construction methods—yet most fire service curriculums present an overly simple system for classifying buildings (the five classic types). We feel this is a trap that can lead to an error in reading a building. To address the above influences and better classify a building, firefighters should recognize several classification methods to help them better read a building:

- NFPA 220 (the five classic types)
- Building era (the construction methods used during a given historical time period)
- Building occupant use (its intended use)
- Building size (external and internal)

Some may argue that the five types listed in NFPA 220 are actually tied to occupancy use and building size. In fact, that is mostly true for a building—when it's originally built or when a

building is renovated or altered using a code/permit process. This interplay of size/type/use has evolved into a very complex body of codes and standards that tell a contractor how a building should be built from an occupant life-safety perspective. The size/type/use code structure has changed over time—meaning that the era or time period during which a building was built must be factored. We hope it is obvious that a 1930s Type III building is significantly different from a 1990s Type III. Unfortunately, the street firefighter is rarely educated in the interplay of size/type/era/use (a massive body of work). Further, the building classifications that are presented to firefighters are too often oversimplified. Most firefighters are taught the NFPA classification system to help them understand building construction methods. This system has served the fire service well for many decades. Having said that, we believe that the oversimplified use of the five types to classify a building at a 2 a.m. fire can be dangerous!

To substantiate, consider a fire in an NFPA 220, Type V, wood frame building (the most common type). Is it 1,100 sq ft or 6,500 sq ft? Is it 1930s or 2010s wood frame? Is it a single-family dwelling or a budget chain hotel? As an arriving fire officer at a structure fire, the specific size, era, and use of the Type V building is much more meaningful than the fact that it is a Type V. There are other issues with the NFPA 220 system. For example, buildings can be built that are actually multiple NFPA 220 types in the same structure, and there are buildings that don't really fit into any of the five types. See the dangerous trap?

The intent here is to not overwhelm you with considerations—our whole goal is to show how the evolution of buildings has created the need to look at them with a better tactical eye. In fact, we've created some Rapid Street-Read Guides to help you better classify buildings (in the last section of the book). In essence, what we're proposing here is that firefighters making strategic/tactical decisions at structure fires need to consider the building as their host—that is, the building is hosting a party and you are the guest. Some hosts are gracious and accommodating and some not so much. In context, the host was built a certain way for a certain use and size during a certain era. Look at figure 4–2. Are you thinking it's just a simple house? You're right. You could even say it's a Type V wood frame house. However, the good fire officer should view it as a small, Type V, late-1800s (balloon) wood frame, Cape Cod style, single-family dwelling. Don't get us wrong, you wouldn't want to say all that for your on-scene radio report! Yet by classifying the house more completely (size/type/era/use), you can engage a given incident with an understanding of the issues that might present themselves.

Fig. 4–2. A typical small, Type V, late-1800s wood frame, Cape Cod style residence

To better classify buildings and maximize the use of the street guides included in section 3, you need to have a further understanding of the whole size/type/era/use relationship. This chapter covers the NFPA 220 classification system. Chapter 5 discusses the other important classification considerations. Trust us, the investment you make in the balance of this chapter and chapter 5 will help make the Rapid Street-Read Guides very useful aids in learning to read buildings.

NFPA 220 OVERVIEW: THE FIVE CLASSIC BUILDING TYPES

As a starting point to reading a building, *NFPA 220: Standard on Types of Building Construction* provides five major types of building construction classifications that are directly related to "the combustibility and the fire resistance rating of a building's structural elements. Fire walls, nonbearing exterior walls, nonbearing interior partitions, fire barrier walls, shaft enclosures, and openings in walls-partitions-floors and roofs are not related to the types of building construction and are regulated by other standards and codes, where appropriate."

Before we look at the five NFPA 220 building classifications, let's briefly consider several definitions that are commonly used in this area of building construction classifications that are based on a combination of combustibility and fire resistance, and then consider how NFPA 220 designations are formulated.

Definitions

Combustible. Will burn, flammable.

Limited-combustible. Materials that have about one-half the heat potential of wood, or not over 3,500 Btu/lb. For comparative purposes, Douglas fir equals 8,400 Btu/lb.

Noncombustible. Materials that will not ignite, burn, support combustion, or release flammable vapors when heated.

Fire resistance rating (FRR). The length of time to burn *through* a given material—rated in minutes or hours. It is important to remember that this rating is derived in a laboratory setting with a predetermined heat/flame source. However, remember that actual fires can burn with less or greater heat and/or flame sources, which can dramatically change the fire resistance rating.

Flame spread rating (FSR). The length of time it takes to burn *across* the surface of a given material—rated in minutes or hours.

Protected. Having a fire resistance rating of at least one hour based on its structural elements or protective envelope for the structural elements.

Unprotected. A material that when exposed (or can be exposed) in its natural state to the effects of heat and/or fire will cause a degradation of its structural integrity.

Numerical designations

NPPA 220 uses a combination of Roman and Arabic numerals to define five types of building construction and the fire resistance of their primary structural components, as shown in figure 4–3.

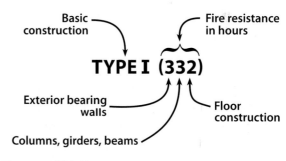

Fig. 4–3. NFPA 220 uses a system of numerals to define five types of building construction.

- The first designation is a Roman numeral that refers to the type of construction (i.e., Type I, Type II, Type III, Type IV, and Type V). This is the most common illustration of the system and is detailed in the next section.
- Following the Roman numerals are three Arabic numbers in parentheses that indicate the fire resistance of the exterior bearing walls, structural frame, and floor construction.

- In figure 4–3, Type I refers to fire resistive construction. The first number (3) indicates a fire resistance of 3 hours for the exterior bearing walls, the second number (3) indicates a fire resistance of 3 hours for the structural frame, and the third number (2) indicates a fire resistance of the floor construction of 2 hours.

- Normally, as the Roman numerals increase, the hourly ratings decrease, and as the Arabic numbers decrease (from 4 to 0), so does the fire resistance.

- The following numerical combinations are normally linked with the Roman numerals:

    - Type I (442 or 332)
    - Type II (222, 111, or 000)
    - Type III (211 or 200)
    - Type IV (2HH, where H indicates heavy timber)
    - Type V (111 or 000)

Remember that any system will have its strengths and areas that need improvement, which specifically applies to this system. As an example, one strength of NFPA 220 is that the system provides a simple way to classify buildings in five areas of consideration. It can also provide a structured outline that can be used for prefire planning operations to familiarize personnel with a building and its structural strengths and weaknesses, areas of construction that enhance the extension of fire and smoke, and so on. One noteworthy area of weakness that should be remembered is that while Type III through Type V buildings can normally be easily identified from the street, Type I and Type II buildings can only be identified from the interior of a building, if at all. Now, with the preceding overview of how NFPA 220 works, let's look at the five types of construction.

Type I (442 or 332): Fire resistive construction

Shall be those types in which the fire walls, structural elements, walls, arches, floors, and roofs are of approved noncombustible or limited-combustible materials.

Of the five types of construction, Type I can be considered as the most fire resistive type because all of the structural members must be constructed from noncombustible materials such as reinforced concrete and protected steel. Reinforced concrete is a noncombustible material and provides thermal protection to the steel reinforcing rebar within the concrete. Steel framing, however, must be protected from heat to yield the appropriate fire resistive qualities that are mandatory within this classification. Protection of steel structural members can be achieved by four methods, as listed below (fig. 4–4).

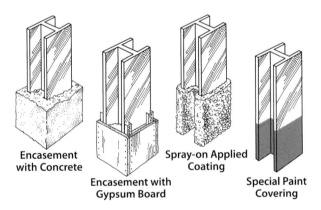

Fig. 4–4. Protection of steel structural members can be achieved by four methods.

1. Steel is enclosed in fire-resistive concrete.

2. Steel is enclosed in gypsum board.

3. A coating material (e.g., vermiculite with a cement binder, etc.) is sprayed on the steel. This is a very common method and is illustrated in figure 4–5.

4. An intumescent fire resistant coating is applied to the steel. These coatings are applied like paint, are about 0.5 in. thick, will expand about

15 to 30 times during a fire, and are designed to provide insulation to steel in the event of a fire. This coating has become a popular fire resistive option when steel structural members must be exposed and concrete and/or sprayed cementitious materials are not viable options.

Fig. 4–5. Type I classification requires steel structural beams are protected with some type of fire protection measure such as a spray-on coating.

This type of construction can be often found in conventional and modern mid- to high-rise buildings (commercial and residential), hospitals, schools, shopping centers, airports, and other similar types of large buildings that often are associated with public assembly. As an example, figure 4–6 illustrates a modern multistory office building that was constructed with a protected steel structural frame and noncombustible walls, floors, and roof structural members (although the roof is commonly covered with combustible materials).

Fig. 4–6. A modern multistory office building of Type I construction

Primary hazard. Compared to the other four classifications of NFPA buildings, Type I buildings are the most resistive to fire and collapse due to their masonry and protected steel structural members. The primary hazard is in the contents within the building. Therefore, the type of building—residential or commercial—will begin to assist the incident commander to establish initial priorities. Nevertheless, there are certain inherent attributes that will enhance and spread the extension of smoke, heat, and fire within these types of buildings.

Open areas. Open areas are the enemy of compartmentalization as they allow minimal restrictions to the extension of fire and its by-products. Normally, the type of building is a good indicator of a particular floor plan. As an example, hotels and other types of residential occupancies are partitioned for each unit, which will tend to confine fire as opposed to open areas as often found in some commercial (e.g., warehouse) occupancies. Additionally, older buildings tend to be more open than newer buildings, which often favor compartmentalization.

Concrete spalling and heat retention. Depending on the length and severity of fire, concrete can offer several significant challenges. First, concrete can retain a significant amount of heat and then radiate that heat back into the structure, making tenability difficult for attack and overhaul operations. Secondly, as there is some moisture in concrete, heat from a fire can cause the moisture to expand and rupture the concrete, and in some cases expose the rebar within the concrete causing a further weakening. This is noted in the case study of the Central Library fire in Los Angeles presented later in this chapter.

Heating, ventilation, and air conditioning systems (HVAC). Although heating, ventilation, and air conditioning systems are primarily responsible for keeping the interior of these buildings comfortable in various types of weather, they can also be a significant hazard as the ducting travels throughout numerous areas, including walls, multiple floors, and other similar areas. Once this ducting (which

can be quite large depending on the system) has been breached with fire and/or heat and smoke, the ducting can easily allow extension to other parts of a building that would otherwise not be an immediate concern. This is noted in the following case study of the First Interstate fire in Los Angeles. This is a primary reason why HVAC systems should be turned off in the initial stages of an incident in these buildings. Remember that some buildings such as motels and hotels can use individual air conditioners/heaters for each occupancy, and thus do not require a large, comprehensive HVAC system that travels throughout a building in order to maintain the interior environment. Common hallways and lobby areas, however, may be serviced by a single system.

Interstitial areas. For this discussion, interstitial areas (see fig. 7–12) are created by false floors and suspended ceilings that create voids above and/or below personnel, and greatly increase the potential extension of fire and its by-products. These areas are not often found in older buildings (unless they have been renovated), but are common in modern buildings. False floors are often found in commercial buildings that house electronic equipment and computers. A false floor is one that is constructed above the structural floor, usually to allow wires and cables to be out of sight. Obviously, these voids can enhance the extension of fire, and fires in this type of concealed space are capable of producing a significant amount of smoke from plastics (electrical insulation).

More familiar to firefighters are the ever present suspended ceilings that are used to create the desired ceiling height and an attractive appearance. This also creates a void above the visible ceiling that can hide HVAC ducting and wiring, and enhance the extension of fire and its by-products. This hazard can also be quickly magnified by the ceiling collapsing in a fire, exposing the wires that support the ceiling that can easily entangle fire suppression personnel.

Curtain wall construction. Curtain wall construction is a relatively new method of construction that allows the exterior of a building to be completed in a rapid manner and accomplishes two goals. One, the exterior of the building has a finished appearance, and two, the exterior now provides protection to workers on the interior of the building. Typically, the primary steel structural beams are erected, and if a Type I classification is desired, the beams are protected by one of the four previously mentioned methods (see fig. 4–4). Then, metal hangars are attached to the exterior of the steel structural beams. Lastly, glass, tile, aluminum panels, slate, concrete, or other similar decorative panels are attached to the metal hangars. As a result, there is a 3 in. to 4 in. gap between the metal hangars and steel structural members (unless sealed) that not only allows the upward travel of air within the building (assisting the HVAC system), but also allows fire and its by-products to extend vertically.

Center core/center hallway floor plans. To be able to evaluate potential access and egress routes for firefighters and occupants, it is important to know the primary floor plan of a building. Type I buildings normally fall into three categories: center hallway, center core, or a combination of center core/center hallway configurations.

Center hallway configurations have a stair shaft at either end of the building that is common to the hallways in the building (visible on the end of the multiple-story building in figure 4–6).

Conversely, center core configurations will place the elevators, stairs, and other necessary essentials in a central core in the middle of a building, with the hallways on each floor arranged around the central core (these are referred to as circuit hallways). Combination configurations are often found in larger buildings and can have a center core with stairways, and hallways that are common to stair shafts at the ends of a building.

Always be familiar with or able to quickly determine the best points for access and egress of personnel and occupants, as well as any vertical openings that can enhance the extension of fire and its by-products.

Shafts. Always be aware of any shafts such as elevator shafts, zero-clearance fireplaces, pipe alleys, poke-through construction, dumbwaiters, and trash chutes that can enhance extension of fire and its by-products. When evaluating vertical stair shafts, determine whether or not the shaft accesses the roof. Stair shafts that do not access a roof can be difficult to ventilate and keep clear of smoke and heat!

Forcible entry. Depending on the type of building, access throughout these buildings can be challenging. As an example, older buildings use wood as the primary material for doors, door jambs, and door frames that in most cases can be adequately forced with common forcible entry tools. Newer buildings routinely use metal doors in metal frames that can be more challenging with conventional forcible entry methods. Some of the most challenging buildings are commercial residential buildings (e.g., hotels and motels) that are compartmentalized, where each unit has a fire-rated door of heavy construction and each door is equipped with multiple locks. Unless a master key

CASE STUDIES: TYPE I BUILDINGS

FIRST INTERSTATE BUILDING—LOS ANGELES, CA

In 1988, first arriving companies found the 12th and 13th floors fully involved in the 62-story, First Interstate high-rise building, with fire extending upward at a rapid rate. The fire ultimately consumed floors 12 to 16, took over 3½ hours to control, and burned about 45 minutes per floor with temperatures estimated at 2,000°F. After the fire was extinguished, the following factors were noted (in no particular order):

- The structural steel had been coated with a vermiculite type material, which was successful in protecting the coated steel. After the fire, the interior of the building was cleared of all debris, repaired, and the exterior aluminum decorative material replaced. The building then reopened for business.

- Fire had extended up a portion of the HVAC system and started a fire in a storeroom on the 27th floor. Fortunately, the fire self-extinguished due to a lack of oxygen.

- Fire extended vertically from five different avenues: (1) lapping (autoexposure), (2) curtain wall construction, (3) HVAC system, (4) poke-through construction, and (5) radiated heat through the floors.

- Typical with fires in multistory buildings, the higher the fire above the ground, the more difficult and time intensive it will be to place sufficient resources at the floor(s) of involvement for extinguishment and search operations.

CENTRAL LIBRARY—LOS ANGELES, CA

In 1986, first arriving companies to the Central Library building in downtown Los Angeles initially found only a very slight odor of smoke. Further investigation found a well-developed fire in a room that was in the process of being extended within the building. The building was built in 1926 and was constructed of reinforced concrete walls, floors, and roof. The fire burned for over 7½ hours and was responsible for an immense amount of damage to the contents of the library. After the fire was extinguished, the following factors were noted (in no particular order):

- Due to the length of the fire, there was significant spalling to the concrete that was in direct contact with fire. Although collapse of the building was not a consideration, the significant damage to the concrete did expose the metal rebar.

is available, forcing entry into each unit will at best be time consuming and difficult.

Ventilation. Vertical ventilation in multistory buildings can be demanding unless the fire is on the top floor. Horizontal ventilation, however, can be simplified in multistory buildings by using windows. In older buildings, windows are often openable and made with plate or annealed glass, which can be easily broken. In newer buildings, often referred to as sealed buildings, windows are not openable (natural ventilation is replaced by modern HVAC systems) and are normally tempered glass, which can be difficult to break unless specific methods are used.

Identification. Type I buildings can be a benefit to fireground personnel due to the fire resistive properties that are inherent to this type of building. However, unless responding personnel are familiar with a particular building, it will be virtually impossible to determine from the street if a building that appears to be a Type I or II incorporates protection for its structural members.

- Due to the radiated heat and lack of sufficient ventilation within the building, temperatures were estimated to have exceeded 2,000°F. This resulted in lengthy and cumbersome attack operations and also required the replacement of attack crews every 15 minutes for rehab considerations.
- As there were few windows in the building, vertical ventilation consisted of using jackhammers to open holes in the concrete roof. This was a slow and arduous process but was somewhat successful.
- A fire that should have been extinguished in 2 to 3 hours took over 7½ hours and required a considerable amount of resources!

RAILROAD FREIGHT WAREHOUSE—FORT WORTH, TX

On April 14, 1992, resources from the Forth Worth Fire Department responded to a structure fire in the vacant Texas & Pacific Railroad freight warehouse that resulted in five alarms being utilized to extinguish the fire. The eight-story building with a basement covered over three city blocks, was constructed in 1905, and was constructed of reinforced slab floors supported by concrete columns and masonry nonbearing exterior walls. The following is a summary of the lessons learned or reinforced:

- The large fire was primarily a result of burning debris that had accumulated in the building. This fire never posed a significant risk of structural collapse despite the heavy fire conditions on arrival. The greater danger to firefighters came from spalling concrete and radiated heat.
- When buildings become unsecured, all floors must be searched during a fire for possible trapped occupants. Search and rescue operations were time consuming due to the size of the structure.
- Open elevator doors on different floors posed a severe threat to firefighting and rescue operations, particularly in minimal visibility.
- This fire clearly demonstrated the importance of protecting vertical openings in a multistory building. On the sixth floor and especially on the eighth floor, where elevator doors were open, heavy smoke quickly entered these floors and made conditions untenable for any occupants. Also, on the eighth floor, an open door to a stairwell allowed smoke to enter the stairwell and make conditions in the upper portions of the stairwell untenable.

Quick summary

- Firefighters must be familiar with the buildings in their area of responsibility in order to achieve a relevant building size-up at structural incidents.

- The ability to read a building can be enhanced by a working knowledge of NFPA 220 and the era/use/size/type classification methods.

- In this chapter, the five building classifications of NFPA 220 that are summarized are Type I, II, III, IV, and V. Era, use, type, and size are covered in chapter 5.

- Type I construction is defined as fire resistive construction and is the most fire resistive type of construction.

- Type I consists of protected steel frame and noncombustible walls, floors, and roof structural members.

- The primary hazard is the interior contents.

- Common hazards are open areas, spalling of concrete, HVAC systems, interstitial areas, center core/center hallway floor plans, interior shafts, and curtain wall construction.

- Similar to Type II buildings, it can be difficult to identify the presence of protected structural members in a Type I building.

Type II (222, 111, or 000): Noncombustible construction

Shall be those types in which the fire walls, structural elements, walls, arches, floors, and roofs are of approved noncombustible or limited-combustible materials.

As a starting point for this classification, notice this NFPA definition is the same as the Type I NFPA definition except the fire resistance requirements are far less stringent as evidenced by the lower numbers highlighting the definition. This is a direct result of the steel structural components being unprotected. As opposed to Type I construction, where protected steel can be a benefit during a fire, unprotected steel in Type II buildings can be a significant detriment during a fire, as unprotected steel that is exposed to heat can soften, expand, and fail even though the steel members can be quite large. From a structural perspective, this correlates to faster failure rates for unprotected structural components when they are exposed to fire and/or heat as compared to protected steel components.

Typically, these buildings are characterized by (1) concrete tilt-up slabs or concrete block walls that support an unprotected metal roof structure, (2) unprotected steel frame components that are enclosed by concrete tilt-up slabs or block walls, (3) an unprotected steel frame that is enclosed or surrounded by metal exterior walls, and (4) other similar variations. The steel structural components are characterized by unprotected steel beams and girders that also typically support unprotected open web bar joists that are covered by an unprotected metal deck built-up roof (fig. 4–7). The roof may be covered with combustible materials such as tar and felt/tar paper. Although the unprotected steel structural components are a significant hazard during a fire, the combustible materials that comprise the roof in combination with the contents within the building are the principal hazards.

Fig. 4–7. Type II construction is characterized by unprotected steel beams and girders supporting an unprotected metal deck built-up roof.

As Type I and Type II buildings can be very similar except for the presence or absence of protection for primary steel structural members and roof systems, the previous hazards delineated under Type I buildings can also apply to Type II buildings. However, there are several noteworthy hazards that should be considered when confronted by Type II buildings.

Steel. Although unprotected steel is a strong structural material that has widespread use by the building industry for its ability to support a significant amount of weight, it also has a consistently poor record in fires as it can twist, sag, and/or collapse when exposed to heat.

Wall collapse. Typically, exterior walls fall into three general classifications: unreinforced masonry, reinforced masonry, and tilt-up or precast.

- Unreinforced masonry (URM) construction that is comprised only of bricks and lime and sand mortar is not a sound structural type of construction. As delineated in following chapters, this type of construction is no longer used and for good reason, as it is prone to collapse in a fire. Of the three types of exterior walls in this discussion, URM construction is the most susceptible to collapse from expanding and failing steel structural members, and can easily collapse outward twice its height.

- Modern masonry construction is not as willing a candidate for inward or outward collapse as URM. Modern masonry is comprised of either concrete cinder block (reinforced and with cells filled with concrete) construction or post-1935 brick construction that utilizes rebar, steel, and Portland cement as mortar.

- Tilt-up concrete slab or precast wall construction that is comprised of solid concrete slabs can present an interesting dilemma and is summarized by two perspectives. The first perspective is that concrete is strong and can resist the effects of fire longer than most other building materials, and therefore is used in exterior walls for its strength and durability. The second perspective is that when concrete is used in a tilt-up or precast configuration, it can present a considerable barrier to expanding steel structural members. Therefore, when steel structural members are embedded into the concrete and are exposed to heat from a fire, they expand outward, forcing the concrete slab(s) outward. Although collapse of concrete tilt-up/precast wall construction is not a common occurrence, it has happened as discussed in the following case studies.

Contents. The primary hazard to the unprotected steel in these buildings is the contents within the building and their proximity to the steel structural members (fig. 4–8). It is no great revelation to modern firefighters that the contents within modern buildings normally consist of synthetic materials (petrochemical-based compounds) that burn hotter and faster than the conventional materials of yesterday (cellulose-based materials). An initial concern should be the type of occupancy and/or the type of content it houses.

Fig. 4–8. Building contents that are in close proximity to steel structural members can create a significant hazard to unprotected structural members.

CASE STUDIES: TYPE II BUILDINGS

HYDRA-MATIC FIRE—LIVONIA, MI

Although there have been many fires in Type II buildings that have been primarily lost as a result of a rapidly extending overhead fire in a metal deck built-up roof, the most noteworthy is the legendary Hydra-Matic fire in Livonia, Michigan that completely destroyed the plant that produced Hydra-Matic transmissions for General Motors. To this day, the fire is still characterized as the worst industrial fire in American history, and the worst dollar loss in the history of the auto industry. In 1953, a small fire started outside the plant in a flammable rust inhibitor that quickly spread to the interior and the roof with devastating speed and disastrous results. Once the fire extended to the combustible roof, it quickly traveled across the roof in all directions and also enhanced the interior fire, engulfing the entire 1.5 million sq ft plant in minutes. Numerous fire departments were called to the fire, but to no avail as the combination of the interior and exterior fire were too much for resources. The primary hazards with this type of roof are the ability of fire to travel and the possibility of collapse. Fire can readily extend between the corrugated metal decking and the insulation/tar paper layers, and the metal bar joist roof is susceptible to weakening and collapse from the roof fire.

MGM GRAND FIRE—LAS VEGAS STRIP, NV

In 1980, a fire at the MGM Grand Hotel and Casino in Las Vegas, a 26-story luxury resort, was responsible for the deaths of 85 people, primarily from smoke inhalation. An electrical fire in a restaurant on the second floor started at 7 a.m. and quickly spread throughout the second floor. However, the primary hazard was the extension of smoke to upper floors. Shafts such as elevators, stairwells, seismic joints, and faulty smoke dampers within the HVAC system and associated ducting allowed the smoke to rapidly extend to the top floor of the building. Interestingly, only one person died from burns from the fire, the rest

A residential or commercial residential occupancy will have a reduced fire load and more compartmentalization as compared to a commercial business type occupancy that will likely have more open spaces and a higher fire load. Remember that the name on the exterior of a commercial building is a great indicator of the contents within a building. Moreover, always try to visualize the distance from the grade floor to the ceiling of the building, as higher ceilings/roofs will be farther away from the burnable contents (also consider the increase of contents during special holidays such as Christmas, etc.). Lastly, remember that the type of contents, their flammability, and their distance from the steel structural components and supports also applies to racked or tiered storage, which can be a significant hazard to attack personnel and overhaul operations!

Extension of fire and its by-products. Although the extension of fire and its by-products is always a consideration in single-story buildings, it should be a primary consideration in multistory buildings, particularly residential and commercial residential buildings. Although fire and heat can weaken and cause collapse of unprotected structural members, it is smoke that is predominantly responsible for the deaths of occupants within these buildings. Remember from the previous discussion in Type I buildings that extension of smoke is enhanced by poke-through construction, vertical shafts, curtain wall construction, and HVAC systems. Some examples of this dilemma are the fires that have occurred in the numerous high-rise hotels in Las Vegas and are detailed in these case studies.

were from smoke inhalation, many of them in their sleep. There have been other similar fires in Las Vegas hotels (e.g., Sahara fire—1964, Caesars Palace fire—1981, Hilton fire—1981, and Bellagio fire—2008) that were also complicated by the vertical extension of smoke. This illustrates the importance of evaluating the upward extension of smoke and the status of HVAC systems.

VACANT WAREHOUSE FIRE—CLARK COUNTY, NV

In 2007, the Clark County Fire Department (NV) responded to "smoke showing" from a large vacant warehouse building. While initial resources were involved in forcible entry operations, the fire quickly accelerated to the point that fire was now showing from the roof. About this time, a firefighter noticed a bow in the north wall that resulted in an immediate withdrawal of personnel away from the building. Within seconds the roof collapsed, pulling the north wall into the building (this collapse happened less than 15 minutes from the time of the alarm). Seconds later, part of the west wall collapsed outward.

The following is an overview of lessons learned as reported by the Clark County Fire Department.

Tilt-slab construction is dangerous in our view, because it relies on intact building components for stability. Generally speaking, wall sections that are attached to the roofing systems usually fall inward as the roof system fails and the trusses pull the sections in. Walls that run parallel or are not attached to the roofing system are a 50/50 guess at best. Never expect these wall panels to fall inward! Three connector plates, all of which failed, attached each of this structure's parallel wall sections to the slab. Three similar connector plates on each side connected each wall section to its adjoining sections. Several of these connections were pulled completely from the point in which they were embedded in the wall system. Any fire in a tilt-up building presents a collapse potential from the onset. This building collapsed in less than 15 minutes, and it involved minimal live-fire loading. The fuel was primarily the building structure in the form of the roofing system, which consisted of wood composite I-beams, sheathing, and general roof-finishing materials.

Quick summary

- Type II construction is defined as noncombustible construction and is a direct result of the steel structural components being unprotected.

- The fire resistance requirements for Type II construction are far less stringent than Type I construction.

- Type I hazards also apply to Type II construction, except this construction will collapse at a faster rate due to unprotected structural members.

- The primary hazards are the presence of unprotected steel structural components and the interior contents.

- Similar to Type I buildings, it can be difficult to determine the presence of protected or unprotected structural members from the exterior.

Type III (211 or 200): Ordinary construction

Shall be that type in which exterior walls and structural elements that are portions of exterior walls are of approved noncombustible or limited-combustible materials, and in which fire walls, interior structural elements, walls, arches, floors, and roofs are entirely or partially of wood of smaller dimensions than required for Type IV construction or are of approved noncombustible, limited-combustible, or other approved combustible materials.

Type III buildings typically consist of exterior walls (load-bearing) that are constructed of noncombustible materials such as concrete block, brick, or a combination of steel studs with an exterior brick veneer. The interior is constructed from combustible wood components (i.e., walls, floors, ceilings) and supporting structural members. A common reference to this type of construction is *Main Street USA* and *brick and joist construction*, which tends to denote a wide variety of older construction such as taxpayer type buildings (fig. 4–9) and multistory residential (hotel) buildings.

Fig. 4–9. Taxpayer type buildings are representative of Type III construction.

However, Type III buildings are also being constructed today, such as the modern strip mall or mini-mall. Although these buildings typically have concrete floors, they also can have concrete block exterior walls (noncombustible) and interior wood walls (combustible), and trussed roof assemblies that also act as ceiling structural members (combustible). Because there are a wide variety of buildings that can fall into this classification, let's consider some structural considerations that are common in Type III buildings.

Construction. When considering the numerous types and methods of construction that have been utilized in these buildings, it must be remembered that they have been built from the 1800s to the present day. Therefore, the era of a building can often give important clues to the type of interior construction. From a simplistic perspective, construction methods that were employed up until the 1940s often used solid concrete exterior walls and a heavy and/or substantial size and grade of lumber for the interior of the building. Exceptions to this generalization are the URM that was used for exterior walls until the mid-1930s and renovations that typically used lightweight components to alter interior floor plans.

Type III buildings constructed after the 1940s and 1950s typically used cinderblock walls, brick masonry that is not classified as URM, and other forms of substantial masonry construction. However, as the age of these buildings decrease (closer to modern times), so does the size and the length of time the interior structural components will last when exposed to fire. In summary, geometry and glue has often replaced size and nails/bolts/steel plates and conventional lumber.

Vertical extension. Although horizontal extension is a real concern in any type of structure, particularly structures with minimal compartmentalization, vertical extension of fire and its by-products is a primary problem in multistory buildings, especially Type III commercial residential buildings (fig. 4–10). Vertical extension has proven to be a major contributor to the loss of life that has

typically been remote from the point of origin of a fire.

Fig. 4–10. Vertical extension can be a significant problem in multistory commercial residential buildings.

As an example, the National Fire Protection Association (NFPA) has documented several dozen hotel fires that date from the 1930s to the 1980s that have been responsible for over 10 deaths per incident. Hotel fires such as the Kerns Hotel (32 deaths), La Salle Hotel (61), Canfield Hotel (19), Winecoff Hotel (119), Hotel Roosevelt (22), Ozark Hotel (20), the Ponet Hotel (19), and others have clearly demonstrated the dangers of vertical extension via open stairways to upper floors in these buildings. The same is true in other types of buildings that incorporate open stairways. This is best illustrated by the following summaries:

- "The fire quickly spread through the highly varnished wood paneling in the lounge and mezzanine balcony before ascending stairwells and shafts." —La Salle Hotel

- "Contributing factors that added to the severity of this incident were delayed alarms, open stairwells, and the presence of combustible materials." —Canfield Hotel

- "The fire started in the lobby and spread through two stairways and all the halls, killing 20 people. The cause of death was smoke inhalation." —Ozark Hotel

- "The hotel suffered a fire which started on the sixth floor and spread rapidly through the hallways and staircases, killing 28 people. Although the hotel was supposed to be fireproof, synthetic carpeting, painted doors and frames and open stairways fueled the spread of fire." —Pioneer International Hotel

In these multistory buildings (most of which were Type III buildings), fires were escalated by vertical extension via shafts, which can be summarized as follows:

- Stairways
- Elevator shafts
- Pipe chases and dumbwaiters
- Light shafts
- Seismic joints

Cocklofts, attics, truss lofts. These three terms are also defined in chapter 8, but we introduce their definitions here for clarity.

- **Cockloft.** A small space that is created when a roof is raised above the level of ceiling joists and rafters to provide a pitch for drainage. Cocklofts are common in older buildings (1800s and early 1900s) and are normally of conventional construction.

- **Attic.** A large space that is created by a steep pitched roof (arch, gable, etc.) for drainage and/or appearance. Depending on the type of construction, attics are normally large enough for storage, can be modified for additional living space, and can contain HVAC equipment.

- **Truss loft.** An attic space created by the open web nature of trusses. The voids that are located between roof rafters and ceiling joists are nothing more than an open area that can allow fire to easily extend in numerous directions.

An advantage of attics and cocklofts that are of older conventional construction is that the size of the structural members (rafters and joists) will resist the effects of fire for longer periods of time as compared to newer lightweight truss

construction. Conversely, when truss lofts are created by modern lightweight construction that is either secured by gang-nail plates or glue, rapid collapse can be expected when these structural members are exposed to fire and/or sufficient heat. In either case, the presence of division walls can effectively compartmentalize these areas if they have not been breached. It should be noted that older division walls stand a greater chance of not being secure due to renovations or modifications made by plumbers, electricians, or other tradespeople. However, newer division walls can also be unsecured for the same reasons.

The preceding three voids can also present a noteworthy concern when they are common to multiple occupancies such as mini-malls, row houses, and so on. Fire that extends into a common attic can easily run the entire attic and expose other occupancies (that are common to the attic) unless fireground personnel pull ceilings ahead of an extending fire and extinguish the fire back to its source point of origin.

Alterations and concealed spaces/voids. Alterations are commonly used to change and/or improve the appearance of the exterior of a building and/or the existing floor plan, particularly in older buildings. However, these modernization techniques can increase the potential of concealed spaces—voids that were not originally present and can also significantly weaken a portion of the original design of a particular building. Examples of alterations include the following:

- If the exterior appearance has been changed (i.e., updated), assume the interior has also been modified.
- If the type of occupancy has been changed (e.g., a commercial warehouse converted into condominiums), expect the interior to be dramatically different and will likely employ lightweight construction.
- If an older building has been enlarged/added on to, expect the use of lightweight construction and nonstandard floor plans.

Remember that alterations typically result in voids and concealed spaces that were not a part of the original building and design criteria. Voids can be created by the following examples:

- Construction designs such as knee walls, like those created when an attic is framed for a room (see fig. 9–28)
- Suspended/dropped ceilings that create a void (often quite large) between the ceiling and the floor/roof above
- Voids created by pipe chases
- Zero-clearance fireplaces
- Soffits

A good example of a pipe chase is one created by *stacked* kitchens and baths. (In multistory buildings, these areas are commonly stacked on top of each other to minimize plumbing considerations.)

Parapet walls. A parapet wall is nothing more than a continuation of a wall above a roofline, and can extend from 1 ft to 8 ft, depending on the occupancy and type of wall. Hazards associated with these type of walls include the following examples.

- Buildings constructed with concrete tilt-up slabs, which are normally higher than the roofline. Although stability of the parapet wall is not a concern, these parapet walls normally encircle the entire building. This is capable of hiding the roofline below the parapet wall (which is necessary to know if personnel are going to the roof) and hiding the type of roof (flat, arch, etc.). In this case, the presence of scuppers will indicate the hidden roofline (see fig. 9–21).
- Modern masonry walls can be constructed from cinderblock/concrete block construction or modern brick construction that incorporates similar concerns as concrete tilt-up slabs. Stability of this type of construction is normally not a concern.
- Unreinforced masonry construction is markedly different from the two previous examples and can present a noteworthy hazard to fireground personnel. The lack of sufficient Portland cement

and rebar plus the deterioration of mortar due to age combine to result in a weak freestanding wall above the roofline that is either an extension of the wall it is on, or is supported by a steel I-beam that is embedded in the wall below the parapet wall. Depending on the age of the building, this type of wall is often capped by coping stones that have lost their adhesive connection through age and weather and are only attached by gravity (see fig. 9–20). These conditions present an extension of an exterior wall (normally the front and side walls) that can readily collapse during fire conditions and/or when struck by heavy streams.

Cornices. Cornices can be categorized by old and new construction (cornices are also described in chapter 9). In new construction, cornices are often nothing more than rigid foam that is glued to a building and plastered over to give an acceptably pleasing appearance at a moderate cost (see fig. 9–19). However, two major problems should be considered:

- They will not support the weight of personnel and aerial devices.
- They can be flammable, as graphically demonstrated in the Monte Carlo Hotel fire in Las Vegas in 2008. At this incident, a fire started on the roof and quickly extended to the foam cornice around the top of the hotel. Due to its flammability, the fire quickly grew in intensity and then extended into the top floor, creating a major fire.

Cornices constructed on older buildings are dramatically different in that they are often an integral part of the construction and made from substantial materials. For example, some cornices made from stone (also known as stone corbelling) rest atop the load-bearing wall or parapet using mortar for a solid connection. Some wood or metal cornices were made by creating "lets" in the exterior wall, inserting short but visible supports for the cornice, and extending roof boards over the exterior wall and false visible supports (fig. 4–11). This type of cornice often used tin to cover the underside of the roof boards for a finished appearance.

Fig. 4–11. Many cornices are not of substantial construction and can create a collapse hazard to fireground personnel.

Age and weather is an enemy to both stone and wood/metal cornices in that it can loosen the stone and degrade the wood or metal. Additionally, the wood cornice can conceal a void area behind or above it that allows the extension of fire to other areas, either on the exterior or to the interior of the building. Any type of cornice should be considered an unstable component of a building and should not be used to support the weight of an aerial device or personnel.

Previous fires. A subtle but not a readily visible consideration is the possible damage done by previous fires, particularly in older buildings, that resulted in repairs of less integrity than the original construction. Although this consideration also applies to floors and walls, roofs are a common example where a fire has destroyed conventional roof rafters of 2×6 in. or larger and then the burned area has been replaced by smaller members or lightweight construction.

As an example, look at figures 4–12 and 4–13. In figure 4–12, the older unreinforced masonry buildings on the left and right originally had bowstring roofs and looked virtually identical from Side A. However, a fire in the left building that destroyed the

original arched roof resulted in a replacement roof of lightweight panelized construction. The lightweight construction on the left looks quite different from the arched roof on the right building. In figure 4–13, notice the different patches of composition roofing that cover lightweight construction used to replace the older conventional construction. Fireground personnel should always be on the lookout for similar examples that have reduced the integrity of the original construction and, when visible, notify appropriate personnel in a timely manner.

Fig. 4–13. Visible modifications to a roof are an indicator of repairs that may not be to the same standard or strength of the original roof.

Facades. Like cornices, facades may conceal voids that can easily allow the extension of fire. For detailed information about facades, see chapter 9.

Fig. 4–12. A style of roof that is not appropriate for the age of a building can be an indicator of a replacement roof.

CASE STUDY: TYPE III BUILDING

CUGEES RESTAURANT—LOS ANGELES, CA

On January 28, 1981, at 3:33 a.m., the Los Angeles Fire Department responded to Cugees restaurant with smoke and fire visible in the interior of the building. The building was constructed in 1935, and had an unreinforced masonry exterior with a conventional wood roof covered with multiple layers of composition roofing. Attack crews entered the building and a four-person roof ventilation team began to vent the roof. While roof ventilation operations were underway, the roof began to collapse and resulted in injury to three firefighters and the death of one firefighter. The cause of the collapsing roof was attributed to the roof construction, which was comprised of wooden beams nailed to wooden shingles that were wedged into cavities (lets) in the masonry walls. This type of construction was further compromised by numerous layers of composition on the roof that was estimated to have provided a significant amount of weight that the original roof was not designed for and was instrumental in an early collapse of the roof. Following the collapse of the roof, the masonry walls of the building also collapsed.

Quick summary

- Type III construction is defined as ordinary construction.
- In this type of construction, buildings have noncombustible exterior walls and the interior is constructed from combustible wood components.
- Taxpayer and mini-mall buildings are common examples of Type III construction.
- The primary hazard is the combustible interior structural components and contents.
- Common hazards associated with these buildings are a combination of old and/or new construction, vertical extension, cocklofts/attics/truss lofts/common attics, alterations and resultant voids, parapet walls, cornices, facades, and remodels after previous fires.

Type IV (2HH): Heavy timber/mill construction

Shall be that type in which fire walls, exterior walls, and interior bearing walls and structural elements that are portions of such walls are of approved noncombustible or limited-combustible materials. Other interior structural elements, arches, floors and roofs shall be of solid or laminated wood without concealed spaces.

Type IV buildings normally consist of exterior masonry walls that surround the heavy wood timbers and construction materials that comprise the interior. From another perspective—*a masonry exterior that encloses an interior lumber yard.* This is a type of construction that was popular in the late 1800s and early 1900s and primarily used for large factories, warehouses (fig. 4–14), churches, lodges, and other similar buildings. These buildings are no longer being built, but can be found in communities large and small. What sets these buildings apart from other types of large buildings is the size of the interior wooden structural members, which is no longer cost effective.

Fig. 4–14. Type IV construction is no longer viable due to the scarcity and cost of lumber required for manufacture.

The American Institute of Timber Construction defines heavy timber construction as a type of construction from which fire resistance is attained by placing limitations on the minimum size of all load-carrying wood members, by avoiding concealed spaces under floors or roofs, and by applying the required degree of fire resistance in exterior and interior walls. These requirements are summarized as follows:

- Wood columns shall be not less than 8 in. thick in any dimension when supporting floor loads.
- Wood columns shall be not less than 6 in. wide and 8 in. deep when supporting roof and ceiling loads.
- Floor beams and girders shall not be less than 6 in. wide and not less than 10 in. deep.
- Floors shall be of tongue-and-groove (T&G) planks not less than 3 in. thick and covered by 1 in. T&G flooring.
- Roof decks shall be no less than 2 in. T&G planks.
- Load-bearing exterior and interior walls shall have a fire resistance rating of not less than 2 hours.
- Floors and decks shall be without concealed spaces, except that building service equipment may be enclosed provided the spaces between the equipment and enclosures are fire-stopped or protected by other acceptable means.

One historical note—heavy timber and mill buildings are often classified together. A true mill building may not meet the requirements above, mainly because it was constructed prior to the establishment of the aforementioned standard. Granted, mill buildings typically used heavy timbers to form a post and beam interior surrounded by exterior load-bearing masonry or stacked stone. The difference is in the dimensional requirements that may or may not be met. With the preceding overview of the minimum standards for this type of construction, it is obvious that these standards are significantly different from modern lightweight construction with gang-nail and/or glued connection points. However, as heavy as the requirement for the structural members in Type IV buildings are and their resultant record of excellent fire resistance, these buildings are capable of multiple significant hazards, which are summarized below.

Renovations. Because of the size of these buildings, their condition, in some cases their location (downtown areas), and heavy construction, they have in many cases become either a candidate for a vacant building (fig. 4–14) that is slowly succumbing to the elements, or a prime candidate for renovations that go beyond their original design criteria. Although a vacant building will have its own set of hazards (see the railroad freight warehouse in the case studies of Type I buildings), renovations present a new set of concerns and hazards as follows:

- Remember that one of the criteria for Type IV buildings is a lack of concealed spaces. With renovations, numerous types of concealed spaces are a result of a modification of the original building. As an example, HVAC systems, suspended ceilings, soffits that conceal plumbing/electrical considerations, and zero-clearance fireplaces for residential units are just a sampling of the construction variations that can conceal and enhance the extension of fire and its by-products.

- The size, inherent structural stability, and often the location of these buildings make conversions to a modern factory, multifamily residential units (condos, apartments, etc.), offices, self-storage units, and even schools an ideal method to transform yesterday's buildings to modern buildings that are not only attractive but an integral part of a community's economy. A good example of a practical renovation of a heavy timber building is the conversion to condominiums when these buildings are located in downtown areas. This affords occupants of these buildings the opportunity to be closer to areas of employment in the downtown area instead of living outside a city and having to commute to work. Another example is the Ford building in downtown Portland, Oregon (fig. 4–15), built in 1915 and used to assemble Model T automobiles, which has been converted to contain small, upscale businesses.

Fig. 4–15. This older heavy timber/mill building has been renovated into an upscale office building. (Photo by Jim Forquer.)

Openings, shafts, and doors. Because these buildings were originally constructed and used for manufacturing purposes, they typically have large open areas with minimal compartmentalization. Although they were constructed with a lack of concealed spaces, openings between the floors for conveyor belts, freight elevator shafts, and to move product are common and available to enhance the spread of fire and its by-products. In some cases, division walls were utilized to separate portions of the open areas and are fitted with fire doors that are activated by fusible links. However, the age and atmospheric contamination of these links has in

most cases rendered them ineffective, which leaves openings in division walls. As a result, extension avenues in these buildings can be numerous. It should also be noted that most of these buildings were constructed prior to sprinkler requirements, so unless sprinklers have been retrofitted, this protection is absent.

Size of fire and collapse areas. Due to the size of the wooden timber construction, a large fire load is present. Although there is a large surface mass for heat to ignite, which does take a noteworthy amount of time (particularly compared to modern construction), these components will burn and create a large hot fire that requires large-bore streams to extinguish. An additional factor that enhances the ignition and spread of fire is the common presence of floor planks that are soaked with petroleum-based compounds from past manufacturing operations. Additionally, many floor beams are designed with fire cuts that allow the beams to pull out of the walls if compromised by fire. In this case, the floors will collapse and leave the walls freestanding, which is also a recipe for collapse. If a fire becomes well advanced, it is difficult to extinguish and normally destroys the entire building, resulting in a massive fire that is hot enough to keep resources away from the building in a defensive mode. In most cases, defensive operations will result in a collapse of the entire building and radiated heat that can project hundreds of feet. It is important that incident commanders position apparatus well away from the potential collapse zone and consider exposure protection if necessary.

CASE STUDIES: TYPE IV BUILDINGS

LATTER-DAY SAINTS TABERNACLE—PROVO, UT

In December 2010, resources responded to a major fire in Provo, Utah, in the Latter-day Saints Tabernacle. The large tabernacle was constructed in the 1880s and utilized URM exterior walls with rough-cut heavy timber construction of up to 10 in. and rough-cut 2 × 12 in. planks held together with iron bolts. The roof decking was comprised of 1 × 8 in. planking, and the floor was 1 in. T&G planks supported by 2 × 12 in. rough-cut floor joists. Numerous renovations were completed between the 1980s and 1990s. At 2:44 a.m., initial companies found visible smoke extending from the structure and soon determined an interior fire had made considerable progress. Interior operations were initiated but it was soon evident that a defensive operation was required as interior operations were unsuccessful. At 3:28 a.m., despite a 4,000 gpm water flow, the roof structure began to collapse, and by 6 a.m., the entire roof had collapsed into the building. The fire was finally declared out at 5:30 p.m.

WOONSOCKET MILL—WOONSOCKET, RI

On June 7, 2011, resources responded to the Woonsocket mill in Woonsocket, Rhode Island. The four-story, 122-year-old building of heavy mill-timber construction had been vacant since 2009 and was formerly used as a factory to produce rubber products and, later, Keds sneakers. The mill was built in 1889 and was more than 217,000 sq ft in size, and was also classified as a historic Woonsocket landmark. Initial companies found a small fire in the mill, but the fire quickly spread to engulf the entire structure, resulting in a tremendous blaze that was responsible for eight alarms and 10 to 15 departments from Rhode Island and Massachusetts being called in to help fight the fire. As the mill was located in an area with no exposure considerations, the size of the fire resulted in a defensive strategy and the ultimate loss of the mill. The cause of the fire was determined to be a welding torch.

Incident scope. As some of these buildings can cover an entire block or can be interconnected to other similar large buildings (for example, by overhead walkways, etc.), the potential of a large conflagration that is capable of exceeding the resources of most fire departments is a reality. Therefore, incident commanders (ICs) must think outside the box and consider such extreme considerations as mutual aid contingencies, water supplies, collapse zones, protection of exposures when radiated heat can make this consideration complex, incident duration, rehabilitation of resources, the possibility of medical challenges, and other similar considerations. Remember that municipalities with these types of structures are candidates for an incident that have the potential to tax even those departments with significant resources. Additionally, just because a building is over 100 years old and has never suffered a fire does not mean that it should ever be considered impervious or ignored.

Quick summary

- Type IV construction generally consists of older buildings made with exterior masonry walls that surround the heavy wood timbers and construction materials that comprise the interior.
- Heavy timber and mill buildings are normally classified together due to their similarities.
- Although these types of buildings have an excellent record of fire resistance, they are capable of large conflagration type fires (depending on the size of a building).
- Common hazards are renovations, openings/shafts/doors, size of fire and resultant exposures, collapse areas, and the scope of the incident involving this type of fire.

Type V (111 or 000): Wood frame construction

Shall be that type in which structural elements, walls, arches, floors, and roofs are entirely or partially of wood or other approved material.

Of all of the preceding building classifications that we have been considering, Type V buildings are those most responsible for the majority of death and injuries to firefighters and civilians alike in this country, and for good reason. The primary components in these buildings, whether it be the walls, floors, or roof, are constructed of flammable wood, plus there are more single-family dwellings than any other type of building. Unfortunately, this dilemma is further complicated by the fact that these buildings (residential and moderate size commercial buildings) have been constructed all the way from the late 1700s up to the present, with a wide variety of construction styles and methods (fig. 4–16).

Fig. 4–16. Wood frame buildings have been constructed from the 1700s to the present and are the most prevalent type of building in this country.

Renovations and various materials that have been used over the years can be a challenge to any firefighter who attempts to read these buildings from the exterior in a minimal amount of time. Additionally, remember that the longer these buildings burn, the weaker they become and thus succumb to gravity. Although the preceding sentence applies to all buildings, it particularly

applies to Type V buildings because all structural components will burn—adding to fire spread severity and collapse potential.

There are many types of all-wood buildings that may or may not meet the requirements of NFPA 220. They include the following:

- Stacked log
- Post and beam
- Balloon frame
- Western platform (conventional, legacy, and engineered wood)

Each of these wood buildings has unique characteristics that ultimately influence fire spread and collapse potentials. Let's briefly describe the differences of each and list some tactical considerations for them.

Stacked log buildings are likely the oldest Type V buildings found in America as they date back to the early Pilgrim days with the timber-rich forests found in the new land. Log buildings use solid logs of various sizes that are stacked and often interlocked together at their ends, forming a strong bond (fig. 4–17). Originally, builders would interlock intersecting wall logs by cutting a notch toward the end of the log, then alternate the two wall stacks. This approach created a very stiff building—but a drafty one. Chinking (a paste-like filler) is used to fill the random voids between stacked logs.

Fig. 4–17. A log frame building can be a very strong type of building.

As the log building evolved, builders started to plane the top and bottom radius of the round log so that horizontally laid logs would sit flat atop the others and eliminate the need for chinking. The ends are still notched using various types of cuts so that the intersecting walls still interlock. Not surprisingly, due to the size of the logs and construction, they are the strongest wood construction—primarily due to the fact that a log building with interlocking logs creates a shell that can support itself and a roof. Although the log walls will withstand the effects of fire for a significant time frame and the structural integrity of the roof under fire conditions can be quite noteworthy, it is dependent on the type of construction that has been used. Older log buildings typically utilized large diameter logs or heavy timbers to form the roof rafters/ridge beam to support roof decking of some sort. Like the walls, this is a tough roof that can resist collapse (the decking is the weak link). The heavy timber or log rafters typically sag before collapse, providing a small window of warning to firefighters.

Newer log buildings are likely to have an engineered roof that includes trusses. A true log truss that uses steel plates and bolts for assembly is still a pretty tough roof. Unfortunately, some log building roofs have the appearance of being true logs when in fact they are lightweight roofs (wood or steel truss) that are faced with a fake log covering. Unless you were there during construction, it is difficult to tell the difference. One visual clue might help—if you see what appear to be log rafter tails extend past the exterior walls only at the corners or only on the gable ends, and the eave soffits are small and finished like a typical wood frame, treat it as an engineered lightweight roof. A true log trussed roof should include multiple rafter tails along the wall length and have tall or deep finished soffits.

The log exterior is normally stained with a protective coating, and the interior can also be stained with a protective coating or covered with drywall or wood paneling to produce a smooth surface. Normally, owner/occupants who want the

look of log construction do not want the appearance of log walls with visible roof rafters that are constructed from lightweight construction. Log construction is more commonly utilized for specific applications such as homes in mountainous areas, restaurants, and other structures that are looking for a unique look, albeit at a high cost in modern times.

Fire spread concerns: Although the exterior log walls and heavy roof members present a heavy fire load, they do not ignite easily. Fire spread is dictated by the interior geometry and content fire load.

Collapse concerns: Wall failure is rarely a firefighter risk in stacked log buildings. Likewise, a solid log roof system (rafters and ridge beams) will resist collapse over time under fire conditions. Localized collapse and burn through of roof decking can be a concern. Many log buildings use interior lofts to maximize usable space (for bedrooms, offices, storage, etc.). These lofts (and most interior partitions) are likely constructed using standard wall and floor framing (2 × 4 in., 2 × 8 in., or even lightweight wood truss floors). Obviously, an interior collapse threat lies in these areas.

Post and beam. Post and beam (or braced frame) construction uses two primary structural components: posts at the corners of the structure (columns) that support horizontal beams (also referred to as *girts* and *ribbon boards*) at each floor level, which are used to support the structural floor components and roof. Prior to the use of truss construction, post and beam construction was used to create an unobstructed floor space. While interior columns may be present in larger buildings, the whole purpose of the posts and beams is to maximize uninterrupted interior space. Post and beam construction is commonly used in barn construction and is sometimes called *pole barn* construction, but was also used in residential structures in the 1800s.

In classic post and beam construction, the connections between the corner posts, horizontal beams, and diagonal bracing supports are secured by mortises and tenons. The mortise and tenon connection is a system that has been used by woodworkers for thousands of years (there is evidence of this system back to 2,600 BC) to join two pieces of wood, particularly at a 90° angle, and without using expensive metal (plates, hangers, nails, etc.).

Fig. 4–18. Here, a mortise and tenon system is used to join diagonal bracing and a vertical post. (Photo by Jerry Knapp.)

In figure 4–18, the mortise and tenon connections can be seen at the junction of the vertical post and diagonal bracing. The mortise is the cavity and the tenon is the projection. A cavity is formed in the corner post and a projection is formed at each end of a beam. Once the mortise and tenon are joined, several horizontal holes are drilled in the post and through the tenon, and then wooden pegs are inserted to form a secure joint (which is visible in the cavity in figure 4–19). As construction methods evolved, the mortise and tenon system was replaced by bolts, plates, and/or hangers to make the column/beam connection.

The vertical corner posts and the horizontal beams form the primary structural skeleton of the building. However, the skeletal frame needs to be made rigid to withstand wind loads and is achieved by adding diagonal bracing, exterior wall enclosures, or both. Exterior wall enclosures are created in numerous ways, although furring strips with wall clapboarding and partition framing (typically 2 × 4s and sheathing) seem to be the most prevalent. Roofs are typically formed using rafters and ridge beams.

Fig. 4–19. A mortise (cavity) with a wooden peg inserted through the mortise

Fire spread concerns: Anyone who has fought a wooden barn fire (or post and beam residential fire) can speak to the rapid fire spread potential of the exposed wood. In many ways, the typical post and beam barn is a wood crib with lots of combustible surface area and very little fire-stopping. The interior arrangement, finish, and content fire load are all factors in the fire spread potential but structurally, the post and beam building was built to provide unobstructed space. If the interior space is not finished out, the structural elements are exposed and the cavernous nature of the interior can create a true fire storm with tremendous heat release rates.

Collapse concerns: Although the primary structural components are strong, the weak portion of this construction is the mortise and tenon joints, particularly the tenon part. As evidenced in figure 4–19, although the mortise cavity reduces the size of the vertical post, the tenon significantly reduces the size of the horizontal beam at the connection point. This connection, and when exposed to sufficient fire, can collapse and also cause a collapse of the floor joists that are supported by a collapsing beam post. This is most pronounced at the lower floors in a multistory building as they are also supporting the weight of upper floors. Post and beam buildings with heavy timbers and bolts/plates collapse differently—once fire consumes the exterior wall and roof partitions/sheathing, rigidity is lost and the building will start to lean subject to lateral or eccentric loads (wind, fire streams, and/or interior content live loads).

Balloon frame. This type of construction, which was employed from the mid-1800s to the late 1940s, is also known as *Chicago construction* and was popular when long lengths of lumber were available and cost effective. Additionally, it was an easier method of building without needing highly skilled carpenters that were often necessary for dovetail joints and mortises and tenons required for post and beam construction. In 1832, a Chicagoan built a balloon framed warehouse that employed the now familiar 2 × 4 in. studs and 16 in. vertical stud spacing. This system was quickly adopted and used until it was replaced by modern platform framing. The defining feature of a balloon frame building is the continuous nature of the exterior wall studs—the stud extends the full height of the wall from the sill plate to the roof plate.

After the foundation was completed, the exterior walls that were comprised of full-length studs would be assembled in a horizontal position and then tilted up to their vertical position. This process was completed in a timely manner and with relative simplicity. Floor joists were attached to a *ribbon board* (a type of ledger) that was nailed or cut into the exterior wall studs and used to support the floor (and ceiling joists) either for single or multistory buildings, and there was a lack of fire blocking in the exterior wall stud cavities. This resulted in open void channels between the stud cavities (foundation to the attic) and access from the stud channels to the attic and the floor joists. These combustible voids create uninterrupted vertical and horizontal pathways for a fire (fig. 4–20).

Fig. 4–20. Balloon frame construction can readily allow the vertical extension of fire into an attic.

The exterior wall framing could be completed with any number of veneers such as brick, wood, stone, stucco, or shingles. Interior walls were typically lathe and plaster—although many have been rehabbed over time (with new electrical wiring and plumbing and finished with drywall). Roof rafters were often constructed from 2 × 4 in. lumber, spaced up to 3 ft on center (referred to as bungalow construction—see figure 8–3), and the roof has likely been covered multiple times without removing the preceding roofing layers, which can result in an overloaded roof. From the street, the following visual clues may help you classify a building as being balloon frame:

- Old wood frame buildings up to about three stories in height
- Older wood shiplap siding or asbestos-type shingles
- The presence of visible 2 × 4 in. rough-cut rafter tails that can be spaced wider than 2 ft on center.
- Windows in multistory buildings that are narrow and line up vertically (fig. 4–21).

Fire spread concerns: This type of construction is renowned for allowing fire to rapidly travel up the continuous void channels into the attic of a building and also to extend into the voids between the floor joists. When confronted with this type of construction, it is important to remember that the void spaces in a balloon frame building are interconnected, which can allow fire to extend undetected throughout an entire building. It is not uncommon for a simple oil-burner furnace fire in the basement or cellar to spread up the walls and into adjoining floors. Next thing you know, fire is blowing out the eaves, the front and rear porches, and the attic window or vents while the occupant spaces remain clear or with "light smoke only." This fire spread potential requires extensive labor to open up each stud channel to check for and ultimately stop the spread of fire. Vertical ventilation is almost mandatory when fires enter the balloon frame voids.

Fig. 4–21. Inline windows in older wood frame buildings can be an indicator of balloon frame construction.

Collapse concerns: The balloon frame building is inherently strong given the era that balloon frame was prevalent. It uses full-dimensional solid lumber (material mass), nails and screws (through connections), and high compartmentalization. Fires that capture the wood voids can burn for some time before strength is lost. Often, the floor and roof joists will sag as wood mass burns away, giving some advance visual warning to firefighters. This is not to say that collapse is easy to predict. Aging and alteration issues can cause a rapid collapse of entire wall segments or sudden dropping of a floor (especially in unfinished basement fires). Additionally, roof collapse can happen quickly due to increased rafter spacing and the likelihood of additional roof loading due to repairs and maintenance. Many believe that the fire spread potential led to the phasing out of the balloon frame construction method. Historically, though, the demise of balloon framing was forced by war effort materials

rationing and better efficiencies in the lumber industry as it became harder and harder to find, mill, package, and ship long/straight wall studs.

> ## Quick summary
>
> - Type V wood frame buildings are the most predominant type of buildings in this country as they have been constructed from the late 1700s to the present. They are also responsible for the majority of injuries and deaths to firefighters.
> - The primary exterior and interior building component is flammable wood, and the overall primary concern is the flammability of the structural components.
> - Stacked log buildings are the oldest Type V buildings and commonly use solid logs of various sizes. This can result in a strong building unless the roof incorporates lightweight construction.
> - Fire spread and collapse concerns are often minimal. The primary concern is the incorporation of lightweight construction in the roof.
> - Post and beam construction uses two primary structural components: corner posts and horizontal beams.
> - Post and beam construction often uses a mortise and tenon system for connection points.
> - Although post and beam construction is flammable, the primary weak portion of this construction is the mortise and tenon joints.
> - Balloon frame construction is defined by the continuous nature of the exterior wall studs that extend from the foundation to the attic area.
> - The floor joists of a balloon frame building are attached to exterior wall studs by a ribbon/ledger board and there is a lack of fire blocking in the exterior wall cavities.
> - Balloon construction results in open void channels between the stud cavities from the foundation to the attic, ensuring a rapid vertical fire spread. Therefore, rapid fire spread is a primary concern.

Western platform. The western platform construction method was born out of the westward expansion of the United States. Frontiersmen erected temporary wood buildings literally by building a wood platform right on the ground that was supported with little or no foundation or wooden posts (fig. 4–22), adding framed walls atop the platform, and capping it off with a roof. They would then cover the frame with anything suitable (clapboarding, animal hides, canvas, etc.). By eastern standards, these were crude, poorly built, and not suitable for long-term use. Rarely was a second floor included. The method continued to evolve as mining and trapper settlements grew (with the addition of foundations, rigid sheathing, utilities, multiple stories, etc.). Jump ahead to the 1940s. World War II was a defining period in the wood construction industry. Also known as the first industrialized war, WWII challenged builders to gain efficiencies in wood harvesting, milling, packaging, shipping, and rapid building. From this challenge, platform wood frame building supplanted balloon frame and became the most common type of wood construction (and continues today).

Fig. 4–22. Western platform construction often includes a minimal type of foundation.

Platform construction differs significantly from balloon construction as it uses a building-block approach that results in each floor of a building being built as a separate unit from the floors below and above it. This also eliminates the void characteristics of balloon construction. After the foundation is completed, 2 × 4 in. or 2 × 6 in. studs

The Art of Reading Buildings

are typically used for the exterior and interior walls and travel from the foundation to a double plate that is used to support the ceiling joists for the first floor ceiling. If a multistory building is being constructed, then each successive floor is constructed in the same manner. This basic cap and stack method allowed builders to use pallets of uniform length/width lumber to build the entire structure (hence the popularity of the 8 ft stud). Obviously, some type of roof is needed to finish off the structure. We'll cover that a bit more with the various forms of platform construction below.

From the firefighting perspective, platform construction provides intrinsic fire stopping from two elements: (1) the fire blocking in the exterior walls between the floor and ceiling joists, and (2) the double plate at the top of each floor (fig. 4–23). The double plate eliminates the void spaces of balloon construction that allow fire to vertically extend upward into the voids between the floor joists of multistory buildings and into an attic. The interior is normally finished with lath and plaster in older construction, and sheets of drywall in newer construction. The exterior can be finished in a wide variety of materials such as wood lap siding, vinyl siding, plaster, brick veneer, and other suitable materials.

The platform wood frame concept allowed designers to diversify as it became much easier to create larger open floor spaces, split-level floors, integral decks and balconies, and other features. This diversity (plus modern material engineering and imagination) has created several subcategories of platform wood frame buildings and subsequent concerns for firefighters. However, universal or official definitions don't really exist for each of these subcategories. Given that, we know that all platform frame buildings are not created equal (from a fire perspective) and that the fire service has developed its own slang or jargon to help differentiate them. So, we've taken liberties to briefly define each and list some tactical considerations for them based on research and firefighting experience. From a firefighters' viewpoint, wood platform buildings can be subdivided into conventional, legacy, and engineered (lightweight) wood styles.

Fig. 4–23. Platform framing uses fire blocking and double plates to restrict the vertical extension of fire as well as provide stronger framing.

Conventional platform. Conventional platform construction was typical from the 1800s through roughly the late 1940s (fig. 4–24). Conventional platform can be defined as the classic wood frame building that employs full-dimensional lumber (a 2 × 6 was really 2 in. deep by 6 in. wide). This full-dimensional lumber was also rough sawn, meaning the surfaces of the wood were not planed smooth. This rough finish, when exposed to flames, quickly chars over and actually slows the burn rate of the wood (all other influences being equal). Conventional platform is also typified by the following:

- Through connections using nails and screws
- Solid wood floor joists and solid wood rafters or joists (no trusses)

- Lathe and plaster interior wall covering (adding structural rigidity and fire resistance), although gypsum (drywall) became popular in the later years
- Tongue-and-groove wood-plank flooring over joists
- Varied roof coverings such as plywood, OSB, shingles/shakes, 1 × 6 in. boards, skip sheathing, tile and slate, and built-up tar and gravel

Fig. 4–24. Conventional platform construction was typical from the 1880s to the late 1940s and used full-dimensional lumber.

Fire concerns: Conventional platform buildings typically confine fires through compartmentalization. The mass of the wood will actually resist fire spread because of surface charring (initially). The greatest fire spread concern in conventional platform buildings is driven by contents and interior geometry (halls, stairs, etc.). Once a fire takes hold of the structural elements, one can expect prolonged burning as a result of so much wood mass.

Collapse concerns: Compartmentalization and the basic nature of fire-stopping in the structural elements make the conventional platform quite resistive to collapse. These were pretty tough buildings considering they were made from wood. In many cases, interior fires are extinguished and a contractor comes in, removes the interior wall boarding, and re-skins the framing. Collapse threats do exist, however. Well-involved fires with long burn times and lack of maintenance (aging, rot, etc.) can precipitate collapse. Exterior brick veneers can fall off the wood frame sheathing. Decks, balconies, overhangs, and the ever-possible alterations can all add collapse threats. In most cases, the structural elements will show signs of sagging or bowing prior to collapse.

Legacy platform. The term *legacy* has been made popular to describe the accomplishments of the generation of Americans who prospered following World War II. Applied to building construction, the category of legacy platform is used to define a cut lumber, solid wood frame building built from roughly the 1950s to the present day. (Yes, some builders are resisting the industry trend toward engineered lightweight wood.) The key characteristic is the use of solid, milled wood lumber for the assembly of wall studs, floor joists, roof trusses, joists, and rafters (as opposed to engineered wood). A true legacy platform building uses solid wood floor and roof joists/rafters. For our purposes, a wood frame building with solid floor joists but with a trussed roof can be classified as legacy as long as *the wood roof trusses are assembled using solid wood lumber and nails or steel staples that penetrate into the centers of the adjoining wood pieces that form the truss* (more on this later).

So how is the legacy platform different from the conventional platform construction? Well, as the greatest generation prospered and built many buildings, they discovered all kinds of new techniques and profit-enhancing efficiencies. While the legacy building still uses solid wood for structural elements, some of the new building techniques and materials add significant structural integrity concerns for firefighters. Examples include the following:

- The 2 × 4s were no longer full-dimension lumber but 1½ × 3½ in. (resulting in loss of fire-resisting mass) (fig. 4–25).
- Rough-sawn lumber gave way to a smooth planed finish. This sounds good, although from a fire spread perspective, a rough-sawn finish actually chars quicker, slowing the wood burn rate.

The Art of Reading Buildings

Fig. 4–25. Modern 2 × 4s are not full-dimensional lumber.

- Nails and screws were replaced with thin galvanized steel hangers (fig. 4–26) and short nails or staples.
- Tongue-and-grove floor decking was replaced with plywood sheathing.
- Solid wood cross bracing for wind loads was replaced with plywood sheets, particularly near exterior corners.

Fig. 4–26. Nails and screws have commonly been replaced by thin galvanized hangers.

Legacy differs from conventional in other realms, namely interior geometry and size. Americans wanted bigger rooms, taller ceilings, more storage, multiple garages, split flooring levels, and added exterior elements such as large decks/balconies and extended eaves.

Fire concerns: As with conventional platform, interior contents and geometry affect fire spread most in the legacy wood building. The legacy platform building is still inherently fire-stopped from floor to floor in the combustible voids. The use of drywall helps protect the wooden elements. Legacy and conventional platforms differ in fire concerns with the advent of the larger spaces, split levels, and the use of more soffits. *Soffits* are false spaces on the underside of eaves and stairways, the space above cabinets (especially kitchen cabinets), and used to hide construction features like plumbing, HVAC, and electrical conveyance systems. Soffit spaces grew as the legacy platform building got bigger, with more features, and more inventive floor plans. Soffit spaces are rarely fire-stopped and only a thin sheathing layer or drywall separates a fire from entering structural spaces.

The split-level home design (fig. 4–27) became a popular variant for legacy platform homes. The staggered arrangement of the split stories means that each level is not fully capped from the other, meaning vertical fire spread is more likely. The interior wall framing where the split levels intersect may include balloon-like studs that allow vertical spread. The finished interior space of a split-level home typically includes wide-open spaces to accommodate the half stair runs to each level. Smoke, heat, and fire spread are accelerated in these areas.

Fig. 4–27. A split-level structure means that each level is not fully capped from the others, which can enhance the spread of fire.

The capped and stacked nature of legacy platform construction presents some challenges for electricians, plumbers, and HVAC installers. To overcome the challenge, the installers simply drill through the inherent fire-stopping (wood plates/sills, studs, and joists). These penetrations or *poke-throughs* for utility chases can create fire spread considerations for firefighters. Most codes require the poke-through to be sealed (fig. 4–28) to minimize the extension of fire and its by-products. However, in some cases the sealant has not been properly applied or is nonexistent (common in renovations). Where present, the status of these areas should be checked (sealed vs. unsealed).

Fig. 4–28. Poke-through construction should be sealed to minimize the extension of fire.

Collapse concerns: We know that increased material mass helps resist fire. The legacy platform wood building has less structural mass than a like conventional one. We are still talking about solid wood, just a higher surface-to-mass ratio. Add to that the switch to joist hangers and connections that do not extend through the full dimension of the material and one can surmise that the legacy building will collapse sooner than the conventional wood frame. Still, the legacy platform resists fire collapse much better than newer engineered lightweight wood frames.

Some legacy platform buildings include a trussed roof. The truss allows the builder to span larger spaces than joists or rafters could. Trusses also offer a simple way to create vaulted and coffered ceilings. The legacy truss is typically *craftsman built*—meaning that only solid wood is used and the connections are driven through the center mass of adjoining pieces (using screws, nails, or deep-pressed staples). As a rule, firefighters should be concerned regarding the rapid collapse potential of trusses. However, a true craftsman-built truss can be just as strong and collapse resistant as a joist or rafter for a similar span.

Engineered wood platform (lightweight). Chapter 2 introduced us to the material known as engineered wood. As you recall, engineered wood is manufactured using various glues to bind wood chips, slivers, veneers, and pulp to form sheathing or wood elements that are used for columns, studs, or beams. The engineered wood lightweight platform building is one that utilizes engineered wood for structural elements. For brevity, we call these lightweight wood platform buildings. Although the concept of platform construction remains the same for lightweight wood, the size, material, and configuration of the wood used is radically different, creating a new subcategory of construction from the firefighter's perspective.

Conventional wood framing typically uses dimensional milled wood lumber that is at minimum 2×4 in. or 2×6 in. for the wall studs, floor joists, ceiling joists, and roof rafters. Legacy brought us reduced lumber dimensions (2×4s are really $1\frac{1}{2} \times 3\frac{1}{2}$), and the use of roof trusses. Legacy and conventional wood buildings typically use nails or screws for connections. Lightweight wood construction relies on members of less than 2×4 in., elements made from glued wood pieces, glue used for connection points (fig. 4–29), geometry in place of mass for structural members, and the concept of assembly building techniques to achieve strength. Further, the lightweight wood building employs a variety of glues and surface hangers for connections.

The Art of Reading Buildings

Fig. 4–29. Glue can be used for structurally bonding pieces of wood together and for the integrity of connection points.

From a historical perspective, the first use of engineered wood can be found in the 1960s—typically in the assembly of trusses or for door/window headers. Lightweight wood construction gradually became more prevalent as the building industry continually looked for ways to construct buildings with a product that was becoming more expensive (wood) and in less time (reducing labor costs). The 1980s saw the biggest shift away from legacy platform to lightweight. Now, lightweight wood construction has been proven to provide acceptable strength while using less dimensional lumber and allowing for quicker assembly. Regrettably, lightweight wood construction has had a momentous negative impact on fireground operations from the perspective of fast failure rates and firefighter injury and death. As an example,

CASE STUDIES: TYPE V BUILDINGS

HOUSE FIRE—MINNEAPOLIS, MN

On July 3, 2010, resources from the Minneapolis Fire Department responded to a three-story residential structure of wood frame/balloon frame construction and wood exterior siding that was built in 1897. Although the structure was licensed as a single-family rental unit, it was actually being used as an unlicensed lodging/boarding house that contained eight sleeping rooms. Initial companies reported smoke showing from the eaves and the top end of the peak. While additional companies were arriving, attack lines were advanced into the structure and up to the third floor. As the knee wall in the third floor was opened, the room exploded with a flashover, seriously burning two firefighters. The cause of the fire was believed to be an electrical malfunction.

HOUSE FIRE—WARWICK, NY

A structure fire in 2003 within the village of Warwick, New York, consumed a two-story, 4,000 sq ft single-family dwelling in a short period of time that can be measured in minutes. At 4 a.m., the Warwick Fire Department received a call from the occupant of the home with smoke alarms audible in the phone call. Approximately 6 minutes later, initial resources arrived to find the structure was fully involved in fire to the point that it was possible to look through the remaining structural skeleton. Due to the immense amount of radiated heat, initial resources were forced to spot to the exposures on Sides B and D of the involved structure.

The home was constructed of 2 × 6 in. lumber for the exterior walls in a platform frame configuration, and sheathed with OSB that was covered with vinyl siding. The floor and ceiling joists and roof rafters were lightweight truss construction. The primary cause of this fast moving fire was the lightweight

the failure rate of typical conventional/legacy floors averages 15–20 minutes when exposed to fire, whereas the catastrophic failure of lightweight construction averages 5–7 minutes when exposed to fire.[1] Aggravating this problem are the structural connection points that are now held together by gusset plates (see fig. 5–13) that are surface attached and/or glue that can liquefy at relatively low temperatures. As a result of lower costs for materials and labor, lightweight trusses and engineered I-joists have become the modern standard for wood frame buildings, hence the category of lightweight wood frame buildings. This is why it is paramount that fireground personnel identify the type of construction they are about to commit to as many wood frame buildings look similar from the exterior but are dramatically different on the interior. We'll talk more about lightweight construction (and some evolving construction techniques) in chapter 5. For now, let's list some tactical considerations for lightweight wood.

Fire spread concerns: The lightweight wood building exacerbates the fire spread concerns listed previously for conventional and legacy framed buildings. The reasons are simple: lower mass, larger combustible voids (trusses for roofs and floors), the use of glues, and the rapid failure of drywall from hotter fires and greater heat release rates. Truss spaces are basically like a wood crib—lots of exposed surfaces and air space that can spread fire. In many ways, the truss space is like horizontal balloon frame, only worse. Instead of separate stud and joist channels, the void space is construction in combination with OSB sheathing and vinyl siding. Interestingly, the exposures on Sides B and D were 70 ft and 150 ft away from the fire structure and received considerable damage to their vinyl siding. As a final note, due to the speed and extent of fire from the initial call to the fire department, initial resources were forced into a defensive operation with a primary search of the structure not a possible consideration.

HOUSE FIRE—GREEN BAY, WI

On August 13, 2006, members of the Green Bay Fire Department responded to a midday basement fire in a 3,500 sq ft (first floor) lightweight wood frame home built in 1999. The home had a 2,200 sq ft partially finished basement. The floor system included engineered wood I-beams and parallel chord open web trusses made from 2 × 4s and gusset plates. The floor trusses and I-beams were covered with OSB sheathing and an in-floor radiant heating system within a poured lightweight concrete and tile finish. The first crew entered the front door of the house and advanced an attack line through to the basement stairway located near the C/D portion of the first floor.

A second crew entered the front door and started a left-hand primary search. The left-hand search led them to an office-like room off the entryway. A floor collapse occurred within minutes, sending the search crew into the flaming basement. A basement partition wall caused the two firefighters to fall in different directions—separating the two by the wall, debris, and fire. One firefighter was seriously injured but was able to declare a Mayday and was rescued through a basement window. Several attempts were made to rescue the other firefighter but fire intensity, debris, and further collapse thwarted the effort. The line-of-duty death (LODD) investigation cited lightweight construction, direct fire impingement on unprotected trusses, and a significant floor dead load as contributing factors.

wide open from truss to truss for fire spread (three-dimensional fire spread).

Remember that poke-through construction also applies to the webbing of engineered wooden I-beams that have been breached to allow plumbing, electrical, and/or HVAC ducting to travel in a horizontal direction (fig. 4–30). From this photo it is obvious that the poke-through openings are significantly larger than the ducting and will easily allow the extension of fire and its by-products. Today's typical interior fire is fueled by synthetic contents (fire load). Synthetic materials burn hotter and release heat at an extraordinary rate (megawatts per second!). Engineered lightweight construction relies on drywall to protect structural elements. However, the high heat release rate of plastics accelerates the calcination of drywall, and hence, faster access to combustible voids.

Fig. 4–30. Wooden I-beams are often breached to allow the horizontal travel of plumbing, HVAC ducting, and other materials.

Collapse concerns: Arguably, lightweight trusses have had the greatest negative impact on structural integrity in lightweight wood buildings. Where the conventional/legacy methodology of platform construction used varying sizes of lumber that relied on mass and nails/plates/bolts for connection points with a somewhat predictable burn and collapse rate, lightweight trusses have introduced a new set of concerns and hazards to fireground personnel. Namely, we've lost interior firefighting time as lightweight buildings collapse very quickly once heat and fire starts to degrade structural elements. Additionally, a recent concern is the impact of "green" technology on building construction, specifically advanced framing techniques that result in less lumber being used for framing. Obviously, less lumber can result in an increased potential of building collapse. Remember that tan or brown smoke coming from structural areas of lightweight wood buildings is a warning that degradation of mass has begun. (We didn't have much mass to begin with!)

Quick summary

- Platform construction has become the most common type of wood construction and continues today.

- If multistory buildings are constructed using platform methods, double plates between floors are used in conjunction with fire blocking in the exterior walls.

- Conventional platform, which predominately uses full-dimensional lumber and conventional construction procedures, dates from the 1800s to the late 1940s.

- Compartmentalization and dimensional lumber is an asset to these buildings.

- Legacy platform construction started in the late 1950s and is still used today. The defining features are the use of milled lumber, a solid wood frame building, and nails are often used for connecting points.

- A primary hazard is the increasing use of smaller-dimension lumber, metal hangers, and plywood/OSB sheathing. These factors can result in the increased spread of fire and reduced collapse times.

- Engineered wood platform construction primarily uses engineered wood products in combination with adhesives to bond structural connections together.

- Lightweight trusses are the staple of this type of construction as well as OSB that is used in exterior sheathing and interior flooring applications.

- The primary hazards with engineered platform construction are numerous but can be summarized by rapid fire spread due to milled and glued lumber, open web lightweight trusses, and the use of OSB. These factors are also responsible for rapid collapse times when this type of construction is exposed to heat and/or fire.

CHAPTER REVIEW EXERCISE

Answer the following:

1. List the four influences that can be used to classify a building.

2. Match the NFPA 220 building type in the left column with the applicable description in the right column.

TYPE I	Heavy timber
	Fire resistive
TYPE II	Post and beam
TYPE III	Wood frame
	Conventional
TYPE IV	Noncombustible
TYPE V	Ordinary

3. List the four ways that steel can be protected from fire and heat.

4. What are the visual clues that can help one differentiate a Type I and Type II building?

5. What type of building features no concealed spaces and typically has large floor-to-floor openings?

6. What materials are used for the load-bearing walls of a Type III building?

7. List the four ways that an all-wood building can be constructed.

8. Which types of construction include combustible voids that can spread fire past the point of origin?

RESOURCES FOR FURTHER STUDY

- Brannigan, Francis L., Glenn P. Corbett, *Brannigan's Building Construction for the Fire Service*, 4th ed., Sudbury, MA: National Fire Protection Association: Jones and Bartlett, 2007.

- Cunliffe, Sarah, Sara Hunt, Jean Loussier, Ed., *Architecture: A Spotter's Guide*, New York, Metro Books, 2010.

- Dunn, Vincent, *Collapse of Burning Buildings: A Guide to Fireground Safety*, 2nd ed., Tulsa, OK: PennWell, 2010.

- http://tiltup.com

- http://www.toolbase.org/

- Morgan, William, *The Abrams Guide to American House Styles*, New York: Abrams, 2008.

- National Fire Protection Association, *NFPA 220: Standard on Types of Building Construction, 2012 Edition*.

- Thallon, Rob, *Graphic Guide to Frame Construction*, 3rd ed., Newtown, CT: Taunton Press, 2008.

NOTE

1 Taken from the "Report on Structural Stability of Engineered Lumber in Fire Conditions," Project Number 07CA42520, Underwriters Laboratories Inc., Northbrook, IL, 2008. Available at www.ul.com.

CLASSIFYING BUILDINGS—HYBRID, ERA, USE, TYPE, AND SIZE CONSIDERATIONS 5

OBJECTIVES

- Define the term "hybrid construction."
- Explain how the building era can impact firefighters.
- List typical building use categories.
- Delineate how reading building types can affect size-up considerations.
- Understand how building size can be an important factor in reading a building.

THINKING BEYOND THE FIVE TYPES

Chapter 4 brought us an understanding of the five basic types of building construction, and hopefully the chapter helped you realize that there are subsets of each of the five types (e.g., Type V can be log, post and beam, balloon frame, or platform) and there are even further divisions for the subsets (a Type V platform can be conventional, legacy, or lightweight). While the basic five types provide a good starting place to understand (and read) a building, it is just that—a starting place.

As stated in chapter 4, there are buildings that combine multiple NFPA types in a single building and there are buildings that don't really fall into the classic five types from the firefighter's perspective. We use the term *hybrid* to label these buildings. As part of this chapter, we highlight various hybrid construction methods.

The street-smart fire officer uses an understanding of the five types and hybrids to form an understanding of the core approach that was used to construct a building. To truly get a good read on a building however, the officer must further classify a building not only by the type of building, but by its era of construction, its intended use or occupancy, and the size of the building. Why?

The answer is tied to lessons learned. Rarely is the type of building cited as a contributing factor in the collapse of a burning building. More likely, collapses are manifested by interweaving type with era, and use with size. The oft-cited Hackensack Ford dealership building collapse is a tragic example:

> *Five firefighters died. The fact that the building was Type III wasn't the only issue. The building was built in an era where cold-rolled steel was replacing heavy timber and iron for bowstring trusses. The truss space was being used as parts storage for an auto repair shop—it was never intended for that. The size (and span) of the open floor area created a weak link. Relative to the building, it wasn't one factor, it was the insidious culmination of four factors.*[1]

We know history has a habit of repeating itself—firefighters have been injured or killed by hauntingly similar situations in buildings. We owe it to the profession to learn from these lessons. Here, we begin to present some tangible era, use, type, and size considerations that have been contributing factors in the hostile fire spread and collapse of burning buildings. By doing so, it is our hope that you will learn those lessons and translate them to better strategic and tactical decisions at your next building fire.

HYBRID BUILDINGS

While there is no official classification known as "hybrid," the term *hybrid building* is becoming more common. In context here, a **hybrid building** refers to a building that combines various NFPA 220 types in one structure *or* a building that is constructed in a manner that, from a firefighter's view, doesn't really fit into any of the NFPA 220 types. It can also describe alternative construction. (As a side note, Webster's Dictionary defines *hybrid* as "anything of mixed origin," which aptly defines alternative construction.) Engineering breakthroughs, material technology advances, and sustainable environmental initiatives are combining to rapidly change the building construction industry. Even as this chapter is written, newer construction methods are being tested and accepted for buildings. Regardless, we will address the hybrid topic by introducing the concept of NFPA 220 combinations and non-NFPA 220 alternative construction methods/types.

Combining NFPA 220 building types

Earlier we mentioned that codes and standards are designed to match building use with a required type of construction that affords life-safety benefits. We know that NFPA 220 outlines building construction types. *NFPA 5000: Building Construction and Safety Code*[2] is a 55-chapter publication that details how a building's intended use and resulting required construction type are matched. The *International Building Code*[3] is also a body of work that matches a building's intended use (known as occupancy) to a required construction classification. Both documents address how multiple construction types can come together in one building based on use (occupancy) intentions.

In an earlier time, if architects wanted to build a mixed-use building, they would have to build the whole building to the most stringent requirement. For example, a small retail store by itself might only be required to be built using Type V methods. A separate production factory might have to be built as Type II. However, if you combined the two—a retail storefront with an attached production factory behind—the whole building would have to be built as Type II. As can be imagined, this could be costly. The concept of *occupancy separation* was introduced into code requirements to accommodate the desire to build a single structure for multiple uses. Occupancy separation allows more architect/builder options—creating flexible floor plans, money savings, and increased profits while still maintaining some life-safety/fire stop benefits. Occupancy separation is usually accomplished

using an approved fire barrier that meets a given time requirement (like a fire door and fire wall) in order to mix various construction types in the same building. In some cases, mixed NFPA types are allowed without separation barriers if an approved fire protection sprinkler is installed.

An example of a mixed NFPA type hybrid can be found in one Colorado ski resort structure that includes a multilevel parking garage, conference center with hotel space, office/retail space, and individually-owned condominiums (fig. 5–1). The parking garage is built as a Type I monolith (poured-in-place concrete over steel rebar). Stacked on that is the conference center/hotel built as a Type I, using spray-on protection for steel posts and beams. The office/retail space surrounding the conference center is built as a Type II steel building. Individually owned condos are stacked on the office/retail space and are built as a Type V engineered (lightweight) wood structure. Visually, the whole structure is finished in a similar fashion, making it difficult for a fire officer to size up the construction classification—a true hybrid building.

Fig. 5–2. An example of a combination of Type II and Type V construction in a single building

Fig. 5–1. An example of a mixed NFPA type hybrid building

Figure 5–2 is another example of a hybrid. This structure is a restaurant that is built as a Type II steel building with a large Type V wood frame roof structure that hides rooftop HVACs and cooking ventilation hoods. The square footage of the Type V roof structure exceeds that of most homes! These separate types will behave quite differently when exposed to fire. Good building familiarization tours and prefire planning can help the local fire department make better decisions when combating a fire in this building.

Alternative construction methods

From a firefighting perspective, there are buildings that don't fit in any of the NFPA classifications, combined or otherwise. Sure, these buildings may be classified as a given NFPA type for code/permit approval purposes, but they really aren't built as the code-designated type suggests. For discussion sake, we'll label these non-NFPA 220 types as *alternative construction methods*. For example, a single-family home can be built using stacked straw bales and covered with a panelized roof made from expanded polystyrene and OSB. That home might be classified as Type V wood frame for permit purposes, but it really isn't wood frame. A straw bale house may seem retro—like the sod homes that became prevalent during the great homesteading period of the 1800s. While similar in idea, the new straw bale house is thoughtfully engineered and is quite simple to build for a long-lasting, environmentally-friendly building (see straw bale construction in chapter 6).

The list of alternative construction methods that are in use today can be quite lengthy, so we have chosen to dedicate chapter 6 to those (we also cover evolving methods in chapter 6). Generally speaking, though, the fire service has very little research information on the stability and fire spread characteristics of alternative buildings during actual fires. Sure, there have been numerous laboratory tests of assembly components for code purposes, but real-time, full-fuel package fires in hybrid buildings is not well documented. An example of a modern hybrid building with a lack of documented fire experience is laminated strand lumber (LST) and cross-laminated timber (CLT), which are being used instead of steel in concert with modular construction techniques (this methodology is also described in chapter 6). Given that, the street-smart fire officer can look at the low-mass, high surface-to-mass, synthetic, glued, and fuel-rich nature of alternative materials and make a judgment on the rapid fire spread and collapse nature of these buildings.

Quick summary

- There is no official building classification for the term *hybrid*.
- A hybrid building can be defined as a building that combines various NFPA 220 types into one structure, or a building that does not fit into any NFPA classification types.
- An example of a mixed-use or hybrid building would be a building that incorporates NFPA Type II and Type V methodologies within a single building.
- Mixing several construction types in a single building is allowed provided approved occupancy separation features are employed.
- Buildings that don't really fit into any NFPA 220 types will be labeled *alternative construction methods* in this book.

CLASSIFYING BUILDINGS BY ERA, USE, TYPE, AND SIZE

Earlier this chapter we discussed thinking beyond the five types of buildings as outlined by NFPA 220 as there are an increasing number of hybrid buildings being built. Further, the street-smart fire officer realizes that the construction of a building has been influenced by numerous factors. These factors need to be considered when making risk and tactical decisions during suppression operations. The factors include:

- Era—the historical time period during which the structure was built and/or altered
- Use—what the building is being used for (occupancy)
- Type—one of the five NFPA types or a hybrid
- Size—the relative footprint and height of the building

Classifying a building is a key step in the initial and ongoing size-up of a building during fires. Using the *era/use/type/size* method of classifying any given building is like using radar to see the threats that are looming—you may not have all the details, but you can at least see what is coming. This approach is especially helpful if you know how to interpret the radar signatures of era/use/type/size.

Era considerations

Basic human needs have always centered on food, water, and shelter. Some may argue that the list should include fire and companionship, but we all seem to agree that the evolution of humankind has been driven by the needs of the first three. Since prehistoric times, the need for shelter has, more than any other basic need, defined progress. The past 200 years have debatably produced the most significant changes in the shelters we call buildings.

The founding and growth of North America are most responsible for those changes. Think about it: The first settlers couldn't make buildings like the ones they sailed from—they arrived with only hand tools! It took some time to develop the quarries, foundries, and mills to process raw materials needed to build the classic European building. In addition to the hand tools, those early settlers brought ingenuity, resourcefulness, and a freedom of spirit that forever changed buildings.

As it relates to classifying buildings, we know that there are certain historic time periods that defined how a building was built using the materials, practices, and discoveries that were common during that period. We will call the historic time period during which a building was constructed the **era**. Some use building *age* to say the same thing. We feel that building age deals with the deterioration of a building over time—rusting, rotting, lack of maintenance, sagging, and other detrimental effects (fig. 5–3), so we prefer to use the term era. In context, we look at the major building era shifts that took place as part of North America's growth. More importantly, we attempt to classify the firefighter's perspective of various building eras as it relates to fire spread, structural integrity, and collapse potential.

Fig. 5–3. The age of a building can refer to deterioration of a building over time.

In simpler times, a firefighter didn't really have to categorize the building era. Sure, there were concerns of age deterioration, irregularities, and traps, but by and large, the buildings were high-mass, highly compartmentalized, and tough. Firefighters were less concerned about the era and more concerned with hidden voids, minimal access/egress options, alterations, and interior fire loads as they performed suppression, search/rescue, and ventilation operations. In today's firefighting world, a firefighter must consider the building era because it directly impacts those operations. Let's talk about several prevailing eras and how they impact firefighters.

While no official era separation or title exists, we know, mainly through firefighters' experiences, that some general lines can be drawn that can help us understand the building we are dealing with. Predominate eras include the historical (pre-WWI), industrial (pre-WWII), legacy (post-WWII), and the engineered lightweight buildings.[4]

Pre-WWI (pre-1914): The historic era. Buildings built before World War I that are still standing are a testament to the toughness, ingenuity, and materials used to create longevity. The earliest settlers built crude wood buildings that didn't last very long, but they served a use until more permanent structures could be built. Once a proper, permanent wood building was built, founding Americans experienced what so many did before them—the ravages of fire. It wasn't long before society demanded a building that would contain an interior fire (like the ones they had back in Europe). Hence, the cut stone or load-bearing brick-walled building became commonplace in the more dense population centers.

From the 1700s up to WWI, building construction techniques and innovations gradually improved—especially in the area of convenience. Systems for running water, sanitation, and heating all matured, and innovations such as piped gas and electric utilities were incorporated. Larger floor spaces and multistory configurations improved. Large warehouses, mills, and manufacturing buildings were erected (fig. 5–4). Mass-produced iron (made from iron ore and carbon) was an affordable building material, although it lacked the strength and ductility that modern steel has (modern steel

ties and anchors on the outside of a building. A spreader, which is an exterior wall plate, indicates the use of ties within the building to help keep the walls true and tight to floors. Basically, a steel rod has been used to connect opposite load-bearing exterior walls together. The steel rod spans the distance of the interior floor from wall to wall. Typically, the rod has a turnbuckle to make adjustments as the walls (and floor beams) sag over time. Spreaders are needed to distribute loads over more surface area, and they are either over-sized flat washers or more elaborate stars, swirls, and other decorative shapes (fig. 5–6, top photo).

The presence of spreaders on an outside wall can serve as a warning to firefighters. The steel rods inside a building can be easily deformed by hostile fires—causing damage to the wall and/or rapid sagging of the floor. Either way, catastrophic collapse can ensue. Anchors are also visible on the exterior of load-bearing walls (fig. 5–6, bottom photo). Anchors are typically smaller than spreaders and serve a different purpose. Anchors are simply designed to help hold a wood floor beam in a wall pocket. Left unchecked, wood floor beams can shrink over time and slide out of their gravity wall pocket. Anchors help prevent this. Most anchors are nothing more than a lag screw or L-shaped hook with a flat washer or small square plate. If the washer or plate has a screw head only, it is a lag screw. If the washer or plate has an adjustable nut (pictured), it is likely to be an L-shaped hook that is dogged into the wood joist. Anchors are often called joist or rafter tie plates. Fortunately or unfortunately, anchored floor beams negate the purpose of fire cut beams. A sagging floor could bring down the walls (remember the structural hierarchy from chapter 3).

Fig. 5–6. (Top photo) The tie-rod and "star" spreader on this building need to be adjusted—it clearly shows the tie rod that spans the building depth. (Bottom photo) Anchors are typically smaller than spreaders and indicate a lag screw or L-shaped anchor to hold the ends of wood joists or rafters in the gravity wall pocket.

Quick summary

- Historic time periods that define how buildings were built can be characterized by eras.
- A general timeline for building eras is
  - Pre-World War I: historical
  - Pre-World War II: industrial
  - Post-World War II: legacy
  - Engineered lightweight buildings
- Pre-World War I buildings can generally be characterized by solid mass beams (or large rough-cut lumber) for floors and roofs and high-mass materials for the walls, open hallways, and stairways.
- Significant exposure considerations that could lead to fire spread issues were the result of closely spaced wood buildings.

- Although high-mass materials take longer to fail from the effects of fire (as compared to modern construction), gravity connections, the use of iron structural members, sand and lime mortar, alterations, added exterior elements, and age are some of the contributing factors that led to collapse of these buildings.

Pre-WWII (pre-1939): The industrial era. The time period between World War I and World II (1918 to 1939) was one of incredible achievement—in spite of the Great Depression and Prohibition. The great industrial machine started to crank out automobiles, airplanes, huge cargo ships, and more powerful locomotives. Electrical generation facilities expanded, water delivery systems improved, and steel foundries and stamping plants flourished. Fuel-powered machinery mutated. With all these improvements came the need for bigger buildings to house assembly lines and massive machinery. The need for larger buildings led to wood, stone, and bricks being replaced with concrete block, reinforced poured concrete, and alloy-strengthened steel as the primary building materials. Wood was (and still is) the material of choice for most residential properties. The classic Sears Catalog mail-order home was offered by the Sears and Roebuck Company between 1908 and 1940 and became commonplace in residential areas that surrounded industrial complexes (fig. 5–7). These homes are often described as story-and-a-half bungalows that can be closely spaced. They were easy to build, quite stout, and relatively inexpensive.

Two other forces came to play in the industrial development era that also changed the way buildings were built: engineering and standardized building codes. This is not say either didn't exist prior—they simply ramped up and became primary influences in how buildings were built. From a code perspective, efforts were made to make building systems more reliable, such as the standards for plumbing, electrical, and heating. Engineers designed buildings to be larger with attention to strength and durability. Fire safety was considered by both the code developers and engineers, although many lessons were yet to be learned.

Firefighters engaged in suppression efforts of these pre-WWII buildings discovered several issues that impacted fire spread and collapse potential: building size and the use of steel.

Fire spread issues: The defining characteristic of the industrial-era building was its sheer size. It's almost as if the architect and engineering communities were daring each other to design and build bigger buildings. Factories grew from city block size to those that covered acres. Apartments went from dozens of units to hundreds, such as tenement and brownstone type buildings (fig. 5–8). High-rise buildings reached new heights. (In 1931, New York City's Empire State Building became the tallest, at 102 floors.) Obviously, fires grew with building size.

Fig. 5–8. Brownstones were often large enough to contain many units.

Fig. 5–7. An example of a Sears Catalog bungalow

Fire spread issues associated with the industrial era include the following:

- Open stairwells and central hallways. Many large loss-of-life fires occurred in industrial-era buildings such as Our Lady of Angels, Cocoanut Grove, and the Triangle Shirtwaist Factory are notable examples. Fire and smoke spread was virtually unchecked and presented firefighters with serious challenges.

- Utility chases. Many historical-era buildings were built with very simple and minimalistic utilities. During the industrial era, complex (and modern) utilities were engineered right into the original construction. Attention was made to hide unsightly pipes, wires, and ductwork. This created numerous shafts, chases, and dropped ceilings—and fire spread paths. Insulating materials for wiring, pipes, and waterproofing were typically flammable in this era.

- Void spaces. Engineering advances of this era lead to larger and tighter void spaces, which increased the incidences of backdrafts and cold-smoke explosions. Dropped ceiling voids and spaces behind knee walls (such as those used in finished attic spaces) are especially susceptible.

Collapse issues: Perhaps the greatest building achievement of the industrial era was that of roof engineering. Larger spans between walls and post columns could be achieved using steel and the truss concept. The truss concept was not new, it was just that previous trusses (used for railroad bridges, bowstrings roofs, etc.) were self-limiting due to the weight of the wood. The use of steel, rivets, and bolts allowed the truss to achieve strengths and spans that were previously unobtainable. The unusable space created by the truss roof structure also led to poorly planned reuse by building occupants—they converted the truss open spaces into storage areas, which added dead and live loads that were not designed into the original load capacity of the roof system. Unsupported masonry (load-bearing brick) was still used extensively in Type II and III industrial-era buildings, although the concrete masonry unit (CMU) was gaining popularity. Portland cement was being used for mortar.

Quick summary

- Pre-World War II buildings began to be constructed to conform to enhanced engineering and standardized building codes. As a result, buildings became bigger and began to use steel as a replacement for iron and heavy timbers.

- The numbers of single-family dwellings dramatically increased and were characterized by the Sears Catalog mail-order homes.

- High-rise buildings were built in greater numbers and reached new heights.

- Conventional type construction that often used high-mass materials was still the standard building methodology.

- Balloon frame construction, open stairwells and central hallways in multistory buildings, utility chases, and void spaces are some of the hazards that can be expected.

- Portland cement began to replace sand and lime mortar and masonry construction dramatically improved as a result.

Post-WWII (after 1945): The legacy era. As previously mentioned, the industrial era changed the face of developed countries—people could travel further, faster, and with more comforts. Our buildings, machines, and supply delivery systems became huge. Materials used for construction were almost limitless. Then World War II hit. Nations fought to preserve freedoms by creating and building more planes, ships, munitions, fuels, and supply management systems designed to deploy resources around the world. The war effort challenged Americans (and others) to ration raw material, fuel, and food. Manufacturing of wartime equipment and materials was left to a labor force that included women. The building construction industry was challenged to create systems that used minimal materials and quicker/easier construction methods. All of these challenges spawned innovation. What separated the American war effort from other countries was the attention given to

engineering, reliability, and durability—along with material efficiency and production speed. Arguably, these innovations won the war.

When the war ended, returning service men and women came back to an America that had changed. That change included a stronger "can-do" attitude and a lust for freedom. Prosperity ensued. People wanted a place of their own and a big family—the start of the baby boomer generation. To feed this desire, buildings had to be built—and build they did. Entire residential subdivisions popped up, seemingly overnight, along with the supporting infrastructure. Suburban America was born. To support the rapidly growing population, hospitals, schools, churches, shopping malls, and huge stadiums were built. High-rise buildings multiplied. Aircraft supplanted rail as the long-distance travel method of choice, so airport buildings and hangars grew in size and numbers. The economy grew and Americans became materialistic and needed more space for their possessions—more furniture, larger cars, recreational gear, and collectables. Thus was born the greatest generation—the legacy era—and all those buildings that presented firefighters new challenges.

The legacy-era building obviously includes the legacy platform wood frame building discussed in chapter 4. Frame buildings of materials other than wood built after WWII also changed. Lessons from the war (build fast with less material and more engineering) and lessons learned from tragic fires (adequate exiting, sprinkler systems, and more stringent building codes) became defining features of the legacy-era building. Let's look at more specific details of the legacy-era building.

Type I legacy: Stadiums, arenas, and other places of public assembly grew to enormous proportions. Fortunate for firefighters, these structures were built following the most restrictive codes to date and employed lots of steel-reinforced concrete. High-rise buildings changed significantly, as concrete core with steel web and glass curtain walls became the construction method of choice (fig. 5–9). The concrete core contained elevators, stairwells, HVAC ducting, and electrical and plumbing features. The exterior was typically finished with glass curtain walls. These steel-web high-rises used pan floors of sheet steel to support a lightweight concrete floor finish.

Fig. 5–9. A concrete core enclosure surrounded by steel floor beams became the construction method of choice for high-rise and other large buildings in the legacy era.

Type II legacy: Prior to WWII, buildings used for manufacturing, warehousing, and institutional-like needs (such as schools and hospitals) utilized soft brick, cut block, and concrete for load-bearing walls and columns. Beams were either steel or heavy timber (Type IV). After WWII, the concrete masonry unit (CMU) was used almost exclusively for load-bearing walls. Floors and roofs were supported with steel I-beams or heavy (forged) steel trusses that were bolted or riveted. Many Type II legacy buildings used no masonry at all—they were entirely steel (fig. 5–10). These all-steel legacy buildings were built in a post and beam fashion and employed galvanized steel panels or aluminum for wall and roof sheathing.

Type III legacy: Like the Type II legacy building, the post-WWII Type III buildings have CMUs for load-bearing walls. Floors and roofs utilize solid wood joists nailed or screwed into ledger boards attached to the CMUs. The classic suburban strip mall is an example of a Type III legacy building (fig. 5–11).

Fig. 5–10. A Type II legacy all-steel building

Fig. 5–11. A classic Type III legacy building

Type IV legacy: There is no such thing! All humor aside, the use of heavy timbers for floor and roof beams had all but disappeared after WWII. Still, some legacy-era buildings are built as Type IV for those who can afford it—most notably, churches and resorts. As can be imagined, the fire spread and collapse concerns with each of the legacy types above can be varied and numerous. Generally speaking, however, some collective issues can apply to the legacy-era building.

Fire spread issues: The application of complex building, fire, and life-safety codes actually helps the firefighter in legacy-era buildings. The tenet of aggressive interior firefighting became commonplace in America because of the inherent strength and protections built into a code-compliant legacy building. The legacy era brought fire protection systems that were very helpful to firefighters: better automated suppression systems, detection systems, occupant alerting devices, standpipes, victim refuge areas, well-defined exit/egress paths, and smoke control systems. The real issue was having a water delivery system that could meet the demands of fires in large and/or numerous interior spaces, and having enough firefighters to move that system into the building.

Granted, not all legacy buildings had all these features, but codes required their inclusion where history showed they could prevent disaster. Overall, we paint a pretty rosy picture of minimal fire spread issues in the legacy-era building. However, there are some serious issues that should be considered:

- Modern contents. The legacy building was built in the era of hardwood and steel furniture/appliances, typewriters, cotton and wool fabrics, and paper communications. Today's typical fire loading includes an abundance of synthetics, electronics, and interconnected communications systems. Explosive smoke environments are created when these materials pyrolize. Protective systems and devices that were built into the legacy building can be simply overwhelmed by today's fire load.

- Glass. The use of more windows and steel/glass curtain walls allow faster floor-to-floor fire spread in the legacy building as compared to previous building eras. Additionally, with the influx of tempered and double-pane glass, windows are more difficult to break for ventilation operations. They are also capable of lasting an additional amount of time when exposed to heat/fire, which can easily increase the potential of an interior flashover prior to their failure.

- Technology updates. Most legacy buildings had simple electrical and telephone communication systems built into the walls (and protected chaseways). The digital age brought the need for more electrical power capability, data transmission systems, and backup storage and power centers. The legacy building had to be rewired for all this. Tearing into walls to rewire a building is quite costly, so to update the legacy building for modern data and communication systems, many

use cable trays attached to the interior finish side of walls, floors, and ceilings. The addition of exposed power and data trays can be found in all manner of legacy buildings. A simple data cable tray fire can cause fire spread from room to room and floor to floor. The smoke from heated or burning cables can be very poisonous and flammable. Uninterruptable power supply equipment and battery rooms (and their associated fire issues) may also have been added to the legacy building.

Collapse issues: The legacy building is arguably the most firefighter-friendly building from a collapse potential perspective. Lessons learned from fires and earthquakes, material ingenuity, engineering, code development, and durability demands seemed to all come together for the legacy building. Having said that, it is also important to note that all buildings will eventually fail/fall during uncontrolled fires. (Gravity is always present!)

Likewise, the legacy era also saw the gradual shift to truss roof construction. Firefighters should be well warned of the dangers of the truss. Looking back, however, the legacy truss was a higher mass (and well-connected) structural element compared to the modern lightweight truss. There are many examples of legacy trusses still intact after fire has totally consumed roof sheathing (fig. 5–12). Interestingly, the conventional flat roof on the right side of the building in the photo has *totally* collapsed.

Fig. 5–12. Trusses in a legacy type building can still be intact after a major fire.

Although firefighter-friendly, the legacy building still has some collapse issues due to a number of factors. We can summarize these as follows:

- Open spans. The use of steel really opened up interior spaces. Open or wide beam spans are always a collapse issue for any building.

- Occupancy shift. The original legacy building was built for a specific use/purpose, following a prescribed code. As these buildings were bought and sold, areas redeveloped and demographic shifting took place, and the original legacy building may now be used in a manner that was never intended. The classic example is the suburban retail strip mall that is now being used as a church or community center. All manner of alterations to fit the new occupancy can alter the collapse resistance that was found in the original building.

- Overloading. The legacy era also brought the materialistic society. Likewise, memories of rationing during war and the Great Depression were still present, which led to overstocking, hoarding, collecting, and storing of tremendous volumes of goods. This added weight may not have been designed into the engineering of the building, and thus accelerating collapse potential. Air conditioning was still considered a luxury in the legacy era. Many of these buildings have since added A/C units as society has come to expect greater comfort. The added weight of rooftop A/C units may have fit into the design strength of a roof, but it does subtract integrity time during fires. Overloading can also come through the cumulative nature of building improvements. For example, a new A/C unit plus a new roof *plus* a new facade *plus* a new elevator and ramps to meet the Americans with Disabilities Act of 1990 (ADA) requirements *plus* high-rack storage for archives *plus* a data/server room for computers and backup *plus* enhanced lighting fixtures *plus* new solar panels *plus* . . . well, you get the picture. Some of these additions may have been installed with thoughtful engineering, while some may have been added in dubious fashion. The catastrophic

The Art of Reading Buildings

collapse occurs when the cumulative dead load is suddenly stressed. For example, when the building catches fire or is subjected to the extreme weight of a 100-year rain or snowstorm event!

- Aging. Steel rusts, wood decays, mortar cracks, and roofs leak. The legacy building is getting old and does require maintenance to keep it sound. If the legacy has an Achilles' heel, it would have to be the roof. Many legacy-era buildings have had to have their roofs repaired, altered, or replaced. In most cases, the various types of roof repairs work against firefighters. Examples are multiple roofing layers, addition of weight (tile, HVAC, etc.), the use of lightweight construction to replace aging structural members, and so on.

Quick summary

- Innovation highlighted the era that followed World War II and was heightened by the baby boomer generation.
- This era was instrumental in the implementation of what is now referred to as the legacy generation.
- The legacy era can be characterized by Type I, II, III, and IV buildings as follows:
  - Type I: Much larger buildings, increased regulations from restrictive building codes, high-rise buildings used steel-reinforced masonry and concrete, elevators and stairwells, increased use of HVAC systems, lightweight concrete for floors, and exterior glass curtain walls.
  - Type II: The use of reinforced concrete masonry units (concrete cinder blocks) for load-bearing walls and the use of steel for primary structural members and trusses. The all-steel building was also introduced.
  - Type III: Utilized reinforced concrete masonry units for bearing walls and solid wood floor and roof joists were nailed into ledger boards that were attached to the CMUs. Rigid wood trusses became prevalent.
  - Type IV: The use of heavy timbers for floor and roof construction was diminishing. New Type IV buildings were limited to expensive examples like churches and resorts.
- Interior contents were typically comprised of conventional materials instead of synthetic materials.
- Many legacy buildings have been modified to incorporate the requirements for modern communications technology.
- Although these types of buildings are more resistant to fire and collapse than modern buildings, they are being replaced (in increasing frequency) by lightweight construction/trusses.
- Due to the availability of these buildings, many have been converted into an occupancy that is significantly different from the original intent of the construction.
- Aging (especially roofs) may be the principal detriment found in the legacy buildings.

Engineered lightweight era. There is no clear dividing line between the legacy-era building and the engineered lightweight-era building—that is, there is no event (like a world war) that triggered the era shift. Some may point to the energy crisis of the 1970s as the trigger to maximize raw building materials through engineering innovation. Actually, fire officers started to see a shift toward lower mass building materials (and faster collapse during fires) starting in the 1960s, as the use of wood trusses held by gusset plates first appeared at that time. Some of the first composite trusses (stamped, cold-rolled steel and wood) were also being used in the 1960s. It can be said, though, that engineered lightweight components were slowly being added to what was otherwise a legacy building through the 1960s and 1970s (mostly in roof assemblies).

One reason the engineered lightweight building (ELB) was slow to evolve had to do with the language that was used to write codes—they specified exact methods to create something. For instance, to meet code, a given size solid wood beam had to be nailed to a given size load-bearing wall using a specific size nail with a specific nailing pattern. Trusses had to be nailed or screwed together following a prescribed schedule. Through

the 1980s, the building industry made the case that codes should allow more innovation because there was more than one way to build the same thing. Thus began the engineered lightweight era in earnest.

Using the short history lesson above, a fire officer may conclude that most anything built after 1990 is likely to be an ELB. This is a safe conclusion, although it is important to note that there are still some architects/builders who insist on using methods that include quality high-mass materials, solid connections, fire resistance, and "over-built" strength.

The defining descriptor for the engineered lightweight era is *lightweight*. The use of low-mass or high surface-to-mass materials, glues, staples, engineered shapes, and assembly-built systems all combine to form the ELB (fig. 5–13). Of particular note is the concept of *assembly-built*. As mentioned in chapter 3, the assembly-built concept relies on each and every component to interact with others to form a structurally sound building. The assembly-built approach accounts for redundancies and safe tolerances for durability and strength for extremes such as earthquakes, winds, snow, and realistic live loads. Engineers have shown that this approach yields a stronger building that uses less material.

Fig. 5–13. An example of a modern engineered lightweight building that uses low-mass materials.

Unfortunately for firefighters, the assembly-built ELB comes up short on duration and structural integrity when attacked by fire.

Chapter 4 discusses the fire and collapse issues associated with Type V lightweight buildings, and chapter 6 covers some of the fire and collapse issues that can be expected in other types of ELB buildings. For now, though, we can summarize the ELB fire and collapse issues by saying that we have hotter fires in lower mass buildings—a dangerous combination for firefighters. Interior firefighting time has been lost in the ELB.

Quick summary

- In the engineered lightweight era, conventional framing techniques and full-dimensional lumber were replaced by truss and lightweight construction that could be comprised of wood, metal, or a combination of both.

- Lightweight and/or truss construction has proven to fail at a significantly faster rate than conventional full-dimensional lumber construction.

- The recent implementation of adhesives to replace rivets, bolts, and gang-nail plates for connection points has further resulted in faster collapse time frames!

- In some cases, interior firefighting time in this type of construction has been degraded to the point that it is considered unsafe and high-risk.

Use considerations

The first sections in chapter 4 made the case that firefighters need to classify buildings by *era/use/type/size*. We also make the point that the various codes used for the design and construction of buildings are written to weave together the intended building use with construction type requirements. In this section, we begin to discuss the building use categories that help define how a building is supposed to be built.

Even though *use* is listed second in the building classification consideration, it is the *first* consideration made when someone wants to build a structure. The intended use or purpose of a building is called the **occupancy**. Once the occupancy is determined, architects factor in the size that the building needs to be for the desired number of people, equipment, storage, and/or processes that will be hosted by the building. These determinations then get referenced to an applicable code that directs the type of construction (and mandatory features) that must be incorporated in the construction of the building. Obviously, these requirements have evolved over time, leading to the various eras we covered earlier. The evolution of the occupancy construction method requirements have been modified/updated by many factors, but none as influencing as the killing power of fire.

Occupancy requirements are designed to give building occupants a reasonable chance to escape a building when a fire breaks out. Notice that we *didn't* say that code requirements were written to give firefighters a reasonable chance to enter a building and fight a fire. As a rule, newer buildings have more life-safety design features than older ones for a given occupancy use. This rule assumes that the building is being used for the occupancy that was intended when it was first built. Life happens. Buildings are commodities that get bought and sold, swapped, altered, razed, renovated, and revitalized based on economics, demographics, and/or owner desires and needs.

The reality is that firefighters fight fires in buildings that were built for one use, but, for many reasons, are now serving another purpose. You can imagine the fire and collapse issues that are created when a building built as a simple retail shoe store is now being used as an auto body shop. Or the classic occupancy switch of a single-family home that is now a law office (fig. 5–14). The home-turned-office will likely have walls removed, data systems installed, tremendous weight of books/files added, and commercial-sized HVAC systems incorporated.

Fig. 5–14. A single-family home that has been converted into a law office.

To combat this very real concern, many local jurisdictions adopt laws or ordinances that require occupancy changes to follow a permit system to ensure that a building is fit to meet the fire and dead/live load concerns for the new occupancy use. Unfortunately, building owners and occupants make undocumented changes to buildings that impact the life-safety and structural integrity characteristics of their building. Following are some common traps that can be expected to be encountered:

- Changing the occupancy use
- Changing the building size (fig. 5–15)
- Adding square footage, subdividing existing space, and altering floor plans
- Lightweight construction that has replaced conventional construction

Fig. 5–15. Changing the size of a building can alter the physical dimensions as well as the interior floor plan considerations. This is a single dwelling that is comprised of five separate additions that are all interconnected!

Defining occupancy uses. Most fire departments require first-due officers to give a brief building description as part of their on-scene radio report (in addition to smoke/fire conditions, actions being taken, and command declarations). Because this radio report needs to be brief, we tend to oversimplify the building occupancy use portion of the report. As an example, residential or commercial seems to be the most common. This approach may be fine for a rapid radio report, but for truly reading a building, it is woefully inadequate. The great fire officer realizes that there are many types of occupancy classifications that are tied to certain features that can help them better understand the fire spread, collapse, and hazardous tactical issues that might be expected for a given building. As an example, there is a significant difference between a small commercial building that houses a barbershop and a large commercial warehouse building.

As a starting place, there are several code-developing entities that have defined occupancy uses for their various audiences. Namely, the *NFPA 5000: Building Construction and Safety Code* developed by the National Fire Protection Association and the *International Building Code* (*IBC*) developed by the International Code Council. These codes are very similar in the groupings of occupancy uses. The two diverge a bit in the subgroupings that are found in each. The point here is not to memorize the code groupings, but to be aware of these codes and differentiate the occupancy uses.

Note: Be aware the previous IBC and NFPA codes refer to occupancy use as types. However, we feel that the term "use" is a better representation in the era/use/type/size building size-up model that we present in this chapter and throughout the balance of this book.

The following descriptions are the primary groupings that are paraphrased (for simplicity) from *NFPA 5000* and the *IBC*. (The following groupings are expanded into more common and definitive fireground buildings in chapter 10.) Just know that the codes have very specific wording, examples, subgroups, and tables that more specifically define occupancy uses.

Assembly. Buildings used for the assembly of people for civic, social, religious, recreational, food or drink consumption, or awaiting transportation.

Business. Buildings used for offices, professional or service transactions, storage of records, or ambulatory health care.

Day care. Buildings used for the supervised care of children with no overnight care. (NFPA recognizes day care facilities separate from educational.)

Educational. Buildings used for educational purposes of six or more people (at any one time) through the 12th grade and some day care facilities. (ICB puts day cares in an E classification.)

Factory. Buildings used for assembling, disassembling, fabricating, finishing, manufacturing, packaging, or repairing operations that are not classified as Group H (hazardous) or Group S (storage).

Hazardous. Buildings used for the manufacturing, processing, generation, or storage of materials that constitute a physical or health hazard in quantities in excess of allowed control areas.

Institutional. Buildings where people are cared for or live in a supervised environment because of health, age, medical treatment, or those detained for penal or correctional purposes.

Mercantile. Buildings used for the display and sale of merchandise and involving the stocking of goods that are accessible to the public.

Residential. Buildings used for sleeping purposes when not classified as Group I.

Storage. Buildings used for storage that is not classified as hazardous.

Utility and miscellaneous. Buildings, and buildings of an accessory character, that are not classified by any other occupancy use. This group

includes structures such as agriculture buildings, carports, sheds, tanks, and towers.

Fire and collapse issues for each of the occupancy groups above are numerous, yet somehow obvious. Specifically, fire spread issues are closely related to the contents that can be found in a given building. The building occupancy grouping can help clue the firefighter into the potential fire load. Likewise, the arrangement and quantity of the fire load (as suggested by the occupancy group) may help a fire officer envision collapse potential.

Quick summary

- Occupancy use is a primary consideration when designing a building. However, when the occupancy use is changed after a building is constructed, the inherent hazards will also change.
- Building code requirements are normally not written for the benefit of fire service personnel.
- In some cases, a brief occupancy use description should be included in an initial size-up report.
- *NFPA 5000* and the *International Building Code* provide a primary basis for the grouping of occupancy uses.
- The occupancy classification of a building can help fire officers envision the potential fire load of a building, which in turn helps them determine fire spread and collapse potentials.

Type considerations

Determining the type (construction) of a building from the street can often be a challenging adventure, as buildings in this country have used various types of construction materials and techniques over time, and those materials and techniques are continuing to change as technology continues to evolve in the building industry. Unfortunately, most firefighters will readily agree that past and present changes that have affected building construction have typically been a detriment to the fire service, and if history is an indicator of what can be expected in the future, this trend is certain to continue. This is a main reason why a primary element of a size-up should focus on the type of building construction, as all buildings are not identically constructed, nor will all buildings react the same and in a similar amount of time when exposed to heat and/or fire.

Without a doubt, the ability to determine the type of construction will aid in determining the following three primary building attributes (fig. 5–16):

- The relevant strength of a building
- The amount of time available for interior operations
- The length of time a building will resist the effects of heat and fire prior to collapse

Fig. 5–16. Determining the type of construction can help to identify building strengths/weaknesses and how a building will resist the effects of heat and fire.

In chapter 4 we discuss *NFPA 220*, which provides a starting point in being able to rapidly classify building construction types into five classifications. Even though we make the point that era and use considerations are important to classifying a building, the five classifications approach provides a quick and widely accepted way to understand and communicate the underpinnings of a building. A brief review follows.

Type I: Fire resistive. This construction consists of monolith (poured concrete on rebar), concrete and coated steel, and protected steel. Modern multistory buildings are common examples that use this type of construction.

Type II: Noncombustible. This construction is similar to Type I construction except steel structural members are unprotected. Modern multistory buildings with unprotected steel components are a common example of buildings that use this type of construction.

Type III: Ordinary. This type of construction consists of exterior load-bearing walls of noncombustible materials such as concrete block or brick and an interior of combustible wood components. Common buildings are older taxpayer and multistory residential (hotel) buildings, and modern strip mall buildings with concrete block walls and lightweight wood truss roof assemblies.

Type IV: Heavy timber/mill. True mill construction consists of brick and timber, block and timber, and block and glulam. Common buildings are older large factories, warehouses, and churches. Some newer applications can be found in buildings that use glulams for primary structural members, such as newer churches.

Type V: Wood frame. The primary building component for walls, floors, and roofs is combustible wood and/or wood products (engineered wood). The predominant example of this construction type is residential homes—the most common type of building in this country.

Although the preceding five construction classifications can be helpful in prefire planning a building, it does not address some aspects of buildings that can become crucial factors when a building is under demolition by heat, fire, and gravity. As an example, it is difficult to discern the difference (from the exterior) between protected and unprotected steel structural components. Likewise, it can be difficult to determine if a building falls under the classification of a hybrid (a nontraditional building or one that doesn't readily fit into one of the five classic types) or if engineered wood products have been incorporated into the building. Again, the era, use, and size considerations must be factored in to ultimately classify a building.

Quick summary

- Determining the type of construction is an essential element of the *era/use/type/size* method of building size-up.
- *NFPA 220* provides a simple and widely-accepted tool for rapidly understanding and communicating the basic construction underpinnings of a building.
- Evaluating the type of construction can enhance one's ability to determine the relevant strength (and weakness) of a building and how long a building will resist the effects of heat and fire prior to collapse.
- Ultimately, the construction type of building must be further defined by era, use, and size factors.

Size considerations

The last (or fourth) element of the *era/use/type/size* method of classifying buildings is that of building size. The size element can present an interesting dichotomy of simplicity and potential frustrations. The simplicity comes from the common fire service terms that are typically used to describe building size: small, medium, large, big-box, mega-box, and high-rise. The frustration comes from the interpretation of these terms. Granted, descriptions like big-box, mega-box, and high-rise may seem obvious. The same cannot be said for small, medium, and large. There is a key difference between an 1,800 sq ft single-story prairie-style home without a basement and a 5,000 sq ft Victorian-type three-story residential home with a basement. Other than square footage, what is the primary difference between these two examples? The answer lies in the potential and quantity of materials available to burn and

the amount of resources and time it will take to complete all suppression operations. Although both of the examples above are residential homes, giving the size of a building during a size-up can make a noticeable difference in a perceived assessment to incoming companies.

In many ways, small, moderate, and large are *relative* terms that can mean different things based on six considerations:

- The perception of the individual making the judgment
- The footprint or single floor square footage of a building (width × length)
- The interior arrangement of walls and contents
- The number of floors above ground and levels below
- The amount of available resources (apparatus and staffing) and the capability of those resources
- Size of the fire

So, in your mental evaluation of a building and/or what is expressed on a radio size-up, what do the terms small, medium, and large mean?

From a street perspective, judging the size of a building is really an impression, like "Whoa, that is a really big house!" Those impressions are usually formed relative to the typical buildings that are found in your response areas and the capabilities you typically bring to a fire event. So again, the question arises, how do we define the differences in building sizes? In an attempt to attach a general definition to the terms small, medium, large, consider using a rule of thumb we call the two-four-six method.

The two-four-six method takes into account the aforementioned six relative considerations used in making building size judgments and can serve as an easy-to-remember communication tool for defining small, medium, and large. Let's show how this can work.

Two = small. A small building is one that

- is *two* stories or less (or a single story with basement);
- is *two* thousand square feet or less;
- requires *two* hundred feet or less of hand line to reach any part of the building from the main access door;
- needs *two* hand lines or less to handle an offensive operation (1¾ in., 2 in., or 2½ in.).

Four = medium. A medium building is one that

- is *four* stories or less (or three stories with basement);
- is *four* thousand square feet or less;
- requires *four* hundred feet or less of hand line to reach any part of the building from the main access point;
- needs *four* hand lines or less to handle an offensive operation (1¾ in., 2 in., or 2½ in.).

Six = large. A large building is one that

- is *six* stories or less (taller is a high-rise by most definitions);
- is *six* thousand square feet or more (dubious, but it fits the theme);
- requires hose lengths greater than *six* hundred feet to reach portions of the building;
- needs *six* hand lines or more for offensive operations (1¾ in., 2 in., or 2½ in.).

It is important to remember that this is only a rapid size-up tool for making an initial judgment of building size. It should be used in a conservative fashion that leans toward the bigger size. For example, a building may be less than 2,000 sq ft and two stories—a small building. But if the arrangement of that building looks like there will be hose stretches of more than 200 ft, it is prudent to call it a medium-sized building.

Applied, figure 5–17 appears as a single-story dwelling that is roughly 30 ft by 35 ft or 1,050 sq ft, and is well within the reach of a 200 ft preconnect.

Two lines could easily drown this house. Obviously, we should call this a small building.

Fig. 5–17. This dwelling would be considered small using the *two-four-six* rule of thumb for communicating size.

Figure 5–18 is a little trickier. Granted, it is a single-story commercial building that initially appears decent sized. A closer examination would reveal dimensions of 45 ft by 90 ft. That would fit the medium category using the *two-four-six* rule of thumb. Actually, if the building was a bit larger or had a second floor or basement, it could weigh in as a *large* building. We know that some firefighters would scoff at the notion that this building could be classified as large—point taken. But for most fire departments, this could be a large building because of its volume and content potential. The conservative fire officer will plan for worse case. The fire load potential of this building and unknown interior arrangement could require more than four hand lines. Granted, you could open up the boarded-up windows and door and find a wide open and empty space. That makes this a medium building—we'd rather judge bigger and be wrong.

Figure 5–19 is a good example of a large building. It is a two-story URM-constructed building with a basement. The building appears vacant and has only one visible entry. It also appears to be an industrial era manufacturing type building that would present wide-open and/or interconnected interior spaces that will enhance fire spread.

The 6,000 sq ft and restricted access would require multiple lengthy hose lines (most likely 2½ in. handlines) for an interior attack on a working fire.

The previous example brings up a good point—size classifications can be skewed based on the intended use of the building: residential or commercial. Later in this chapter, we look at bridging the size/use issue with some tangible influences. Before we do, let's finish defining the more obvious size classifications of big-box, mega-box, and high-rise.

Fig. 5–18. A common single-story commercial building of about 4,000 sq ft.

Fig. 5–19. An example of a large building that has the potential of presenting a formidable challenge to most fire departments.

Big-box. Big-box buildings are large box-type buildings with noteworthy fire loads that can easily exhaust the resources of many fire departments. Common examples are Lowes, Home Depot, Wal-Mart, Costco, and the anchor stores of large open-air malls. By code, these buildings typically require the installation of fire suppression sprinkler systems. Fires that exceed the capability of the sprinkler system (or an older big-box that is not protected by a system) present major challenges. Other buildings may fit into the big-box category, such as aircraft hangers, warehouses, multiplex theaters, and assembly-line manufacturing buildings. These buildings can require resources (depending on the size of the fire) that most fire departments cannot supply without outside assistance.

Mega-box. Mega-box buildings can be defined as buildings with mammoth proportions. As a result, this definition would refer to those few buildings that are significantly larger than big-box buildings. Some of the largest examples include the MGM Grand and Caesars Palace hotels in Las Vegas as well as the Boeing 747/777/787 assembly plant near Everett, Washington (Payne Field). Although smaller (pun intended), the mega-box size category can be applied to buildings like large commercial airport terminals and mega-malls that have multiple anchor stores plus enclosed concourses and shops (each anchor store is likely to be a big-box unto itself). Like the big-box, mega-box buildings can require resources (depending on the size of the fire) that most fire departments cannot supply without a significant amount of mutual/automatic aid.

High-rise. The previous size categories are basically street slang and defined based on a firefighter's perspective. High-rise buildings are a bit different as there are several nationally-recognized definitions for them. While these definitions vary (minimally), most categorize a building as a high-rise when the vertical height of the building exceeds 75 ft and/or seven to eight floors. The challenges of high-rise fires are well documented and subsequently, fire departments have developed high-rise protocols for their occurrence. Interestingly, the aforementioned big-box and mega-box buildings refer mostly to the horizontal footprint of the building—they are *horizontal high-rises* that should also be guided by protocol.

As promised, we now cover some tangible influences that help bridge the size/use issue.

Residential. Although residential buildings are the most common type of building, do not overlook the fact that these "common" buildings are responsible for the greatest number of civilian and firefighter deaths. Additionally, a residential building is not necessarily a typical single-family dwelling. As an example, there is a substantial difference between a two-and-a-half-story craftsman-type home, a three-story hotel, a 1,600 sq ft single-story home, and an older large residential home that has been converted into a church (fig. 5–20). As a result, when observing and/or giving a size-up regarding a residential structure, it can often be advantageous to delineate the size and the use. Let's consider some of the more common factors that can influence the size evaluation of residential buildings.

Fig. 5–20. A large residential home that has been converted into a church.

- External size is one of the first factors that is viewed and is comprised of length, width, and height—the combined total of the portions of a building above and below grade (i.e., multistory and/or basement). Fortunately, the external size of most residential buildings can be one of the most visible factors in evaluating a residential building (and other types of buildings also).

- The presence of below grade features may not be so visible. As an example, in some areas of this country, cellars and basements are common below the grade level.

- Daylight and walk-out basements are very common to residential-type buildings. It is important to size up the perimeter (do a 360° size-up) to ensure the entire dimension of a building is viewed.

- Some residential buildings are actually commercial-residential. The local Holiday Inn, apartment complex, and tenement can fall into this commercial-residential mix. If the building is a commercial-residential building, it is not only a larger type of building but also has a higher occupant load than common residential dwellings (and at all hours of the day).

- Is there a possibility of a multistory residential home (e.g., Victorian, colonial, etc.) being converted into an illegal multifamily residential building?

- Has the building been converted into a commercial-only building? Doing so often includes the removal of interior walls, which is a double-edged sword. This changes the compartmentalized areas within the building and may help with fire stream penetration, but it increases heat release rates and accelerates collapse potential.

- Some residential homes (typically of custom construction) will have large attic areas that have been primarily constructed to adhere to a particular style. The inherent size of these attics can easily offer the same or greater volume as an additional floor.

- Consideration must be given to the possibility of shared or common attic or cockloft spaces that span several dwelling units.

- Additions that significantly expand the original residential building are rarely visible from the street side. Accuracy in judging building size may require a 360° look at the building.

Commercial. The ability to conduct a rapid size-up of a commercial building can be more difficult due to three primary factors that are not often inherent to common residential-type buildings. These factors are size (exterior and interior), contents (fire load), and a potential lack of people after business hours (search considerations). The factors can have a major impact on the time and resource commitment that is necessary to safely mitigate a structural incident that involves commercial buildings. However, as commercial-type buildings come in a vast variety of sizes, the visible size of these buildings may often be one of the most simplistic factors that can be quickly verified.

Fig. 5–21. Commercial buildings are not always large buildings.

As an example, there is a vast difference between a 30 ft by 50 ft single-story Type III building that is being used as a party supply outlet (fig. 5–21) and a 100 ft by 250 ft single-story concrete tilt-up Type III building that is being used as a warehouse/distribution center for electrical supplies. As you can see, there is a vast difference between these illustrations, yet they are both examples of a commercial building. As a result, this is a primary reason why the size of a building should be considered, and in many cases included in a verbal initial size-up. Let's consider some of the more common factors that can influence the size evaluation of commercial buildings:

- The external size is one of the first factors that is viewed, and includes the length, width, and height—a combined total of portions above and below grade (i.e., multistory and/or basement).

The Art of Reading Buildings

Photo exercise 5–2

Era:

Use:

Type:

Size:

Potential fire spread and collapse issues:

Chapter 5
CLASSIFYING BUILDINGS—HYBRID, ERA, USE, TYPE, AND SIZE CONSIDERATIONS

Photo exercise 5–3

Era:

Use:

Type:

Size:

Potential fire spread and collapse issues:

Photo exercise 5–4

Era:

Use:

Type:

Size:

Potential fire spread and collapse issues:

Photo exercise 5–5

Era:

Use:

Type:

Size:

Potential fire spread and collapse issues:

Photo exercise 5–6

Era:

Use:

Type:

Size:

Potential fire spread and collapse issues:

RESOURCES FOR FURTHER STUDY

- *International Building Code, 2012 Edition*, Washington D.C.: International Code Council, 2012.

- Mintz, S., & McNeil, S. (2013), *Digital History*, retrieved March 5, 2014, from http://www.digitalhistory.uh.edu.

- Morgan, William, *American House Styles*, New York, NY: Henry H. Abrams, Inc., 2008.

- NFPA 5000: *Building Construction and Safety Code, 2012 Edition*, Quincy, MA: National Fire Protection Association, 2012.

NOTES

1. Taken from Dave Dodson's "Incident Safety Officer Academy" lecture series.

2. NFPA 5000: *Building Construction and Safety Code, 2012 Edition*, National Fire Protection Association, Quincy, MA.

3. *International Building Code, 2012 Edition*, International Code Council, Washington D.C.

4. The authors have taken liberty to create these era categories. We feel it is important to teach firefighters how to read a building and understand that era considerations are an important part of that skill. We also know that there are buildings that can defy or negate the era descriptions. Nothing replaces the importance of walking through buildings and becoming familiar with the traps, fire spread, collapse, and tactical issues associated with the buildings in one's jurisdiction. We welcome input to better communicate the concept of *era*.

ALTERNATIVE AND EVOLVING CONSTRUCTION TRENDS 6

OBJECTIVES

- Describe the difference between prescriptive design and performance design.
- List four common alternative building construction methods.
- Define the term "green construction."
- Explain what is meant by advanced framing methods and list several features of AFM.

PERFORMANCE DESIGN HAS NOTHING TO DO WITH FIREFIGHTERS

Chapters 4 and 5 outlined common construction methods and classifications as well as their associated fire and collapse issues. We also introduced the concept of performance-based design construction methods that are found in engineered lightweight construction. In this chapter, we look at alternative and evolving building construction methods. Before we dive in, it is important for firefighters to understand the trends and influences that are changing the way buildings are now being built.

The building construction industry is being tasked to produce buildings that are economically and environmentally friendly. More specifically, the industry must consider the following criteria:

- Building energy use
- Indoor environmental qualities such as light, air, and noise
- Water management
- Waste and pollution reduction

- Material sustainability and durability
- Protection of the surrounding environment during construction and subsequent occupancy

To achieve positive outcomes for these considerations, the building industry (designers, builders, and suppliers) has had to change its general approach to building. Previously, buildings were built using a *prescriptive* approach. To achieve economic and environmental friendliness, the industry has changed to a *performance design* approach. Let's look at the difference.

Prescriptive building approach

The prescriptive approach basically means that a designer (architect and engineer) draws up a plan for how a building is to be built. That design then goes to a contractor to bid the job. The contractor then finds suppliers to deliver materials needed for the project. This approach emphasizes how a building needs to be constructed and is less focused on the end performance of the building. Often, it is strict codes, regulations, and standards that drive the prescriptive approach. This approach places the contractor in a position to make many workaround-type decisions when things don't exactly fit or material requirements are not specified or available. This is not to say that the prescriptive approach is bad, most industrial- and legacy-era buildings were built prescriptively. Unfortunately, some of the prescriptive methods stifle innovation toward environmental friendliness.

Performance design building approach

The origins of performance design construction began in pre-biblical times—the architecture code was basically a performance criteria to "build a house that would not collapse and kill anybody." Through trial and error, newer performance design alternative building methods began to emerge in the 1980s, although examples were few. The 1990s brought a period of great change in code allowances and the subsequent increase in alternative building methods.

The performance design approach to building is more concerned with what a building is required to do and not necessarily with how it is to be constructed. Simply stated, it is an approach that concentrates on the ends and not the means to the end. To make this work, some changes had to take place in the codes/regulations/standards world. Instead of telling someone how something needs to be assembled, the codes had to be reworded to tell someone how something needs to perform. Practically applied, instead of telling builders they must use 2 × 4 wood studs with ⅝ in. drywall on both sides to build a fire separation wall, they now say that a fire separation wall must not lose integrity in a given time frame with a given fire. This performance design approach allows material suppliers to diversify and innovate, and allows building designers to consider alternative materials and methods.

This general switch in building construction approach sets the stage for a whole new set of building construction considerations for firefighters. It is interesting to note that during the research for this chapter, it was never mentioned that a building needs to perform better for firefighters. In today's world, designers, contractors, and material suppliers have a moral (and legal) responsibility to create a building that notifies occupants of a fire and gives them a reasonable time to exit. Firefighting time is not part of the equation. As you'll see, alternative, performance designed buildings utilize lots of low mass (and high surface-to-mass) materials as well as an assembly-built approach that can lead to rapid structural failure during fires.

Quick summary

- The prescriptive building approach focuses on how a building needs to be constructed and not the end performance of the building.
- The performance design building approach focuses on what a building is required to do (end performance) rather than how it needs to be built.
- Firefighting time is not part of the equation for performance design buildings.

ALTERNATIVE BUILDING METHODS

When applied to the fire service and building construction, **alternative building methods** can be defined as building construction materials, assemblies, and systems that are nontraditional, unusually innovative, or don't readily fit into the five classic types. Most firefighters would agree that alternative building methods equates to buildings that are less substantial than their predecessors, which can result in rapid degradation of structural integrity during fires. On the surface, we agree that alternative building methods should be viewed cautiously from a firefighter's perspective even though some alternative building methods are very resistive to collapse under fire conditions (e.g., insulated concrete form flat panel, covered later in this chapter).

In the previous sections on prescriptive and performance design building approaches, we discussed the primary differences between these two methodologies that have resulted in buildings that could best be described as using a "standard" or "conventional" approach toward constructing a building. However, the current and future trend to build buildings faster, higher, and more economically (in many cases, economical can be defined by the fire service as dangerous) has led to alternative and evolving construction methods. These buildings can either incorporate a mix of standard and alternative methods and materials, or methods and/or materials that could best be described as leading edge, innovative, or in some cases, substandard as compared to the conventional construction methods and materials of yesterday.

As we have described in other portions of this book, a classic example of an alternative construction method is the substitution of lightweight materials for conventional materials. The substitution and resulting negative impact on the fire service is no stranger to firefighters. Accepting the reality that new buildings are likely to be lightweight is only a starting place. Firefighters must also accept that the rate of change has reached lightning speeds. That is, the building construction trade has always been evolving, but the rate of evolution was such that the fire service had time to digest the changes and develop training, code language arguments, and/or tactical procedures to deal with the evolution. In today's world, a new construction method or material can be developed, tested, produced, marketed, and applied to a building in less time than it takes to send a firefighter through academy (thanks to performance design codes). This rate of change underscores why it so important for firefighters to be constantly aware of the changes that are occurring in the buildings within their response areas, and how any notable changes can impact applicable structural fire suppression operations—either from a positive or negative viewpoint. Nothing can replace prior familiarity with a particular building as compared to being surprised, caught unaware, or paying the ultimate price from an unknown hazard that should not have been a surprise.

Fig. 6-1. A simple single-family dwelling that incorporates many modern, environmentally-friendly features.

A simple example of being aware of building construction changes in your response area is illustrated in figure 6–1. Although this modest single-story, single-family dwelling appears to be newly constructed and typical of many new homes, it is far from typical. This home incorporates many of the latest "green" characteristics: solar energy, energy-efficient double-pane windows, walls constructed from straw mixed with light clay, and the exterior finished with stucco-type materials

(the same dwelling before the stucco was applied is shown in figure 6–10). Additionally, the roof is constructed from lightweight wood trusses that use glue at their connection points. So, even though this is not a large building, how will it perform during a structure fire? Think about the following considerations:

- When exposed to heat and fire, the roof can catastrophically collapse in a short period of time from failure of glued connection points.

- The solar panels are imposing an added dead load on the roof, in addition to the electrical considerations that are associated with the solar panels.

- The energy-efficient windows will take longer to fail than standard windows, allowing the interior temperature to more easily reach flashover conditions in a shorter period of time.

- Similar to the energy-efficient windows, the enhanced insulation qualities of the light clay and straw walls will also accelerate flashover potential.

- Although the walls are a combination of clay and straw, the clay will not burn but the straw will. This does not present a notable hazard during suppression operations, but it can entail a laborious overhaul operation.

The balance of this chapter explores some alternative construction methods and materials that seem to be catching on. You may note that lightweight or engineered wood buildings are missing from this chapter—we believe that we covered them adequately in chapter 5 (in many ways, the lightweight wood building is now commonplace). We have also omitted some of the bizarre, demonstrative, or one-off methods that are out there. And, in full disclosure, there are likely alternative methods that we've never heard of and new methods and materials that will come out before this book can go through its first printing. For those that are listed, we give a brief description of the building elements and include some fire behavior and collapse issues that may present challenges to firefighters.

Quick summary

- Alternative building methods can be defined as building construction materials, assemblies, and systems that are nontraditional, unusually innovative, or don't readily fit into the five classic types.

- Most alternative construction buildings include lightweight/engineered elements, which equates to reduced integrity and firefighting time under fire conditions.

- The rate of construction evolution is such that firefighters must invest more time exploring new or renovated buildings in their response area.

Lightweight steel

In spite of the advancements in the use of wood in building construction, the current advantages of using lightweight steel as compared to wood are numerous:

- Reduced cost
- Significant weight reduction
- Ease of installation
- Resistant to termites and mold
- Noncombustible material to help satisfy code requirements for certain occupancies

Buildings that use lightweight steel can be a true steel framed style (studs, sills, headers, and joists) or they may be a post and beam style with lightweight steel C-channel studs to help partition the structure. In both cases, OSB sheathing is added as an exterior wall covering that stiffens the structure and adds shear strength for wind loads. Floors can be either metal pan with lightweight concrete or lightweight engineered wood (trusses or I-joists with OSB floor sheathing). Roofs are typically engineered wood trusses or lightweight formed steel trusses (fig. 6–2) with lightweight steel purlins for spacing and sheathing attachment. New lightweight steel buildings resemble wood frame

| Chapter 6 | ALTERNATIVE AND EVOLVING CONSTRUCTION TRENDS |

types when finished. Drywall protects the steel and engineered wood structural elements from interior compartment fires.

Fig. 6–2. Lightweight steel trusses are used for roof structural members. (Photo by Rick A. Haas, Jr.)

Fig. 6–3. This residential building uses lightweight steel for all framing members. (Photo by Rick A. Haas, Jr.)

For the true lightweight steel framed building, C-channel and H-channel shaped studs, sills, headers, and joists are used, as in the residential building in figure 6–3. These light steel components are only ⅛ in. thick, making them very flexible. Strength is achieved by the shape of the components rather than the mass, and the OSB sheathing that makes it stiff. For the post and beam, lightweight steel style, lighter I-beams and H-columns are used to form the external skeleton of the building (often called red-iron because of a red oxide coating) and lightweight C-channel steel studs are used to fill exterior and internal wall sections. Lightweight steel (⅛ in.) is commonly used in truss configurations (open-web bar joist) for structural members in many types of roofs (see fig. 8–17). An additional popular use of lightweight steel is on the exterior of structures built of concrete masonry units to provide decorative enhancements to an otherwise simplistic appearing building (fig. 6–4). Remember that these enhancements will create numerous voids between the steel members and concrete masonry units.

Fig. 6–4. Lightweight steel is used to provide decorative enhancements to the exterior of a concrete masonry unit building.

Fire spread issues: The interior finish of lightweight steel construction uses drywall, which allows fire spread like most other new Type V buildings. Building features like facades, large truss spaces, and open interior geometry can make smoke-induced fire spread a serious concern for firefighters. An additional hazard associated with lightweight steel, particularly C-channel steel studs, is that heat can cause the studs to warp/twist, pulling the attaching screws out of the drywall. This can result in drywall detaching from the steel studs and allow a fire to more easily extend in a horizontal direction. This has been noted in some

office buildings and strip malls where the use of steel studs is commonplace. Similar to open-web wood trusses used for roof structural members, lightweight steel open-web bar joists and similar configurations used for roof structural members also allow the rapid horizontal spread of fire between adjacent trusses. Lastly, remember that steel decorative enhancements on the exterior of concrete masonry unit buildings will conceal voids behind the steel facing.

Collapse issues: High fire loading from contents is a real fire spread threat in almost all new buildings and can deteriorate drywall in rapid fashion. The drywall is an essential part of the collapse-prevention system during fires in lightweight steel buildings. Once fire or hot smoke enters structural element spaces, the lightweight steel will rapidly lose its integrity and start to soften and elongate or buckle, losing the shape that is essential to handle imposed loads. The biggest collapse threat of these buildings comes with a fire in an unfinished basement. Expect a rapid and general collapse of flooring due to the low heat resistance of lightweight steel.

Roof structural elements of lightweight steel typically use pinned truss shapes that are spaced using crosswise steel purlins that support OSB sheathing (refer back to fig. 6–2). This assembly arrangement suggests that the roof will sag before it collapses—providing a slight, but not foolproof warning sign. Additionally, lightweight steel bar joist roof structural members can readily sag and/or collapse when exposed to heat and/or fire. Roofs that use engineered wood trusses and I-joists (in combination with lightweight steel framing members) will collapse quickly after glue breakdown or loss of gusset plates due to heat or fire involvement (look for tan or brownish smoke leaving eaves and roof space vents). Sagging of an engineered wood roof assembly is a late sign—the roof assembly can collapse well before sagging occurs. Firefighters walking on a heat-weakened engineered wood roof can cause the collapse.

Quick summary

- Lightweight steel framing has numerous benefits over lightweight wood construction.
- Lightweight steel buildings can be constructed as a true frame style (load-bearing studs) or as a post and beam style.
- Lightweight steel is being used for the makeup of trusses, facades, and other structural and nonstructural additions in block (CMU) buildings.
- Steel studs (and other structural or framing members) can warp/twist when subjected to heat, causing attached drywall to fail and allowing the horizontal extension of fire.
- Sagging of lightweight steel elements is a proactive warning sign of collapse.

Insulated concrete form (ICF)

Insulated concrete form (ICF) is a newer construction method that was initially developed in Europe after World War II as a cost-effective method to replace/rebuild damaged buildings. **ICF buildings** are those that use permanent expanded polystyrene (EPS) forms for poured concrete and come as panels, planks, or blocks (fig. 6–5). EPS is the most common material for ICF forms, although they can be manufactured from cellular concrete, cement-bonded polystyrene beads, cement-bonded wood fiber (composed of 15% cement and 85% recycled expanded polystyrene), and a combination of approximately 15% cement and 85% mineralized wood chips. The forms are left in place after the concrete cures and serve as a backing skin for interior drywall and exterior stucco, siding, or brick veneer. Obviously, the EPS also provides increased insulation for the wall. As early as the 1960s, ICFs were used as basement foundation walls. In the 1990s, builders started to use the ICF concept to form all exterior walls—as high as three stories. ICF buildings over three stories are likely to be more like a reinforced concrete building with substantial use of steel.

Chapter 6

ALTERNATIVE AND EVOLVING CONSTRUCTION TRENDS

Fig. 6–5. Insulated concrete forms

ICF buildings are becoming quite popular in areas where outside temperature extremes are common, due to their outstanding R-values. Additionally, ICF decking is gaining in popularity as it weighs about 40% less than conventional concrete flooring and also provides good insulation properties. ICF decking can be designed with ICF walls to form a monolithic structure that is joined by rebar. Most ICF buildings utilize engineered wood floors and roof assemblies, although some cases of lightweight steel exist. When finished, the ICF building looks like any other building, although the walls are typically thicker than traditional framed ones. Identifying these buildings is best accomplished during construction and appropriate notations need to be filed in accessible preplan systems. As mentioned previously, there are three principal methods to build an ICF wall: flat panel, post and beam, and grid block.

- **Flat panel.** This system uses interlocking flat panels of EPS to form a mold wherein a solid mass of concrete is poured. The voids between the EPS panels (for the concrete) are spaced using plastic or coated metal plates, which help ensure a uniform thickness of concrete. The plates are also used to support window frames, conduit, and piping runs that are installed prior to the concrete pour. The flat panel system yields a continuous mass of concrete like a conventionally-poured foundation wall. The concrete used to fill the forms can be low-slump (thick) with a high aggregate content—providing a fairly resistive mass for prolonged integrity under fire conditions. The finished flat panel ICF wall is approximately 80% concrete and 20% EPS and steel.

- **Post and beam.** The post and beam ICF system creates just that—a concrete post and beam skeleton (typically with steel rebar). The system is formed by using a combination of EPS planks and panels to form the hollow molds for the concrete posts and beams. The molds are interconnected to solid EPS blocks that create the wall infill (which uses minimal concrete) and serve as the mounting surface for interior drywall and exterior finish. The finished post and beam ICF wall can range from 40% to 60% concrete.

- **Grid block.** An ICF grid block system is one where thick EPS blocks are stacked and interlocked (like LEGO blocks) to create a wall. The EPS blocks are designed with internal cavities that create continuous vertical and horizontal channels for the concrete. Once poured, the concrete forms a lattice or waffle pattern that is encapsulated by the EPS. Steel rebar is typically added to some of the cavities (prior to pour) to give strength in areas around doors, windows, and corners. The concrete used to fill the cavities has to be a high slump (free-flowing) with minimal (or small) aggregate content. When finished, the grid block wall is primarily EPS or form material with as little as 20% to 30% concrete.

123

While all three form styles are categorized as ICF buildings, each reacts differently under fire conditions. The ICF panel system is a friend to firefighters (high-mass walls) although the inclusion of lightweight floors and roofs minimize the gain. The ICF post and beam system has some benefits (protected steel encased in concrete) but not like panel ICF. The grid block form should cause firefighters concern. Additional issues regarding fire spread and collapse are outlined next.

Fire spread issues: The EPS used for ICFs contains a fire-resistive additive that is not supposed to support flaming. Manufacturers of ICF products often cite this feature when questions are asked about fire spread potential. As most firefighters know, flames from other burning materials will cause the EPS to melt and produce additional smoke (pyrolysis). As the EPS smoke becomes airborne and heated, the additive breaks down and the smoke becomes flammable (remember, EPS is a petroleum product). The surface of EPS may resist flaming, but the emitted smoke eventually burns. Expect greater volumes of explosive smoke when the EPS pyrolizes.

Collapse issues: With flat panel ICFs, the firefighters' collapse concerns are mostly tied to the lightweight floors and roof: Once the EPS melts away, there is a huge solid mass of concrete. This is not true with grid block and post and beam ICFs. Expect minimal interior firefighting time and rapid collapse of ICF grid block and post and beam systems. The integrity of the concrete within these two systems is absolutely dependent on the EPS being intact. Once the EPS melts, the concrete becomes unsupported. Even though steel rebar is included in some critical areas, the concrete/steel was never intended (or tested) to be self-supporting. This is another example of the concept of assembly-built. The entire assembly—EPS, concrete, steel, shape, and covering—works together to prevent collapse. The waffle or lattice concrete found in grid block ICF should be especially concerning to firefighters because it is especially fragile without the EPS.

Quick summary

- Insulated concrete form (ICF) uses expanded polystyrene (EPS) molds for concrete, which are left in place after the concrete cures.
- The three primary configurations of ICF are flat panel (mostly concrete), post and beam (half concrete), and grid block (minimal concrete).
- The ICF EPS includes a fire resistive additive. Pyrolysis breaks down the additive, making the smoke produced from EPS explosive (and toxic).
- If polystyrene forms are weakened or removed by fire, the integrity of the concrete is minimized and/or lost.

Structural insulated panels (SIPs)

The **SIPs building** is constructed using panels to form load-bearing walls and the roof. Each panel consists of two outer skins of OSB with an insulating core made from expanded or extruded polystyrene (EPS or XPS, EPS is most common). Panels can range from a common 4 × 8 ft to panels as large as 8 × 28 ft and even 10 × 36 ft. Thickness can range from 4½ in. to 12½ in. The thickness includes the OSB and EPS. Most SIPs use 7/16 in. thick OSB for the skins, although 5/16 in. can be specified. Individually, OSB and EPS lack the strength to form a load-bearing element. When pressure laminated together, a synergy is created that forms a strong panel.

SIPs panels are prefabricated in a factory and then shipped to the job site. There, a small crane is used to assemble the house like the house of playing cards you built as a child (fig. 6–6). Panels are joined using glued OSB splines that fit into slots formed in the EPS (fig. 6–7). The wall panels are topped with a solid wood plate to serve as a receiver for panel screws that secure roof SIPs. Multistory SIPs buildings are typically platform-built using engineered wooden I-beams or wood trusses for floors. SIP floor panels are rare but

exist. SIPs can be built up to three stories without a substructure. Buildings taller than three floors use a steel post and beam skeleton with SIPs infill. Drywall can be screwed right into the OSB skin to create the interior finish. The exterior finish can be myriad: stucco, cement board, lapped wood or cement-fiber siding, and/or stone/brick veneers.

Fig. 6–6. Installing structural insulated panels (SIPs) is fast and easy.

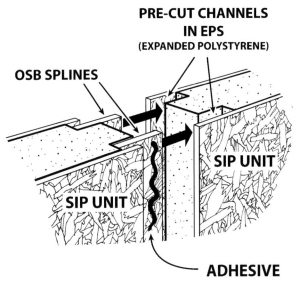

Fig. 6–7. SIP spline connections

Conventional Type III and Type V buildings may be roofed with SIPs. In fact, the original SIPs concept was designed just as a replacement roof for buildings in cold regions to help eliminate heat loss. Likewise, existing structures may have a SIPs addition. Regardless, the SIPs building, roof, or addition is amazingly strong, energy efficient, and economical. That's assuming there is no fire.

Two variants of the SIP concept have evolved in recent years. For those concerned about using plastic foam, the EPS core is replaced with a proprietary fill made from compressed wheat straw or other cellulose material. The second variant replaces the OSB skins with thin steel panels that help provide more impact resistance for buildings in hurricane-prone areas.

Fire spread issues: The finish material applied to the panel's OSB skin is a critical barrier for SIPs fire protection. Once that finish is compromised, the OSB will burn and contribute to rapid fire spread. (Remember that OSB glues are petroleum-based.) Heat alone can cause OSB glues to degrade. Likewise, heat can enter the panel insulation core through utility chase conduits that are formed or cut into the panels (fig. 6–8). The heat will cause EPS to melt, leaving large combustible voids between the OSB skin. Smoke production will be significant. The smoke from OSB and EPS is extraordinarily toxic. Remember also that the OSB lacks the strength to be load bearing without the bond to EPS.

Fig. 6–8. Heat and/or fire can enter utility chase conduits in panel cores and compromise the insulation core.

Because SIPs buildings have high insulating values and few exterior wall voids, smoke and heat will initially be trapped in the building. Be

require specialized equipment and/or tools, has good insulating properties, and is resistant against vermin, rot, and fire.

Fig. 6–10. A straw clay building

The building process consists of the following steps:

- Construct the wooden framework of the building.
- Prepare the clay (earth loam with high clay content) by mixing with water to form slip.
- Mix with dry straw until the straw has a light coating of slip.
- Pack the straw/slip combination into forms that are located in the walls/ceilings and between the wooden framework of the building.
- Remove forms to allow drying that can take one to two months.
- Finish exterior and/or interior with plaster, stucco, applicable sidings, masonry, or other suitable materials.

Fire spread issues: The spread of fire in these types of buildings will primarily come from flammable materials within the interior of a building (e.g., furniture, carpets, clothing, etc.) as the interior contents and floor plans are similar to common residential buildings. Other than the wood framing, clay straw used for walls, ceilings, and roofs is basically not flammable. However, the straw can smolder and will likely present the greatest hazard during overhaul operations.

Collapse issues: Structural collapse is not a primary concern from clay straw materials, but is more a concern for the wood framing that supports and/or encloses clay straw materials. It is easy to see that collapse will be more of a consideration for clay straw used in ceilings and roofs (suspended weight) than walls (supported weight).

Quick summary

- Straw bale buildings use straw and not hay type products as straw is much stronger than hay.
- Straw bale walls can be either load-bearing or non-load-bearing. Load-bearing walls rely on compression of the straw for strength and stability.
- The two methods of compressing load-bearing straw bale walls is through the weight of the roof on a box beam or through compression strapping.
- Overhaul operations in straw bale construction will be complicated as it is necessary to ensure all fire has been eradicated (similar to cellulose insulation).
- Collapse considerations are more prevalent with the roof construction than with the straw bale walls.
- Straw clay construction is comprised of a mixture of earth (clay) and straw.
- Straw clay used for ceilings and roofs presents more of a collapse hazard than straw clay walls.
- Straw clay materials can present challenging considerations during overhaul operations as straw can smolder and spread fire.

EVOLVING BUILDING METHODS/MATERIALS

The alternative construction methods listed above are far from complete—they just tend to be ones that are more prevalent. As this chapter is written, innovation is taking place. One thing is certain: New materials and methods are being tested, marketed, and incorporated into buildings—old and new. Some of these innovations will work well to make a strong, economically-sound, and energy-efficient building. Others will take a few years to be weeded out. Regardless, firefighters have to deal with them. To finish this chapter, we briefly explore some evolving concepts and materials that could influence fire and collapse concerns for firefighters.

Green construction

The *green* movement is alive and well in all parts of the world. In context here, green construction is not a construction type. It is instead an approach to construction with features that include the following:

- Earth-friendly or sustainable materials
- Energy-saving design
- Alternative power sources like solar, geothermal, or wind
- Recycled or repurposed materials
- Health-conscious methods and finish
- Reduction in waste and pollution
- Optimization of building operations and maintenance
- Building life cycle considerations, including renovations and demolition

In the United States, the American Recovery and Reinvestment Act (ARRA) has allocated more than $80 billion dollars to clean energy and/or green technology that is viewed as a resource that has the potential of replacing nonrenewable resources. In partnership with the ARRA, several examples of the growing number of agencies providing oversight in this area are the Green Building Certification Institute, National Association of Home Builders (Green Approved Products Program), and National Green Building Standards (ICC 700). As a result, the perspective of *eco-friendly* has gained a significant amount of popularity both in this country and globally and has the potential of not only impacting how people live their lives but also changing size-up and operational perspectives for structure fires.

One of the many challenges of green construction the fire service will encounter is how to classify a green building or a building that incorporates a notable amount of green attributes. As an example, a Type III building is still a Type III building, but may be built as green using the aforementioned features. As it relates to firefighting, green buildings may bring challenges such as alternative power sources, unusual or increased dead and live loads, low-mass synthetics, and recycled or repurposed materials. Alternative power sources are covered in chapter 9, "Building Features and Concerns." Unusual or increased live and dead loads could include items like a rainwater collection system or a live roof load that is actually a garden with living plants/grasses and an irrigation system. Low-mass synthetics can be petroleum- or cellulosic-based and off-gas flammables with minimal heat application. Recycled or repurposed materials can add a degree of uncertainty to engineering and therefore dubious integrity.

The language used to describe a green building can create some confusion. For example, let's say a new two-and-a-half story, 6,000 sq ft house is being built in your jurisdiction. The house is being constructed using grid-block insulated concrete forms, engineered wooden I-beam floors, a structural insulated panel live roof, and includes numerous solar panels. The contractor's sign calls it *green construction*. The building permit will call it a Type V, residential occupancy. Following the era/use/type/size formula (from chapter 5), the street firefighter should mentally classify it as a *new, single-family, hybrid, and large* building that just so happens to have a significant distributed roof

dead load. Again, it is not suggested that all of this information would be used for an on-scene radio report. However, all of your strategic and tactical decisions should be based on your understanding of the visible and perceived attributes of that building.

Because there are numerous *green* considerations that can affect fireground operations, let's look (in no particular order) at a summary of applicable considerations. (Solar considerations are covered in chapter 9.)

- There are many variations of construction that are different from the traditional wood frame structure. Several examples are rammed earth brick, straw bale, and light clay straw construction. These buildings stay cool in the summer, warm in the winter, and are not overly combustible. However, these materials will confine more heat within a structure fire and enhance the potential of a flashover.

- **Advanced framing methods (AFM)** or **optimum value engineering (OVE)** is a wood framing technique that is designed to use less lumber, minimize waste that is generated in the construction of wood framed structures, and improve the energy efficiency of a structure. This framing method is discussed in more depth later in this chapter.

- **Double-stud wall construction** is not a new framing technique (it can be used in a similar configuration known as offset studs between adjoining occupancies such as condos, etc.) but is beginning to be used to provide an energy-efficient envelope for exterior walls. Two parallel walls are framed with dimensional lumber (2 × 4 in. is common) and configured with either opposing (aligned) or offset (staggered) studs. Only one set of studs is load bearing, and the parallel studs are placed about 3½ in. apart. This allows the gap and interior wall to be completely filled with insulation and can result in an R-value of 40 or more. Unfortunately, the gap also provides an opening that can allow the horizontal extension of fire between the studs.

- A new approach to high-performance wall systems that are capable of providing increased insulation values is known as the **pressure-equalized rain screen insulated structure technique (PERSIST)** that has been developed by the National Research Council in Canada. The method consists of 2 × 4 in. framing, OSB sheathing covered with a peel-and-stick membrane (rubberized asphalt adhesive backed by a layer of high-density cross-laminated polyethylene), and single or multiple layers of rigid extruded polystyrene (XPS) foam insulation. Finally, the building is finished with any type of preferred siding. This combination of materials provides an increased barrier to air and vapor migration. The additional use of synthetic foams and rubberized asphalt adhesives will enhance a fire, provide additional toxic gases when exposed to fire and/or heat, and allow a faster heat buildup within the interior of a building that can lead to a more rapid flashover as well as increased levels that will be encountered by firefighters.

- **Radiant barrier sheathing (RBS)** is a product that has been around for several years but is gaining popularity due to its energy-efficient properties, particularly in energy-efficient buildings, both residential and commercial. Simplistic in design but effective in operation, RBS is nothing more than plywood or OSB with an aluminum type foil affixed to one side that is designed to reflect radiant heat away from the foil. RBS comes in the common sizes of 4 × 8 ft and 4 × 10 ft panels and can be used for wall or roof sheathing.

  When installed as wall sheathing, the foil must be facing outward. When installed as roof sheathing, the foil must be facing inward (toward the attic). For proper operation, a ¾ in. airspace should be maintained in front of the foil, which means that exterior siding needs to be spaced ¾ in. away from the RBS (this can be achieved by furring strips). When properly installed, RBS will lower interior temperatures (summer) and reduce radiant heat loss (winter), thus lowering heating and cooling

costs. However, RBS is capable of enhancing flashover conditions by reducing radiant heat loss from the interior of a building. Additionally, the reflective nature of the RBS will greatly diminish the effectiveness of thermal imaging cameras (TICs) used to find hot spots.

- **Green insulation** uses recycled materials such as cotton and denim in place of fiberglass for insulation. Another common example is cellulose insulation that consists of 80% recycled newspaper, which is superior at preventing airflow as compared to fiberglass. Cellulose insulation is well known for its use in attic insulation *and* its ability to hide a smoldering fire for hours and making detection of fire difficult and time consuming.

- **Cool roofs** are designed to reflect more of the sun's rays than common roofs, such as composition and shingled roofs. Typical construction methods use an increased depth of insulation that is covered with special reflective paints and roof tiles or shingles. Not only do these roofs reflect heat, they also prevent cool or warm air inside a structure from escaping through the roof membrane. Obviously, this same trait will also trap more heat inside a structure from a fire and increase the potential of a flashover.

- **Live green roofs** are often categorized with cool roofs. Although they can also lower the interior temperature of a structure by insulating it from the sun's heat, the actual roof is more complicated than a specific type of paint or roofing shingle. Live roofs can be found on some residential structures, but are primarily found on commercial structures and provide the advantages of lowering heating and cooling costs while maximizing the attractiveness of urban structures. There are two basic types of live roofs:
  - *Extensive* live green roofs typically have a soil depth of 2 in. to 4 in. and are planted with hearty type vegetation that requires minimal maintenance.
  - *Intensive* live green roofs typically have a soil depth of over 1 ft and support a wide range of plants that require regular maintenance. These roof gardens also require substantial structural enhancements to support the additional weight.

A live green roof (or vegetated roof) consists of four basic elements (fig. 6–11) as follows:

- A building with a roof and supporting members that are capable of supporting the additional weight of the materials above it
- Multiple membrane layers to protect the structure from water, a drainage layer (e.g., drainage mat) that will allow water to properly drain from the roof without eroding the soil layer, and a barrier to keep plant roots from penetrating the roof
- A layer of soil for the vegetation that is comprised of specialized materials such as a mixture of vermiculite, lightweight soils, and other appropriate materials
- Planted vegetation

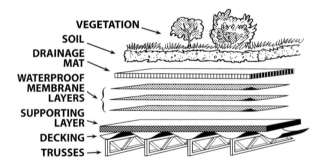

Fig. 6–11. Live green roofs are comprised of four basic elements: supporting members, membrane layers, a layer of soil, and vegetation.

- **Electrochromic smart glass windows** (also known as suspended particle display windows) are primarily designed to allow an occupant to change the amount of light a window reflects. This is accomplished by using tiny transparent electrodes sandwiched between two panes of glass. A small burst of electricity (approximately 1 volt) is used to change the glass from clear to opaque. Interestingly, no electricity is needed to maintain a particular shade once it has been selected. This process is very adept at reducing

the solar radiation passing through windows and can lower the internal temperature of a structure. However, it will be more difficult to break these windows for access and egress and/or ventilation due to their complexity and double panes of glass.

Advanced framing methods (AFM)

Advanced framing methods refers to a variety of wood framing techniques that reduce the amount of lumber (and waste) used to construct a wood frame building. AFM is also known as optimal value engineering (OVE) and is the result of a partnership between the U.S. Department of Housing and Urban Development and the National Association of Home Builders Research Center. The basic premise of AFM is to reduce the amount of *thermal bridging* (the movement of heat through framing members) and the amount of lumber used to construct a wood frame building.

For example, the corner where two wood frame walls intersect is often built using three or four studs. The AFM replaces this with two studs per corner. Other AFM features could include (subject to local code) the following:

- Stud spacing is increased from 16 in. on center to 19.5 in. or 24 in. on center, and the walls use a single 2 × 4 for a top plate (fig. 6–12).

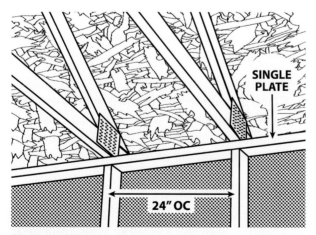

Fig. 6–12. Advanced framing method features

- Butted single 2 × 4 top plates use a metal connector plate, and one or more of the studs are placed only to provide a nailing surface for drywall. This configuration can be replaced by metal clips to receive the drywall sheets (fig. 6–13A).

- A header placed over a window opening that is typically supported by jack studs (double studs) is replaced by single studs, one of which has a backing support for gypsum board/drywall sheets (fig. 6–13B).

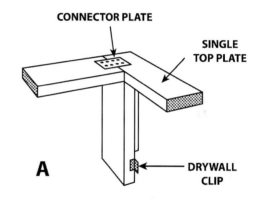

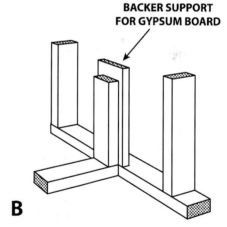

Fig. 6–13. AFM top plate and corner (A) and window header (B) features

- Nailing boards for external siding are 2 × 2 in. instead of 2 × 4 in.

- Metal T-bracing or strapping is used for wind load requirements rather than OSB or plywood stiffeners.

The AFM framing technique is environmentally friendly, uses 10% to 30% less lumber, has a faster implementation time, saves energy, and reduces the framing factor from 15% to 25%. It is easy to see, however, that this new version of in-line or stack framing results in a structure that has less structural components and can burn at a faster rate than a standard framed structure, resulting in faster collapse times when exposed to fire.

Modular panel systems

Like a SIPs building, the concept of factory-created panels assembled on a job site to form load-bearing walls is evolving. The ThermaSteel system uses a galvanized steel frame with carbon-fiber reinforced EPS for panel fill. The Insteel 3-D Panel System is EPS sandwiched between steel wire mesh and sprayed with shotcrete at the building site. Shotcrete is a nozzle-sprayed concrete mortar mix (the water is either premixed or mixed at the nozzle) that is applied on a surface using an air compressor.

Building block systems

Much like CMUs, various alternative materials have been introduced to form a stacked wall. For example, mortarless blocks are like CMUs but have unique engineered internal shapes filled with expanded polystyrene (EPS). The blocks contain perhaps half the concrete as a common CMU. The blocks include tabs and recesses that lock joining blocks together, eliminating the need for mortar. Obviously, the reduced concrete content means faster degradation under fire conditions.

Autoclaved aerated concrete (AAC) blocks are building blocks made from a mixture of sand, Portland cement, gypsum, water, expansion agents, and air that forms a solid block that is one fifth the weight of a similar size concrete block. While the blocks are considered solid, in reality the block is mostly air. These blocks can be formed in unique shapes for almost any architectural wall need. The blocks can be mortared together or glued with special adhesives. Steel rebar is formed within the block for key structural areas (corners, door and window frames, etc.). Using your firefighter knowledge, imagine what a concrete block filled with millions of air bubbles will do in a hostile fire environment.

Engineered wood systems

We presented many different engineered wood products in chapter 2, yet innovation continues. For this discussion, the term *engineered wood systems* does not specifically refer to lightweight truss construction as one might imagine. Instead, it refers to a wide range of derivative engineered wood products (EWPs) and how they can be used in building construction, particularly when used to replace steel and/or concrete materials. Debatably, the evolving concept of EWP has the potential to overshadow other building construction materials, the way they are used, and the resultant type of buildings that can be constructed from this evolving technology. As wood has been the staple of the building construction industry, it is easy to envision how a dramatic change to this product will affect the ways and types of buildings that can and will be constructed. With these thoughts in mind, let's take a summary look at the concept of engineered wood systems in the building construction industry.

Engineered wood can also be referred to as *manufactured board*, *man-made wood*, and *composite wood*, which includes a diverse range of EWPs that are primarily manufactured by binding fibers, strands, particles, or veneers of wood together with adhesives. These form a variety of composite wood based materials that can be used in a variety of residential and commercial applications. Interestingly, EWP can be made from the same softwoods and hardwoods that are used to manufacture dimensional lumber. Sawmill scraps and other types of wood waste can also be used for EWP, as well as some cellulosic products that are derived from vegetable fibers. Most firefighters are aware of the previously mentioned types of EWP such as OSB, LVL, PSL, glue-laminated timber, I-joists, and

finger-jointed lumber. However, the growing variety of new EWPs such as cross-laminated timber and fiber-reinforced products adds a new dimension to the term "engineered wood products," as well as an increased list of hazards that a firefighter will encounter from these products. Let's look again at a few of these new products.

- **Cross-laminated timber (CLT).** A CLT is a panel of wood formed by first gluing together planks of timber to form a thin sheet (½ in. thick and 4 × 8 in. or larger). The thin sheets are then cross laminated with other sheets, usually rotated 90°. In many ways, this is similar to plywood except true timbers are used instead of native wood veneers. Sounds strong, right? Well it is. Now for the alternative spin: These panels can be used to form the load-bearing walls for a high-rise! The surprising capability of CLT in combination with advancing technology is allowing the building industry to begin building multistory buildings out of wood instead of concrete and steel. Several examples are a 9-story apartment building (timber tower) built of wood in nine weeks by four workers in the United Kingdom, a 78 ft bell tower in North Carolina, a multistory commercial building in Montana, a 10-story high-rise building in Australia, and the granddaddy of all wood buildings is the proposed "Big Wood" project in Chicago which will be a mixed-use university complex that will include an all-wood high-rise building! Although the aforementioned buildings/projects are currently referred to as *plyscrapers*, it is easy to see that building construction is not static but dynamic in its potential impact on the fire service.

- **Fiber-reinforced product (FiRP).** A FiRP (pronounced "furp") is a wood beam that has layers of high-strength synthetic fiber material sandwiched and bonded to layers of cut timber or laminated strand lumber (LSL). Carbon graphite strands can also be sandwiched in place of the synthetic fiber material. FiRP beams can carry twice the load of a similar dimension solid wood beam.

Quick summary

- Green construction is not a construction type but rather an approach to building that is earth-friendly, energy saving, and health conscious (as well as other like attributes).

- The various types of green framing techniques will result in different considerations for an incident commander, and are dependent on the type of construction. Due to enhanced insulating qualities, it can be anticipated that interior temperatures will rise at a faster rate, thereby enhancing the potential of flashovers, and will also negatively impact overhaul considerations.

- Advanced framing techniques or optimum value engineering is designed to use less lumber and therefore results in faster collapse times when exposed to fire.

- The advantages of a live green roof are diminished from a firefighter's perspective; the multiple layers of membranes, earth, and vegetation make vertical ventilation operations difficult if not impossible in an acceptable time frame. The additional roof load can cause a more rapid failure under fire conditions.

- Electrochromic smart glass has the potential to hold more heat within a structure, and can be more difficult to break for access, egress, and ventilation operations.

- Engineered wood products have begun to provide a new perspective on the use of wood, especially when used in place of concrete or steel materials. Although EWP offers many advantages over conventional materials, remember they universally use adhesives of various compounds, many of which are capable of readily softening when exposed to heat and fire and emitting toxic gases that are detrimental to fireground personnel.

CASE STUDY: THE FOAM HOME

The Tamarisk House in northwest Scottsdale, Arizona was built using lightweight, sustainable materials consisting mainly of expanded polystyrene coated with a concrete composite skin. This type of construction is officially called the Saebi alternative building system (SABS). The walls are formed in a factory that uses a robot to carve a set of building blocks that may be curved, arched, squared, or angled for a given portion of the home. The exterior wall blocks are 8 in. to 10 in. thick and up to 8 ft in length. Interior walls are closer to 4 in. thick. The building blocks are then shipped to the home site and assembled much like building a child's lock-block toy. The roof and roofing beams are made from the same materials, with beams approaching 30 in. thick to help create large open spans for the interior geometry. A hot-wire foam cutter tool is used to make window, door, or utility openings. Once the blocks are in place, they are shot with the concrete composite.

The SABS technology has undergone years of computer-aided testing and independent laboratory testing to prove its durability. Myriad tests evaluate fire, seismic, aging, water absorption/penetration, and freeze/thaw durability, to name a few. Most tests show that the system is extremely strong and durable. In many ways, the finished SABS product is like a car with unibody construction.

Of interest to firefighters is the fire test. A fire was built in an interior room corner and allowed to develop to 2,000°F in 15 minutes. In that time, the interior concrete skin contributed no smoke or flame spread characteristics, nor did it transfer any heat to the outside wall surface. Obviously, firefighters want to see the performance of the system in an actual, full-fuel load interior fire like the ones firefighters face in homes and businesses. While fire service logic suggests an early failure, it is important to note that builders have a moral and code required obligation to notify occupants of a fire (detectors and alarms) and build the building in such a way that occupants have time to leave (the 15 minute test). There is no obligation to build a building in such a way that firefighters can perform interior fights with relative structural safety.

The only way an arriving fire company would know that a home is built using the SABS technology is if they documented it in preplans. The homes using this system look like typical stucco, Pueblo-styled structures.

TACTICAL POINTS FOR FOAM HOME DISCUSSION:

1. What fire behavior issues do you think the SABS home presents—given a typical household fire load (contents)?

2. Compare and contrast the collapse resistance of a SABS home to a new wood frame home (lightweight, engineered).

3. What forcible entry and ventilation issues does the SABS structure bring? What tools do you think are well suited for SABS penetration?

4. What system does your fire department have to help identify unusual or advanced technology constructed buildings?

The list of evolving materials and methods could go on. We've just included some of the more popular ones here. Buildings constructed with alternative methods and materials may easily pass a series of test lab fires or satisfy the requirements for building codes, regulations, and standards. But how will these buildings react under real-world, full-fuel fire loading, and actual firefighting circumstances? As stated earlier, the fire service has little documentation on the fire spread and collapse issues associated with alternative construction and evolving methods and materials. We must use our collective wisdom to figure out some tactical solutions to these buildings. (Try to answer the tactical discussion questions in the case study on the previous page.) The bottom line is these buildings are being built and they already may exist in your jurisdiction. Get out of the fire house and find them—be curious. If you have a fire in an alternatively constructed home, be sure to document your experiences and share them on our social and technical networks!

CHAPTER REVIEW EXERCISE

Answer the following:

1. How does a performance design building approach differ from prescriptive design approach?
2. What are the collapse issues associated with lightweight steel construction?
3. What is ICF and what are three general types of ICF?
4. How can structural integrity be obtained using the SIPs method of construction?
5. What is the purpose of a toe-up and box-beam in a load-bearing straw bale building?
6. What is "green construction" and what features might you find in a building built green?
7. List three features that might be found in a building built using advanced framing methods.

RESOURCES FOR FURTHER STUDY

- APA—The Engineered Wood Association, www.apawood.org.
- Engineered Wood Products Association (EWPA), www.ewpa.com/.
- Frechette, Leon A., *Build Smarter with Alternative Materials*, Carlsbad, CA: Craftsman Book Company, 1999.
- Gibson, Scott, "High-Performance Walls for Energy-Efficient Building," *Home Power*, April-May 2013.
- Morley, Michael, *Building with Structural Insulated Panels (SIPs): Strength and Energy Efficiency Through Structural Panel Construction*, Newtown, CT: Taunton Press, 2000.
- National Association of Home Builders (www.nahb.org), Green Certified Product program, http://www.homeinnovation.com/greenproducts.
- Spence, William P. and Eva Kultermann, *Construction Materials, Methods and Techniques, Building for a Sustainable Future*, 3rd ed., Clifton Park, NY: Cengage Learning, 2010.
- U.S. Green Building Council (USGBC), www.usgbc.org.
- www.BuildItGreen.org, "Advanced Framing," California Public Utilities Commission, 2008.

SECTION 2
BUILDING COMPONENTS AND FIREFIGHTERS—PRACTICAL LESSONS

FOUNDATIONS, FLOORS, CEILINGS, AND WALLS

7

OBJECTIVES

- Define the basic types of foundations.
- Identify the fundamental types of floors.
- Describe the various types of ceilings.
- List the different types of walls.

THE BOX THAT SURROUNDS YOU

In this chapter, we look at the components that surround you when you engage in interior firefighting operations: the foundation, floors, walls, and ceiling (roofs are covered in the next chapter). In chapter 3, we gave a brief introduction to foundations and walls as part of the discussion on structural elements and the structural hierarchy. This chapter provides more definition and depth to that introduction.

Quite often, when resources arrive on the scene of a structure fire, it is easy to focus on the excitement of the actual fire and perhaps the possibility of trapped occupants within a structure. This situation can also be exaggerated by hysterical occupants/civilians who want the fire put out in the least amount of time and are more than willing to voice their recommendations.

With these thoughts in mind, it is easy to overlook and/or minimize the importance of the very box that will be surrounding you when you enter. From a simplistic perspective, the foundation, floors, walls, and ceiling are cooperatively responsible for contributing to either a partial or total collapse of a structure, or not providing sufficient time and structural integrity to allow the safe and effective extinguishment of a fire in a timely manner. Therefore, it is imperative that fireground personnel take the necessary time to quickly evaluate the primary attributes of a structure that are, in many cases, visibly perceptible if

proper emphasis is placed on their importance for fireground operations. First, let's review the definitions that are used in this chapter.

> ## Definitions:
>
> **Base of operations.** Concept of ensuring the platform you are working on (roof or floor) will safely support you for the duration of your operations.
>
> **Foundation.** The building's anchor to earth and base for all elements built above that anchor.
>
> **Floor.** The platform and substructure that serves as a base for accommodating people movement, furnishings, and fixtures within a building.
>
> **Ceiling.** An interior surface (lining) that covers the top of a room and is not considered a structural element such as walls, floors, and foundations.
>
> **Wall.** A vertical or upright surface designed to enclose or divide a compartment. Walls can be load-bearing (a wall column structural element that supports floor or roof beams) or non-load-bearing (supports its own weight plus anything attached to it).
>
> **Roof.** The top portion of a structure that shelters interior spaces and includes structural supports and coverings (covered in the next chapter).

FOUNDATIONS

The foundation of a structure may be the most obscure portion of a structural size-up, but in many cases its importance cannot be overstated from the perspective of providing and/or clarifying important nuances of a structure. Although foundations do not burn and rarely collapse, they are not only responsible for supporting a building and keeping the wooden parts of a structure above the ground, but they can also give some immediate clues about the era of construction and areas within or under a building that interior personnel should be aware of before entering a building. Although we briefly discussed foundations in chapter 3, let's look a little closer at the common foundation and see why it can be important as a structural size-up consideration regarding the era and type.

Era

In some cases, the general era of building construction can be appraised by looking at the visible foundation. In older residential structures that were built in the 1800s and early 1900s (historic era), field stones or blocks of granite were often used for the foundation. Field stones are defined as easily accessed stones that are common to the area of construction. In some areas of the country, granite is readily available and became the field stone of choice because of its dense, durable nature. Quartz, limestone, and various forms of river rock are other examples of field stones. All of these materials typically predate concrete foundations and can generally be used to date a structure as follows:

- Field stones were used in the 1800s and up until about the 1930s. Unfortunately, the sand-based mortar that was used has often deteriorated over the years, resulting in a substandard bonding agent. In some cases, field stones were also used to form the foundation and walls of a basement or cellar. This type of construction can be considered a forerunner to formed concrete foundations.

- Clay bricks were also used as a foundation material in areas where clay was readily available or other options were not plentiful or cost effective. Obviously, clay bricks and substandard mortar will degrade when exposed to harsh weather conditions and time.

- Granite (and in some cases limestone) used for foundations was cut into blocks and then stacked to form a solid foundation. In some areas,

granite blocks can indicate a general time frame of construction. As an example, in Syracuse, New York, rough-cut blocks were used in the early 1800s, and blocks that were more finely cut (using superior quarrying techniques) were used after the 1840s. Additionally, rough-cut blocks can also denote post and beam residential structures in this area. In figure 7–1, notice the structures appear to be newer, while the foundation blocks are rough-cut.

Fig. 7–1. Foundations can be a visual clue to help determine the era of original construction.

- Concrete gradually became the foundation material of choice during the industrial era. The typical concrete foundation uses a low-slump, high-aggregate blend (less water, more Portland cement and gravel). Steel-reinforced concrete is especially common in foundations used for larger buildings and almost all foundations used to form a basement. More recently, foundation concrete is made using blends that include fibrous filler like fiberglass, plastic, and carbon fiber.

Foundation types

There are three common types of foundations in this country, slab-on-grade, perimeter, and basement/cellar:

Slab-on-grade foundation. In some climates, the ground floor commonly consists of a concrete slab that is poured over a suitable rock base on the ground, and then the walls, floor(s), and roof are erected on top of the slab foundation. Although this is a simplistic approach from a construction and size-up perspective, it is used on a wide variety of buildings ranging from single-family dwellings up to large commercial buildings for big-box stores such as WalMart and Costco. There are several characteristics of this system that can be beneficial to fireground operations:

- In this configuration, concrete can be considered a substantial building component when exposed to heat and/or fire.

- Interior personnel do not have to be concerned with a floor collapse. Their base of operations from this type of grade floor is secure.

- The overall building height will be lower than a similar structure that is constructed on a raised foundation (this is dependent on the height of the slab vs. a raised foundation). This can simplify roof operations with shorter ladders, and can also simplify emergency forcible entry/exit considerations as doors and windows are closer to the ground. From a simplistic viewpoint, one of the most common and basic residential structures is a single-story ranch structure with 8 ft walls/ceilings that is built on a concrete slab and lacks a basement.

- The possibility of basement fires with their associated hazards is eliminated.

- Structures that are constructed on slab foundations are relatively easy to identify.

- The presence of slabs eliminates the need to consider voids under the grade floor.

- Slabs will not have crawl space vents.

Perimeter (deep foundation). Perimeter foundations can be found in many types of climates and are typically constructed from concrete, concrete block, field stone, or clay brick. This configuration will transfer a load through the weak layer of topsoil to the stronger layer of subsoil below and can be used to provide a level floor surface over sloping ground. This type of foundation is found in

two configurations: with a slab poured within the perimeter foundation and with a floor constructed atop the perimeter foundation. The slab-on-grade poured within a perimeter foundation presents minimal hazards to fireground personnel. A floor constructed on top of and supported by a perimeter foundation places the insulated ground floor of a structure above the ground. This results in a crawl space, which can be of various heights, between the ground and floor joists. This crawl space is used for electrical, plumbing, ductwork, insulation, and other similar considerations.

Perimeter foundations will normally have vents in the crawl space that allow air to circulate from the space beneath a building to the exterior to minimize dry rot and other considerations that can be harmful to a building, and also provide a level floor over sloping ground. This is illustrated in figure 7–2 by a perimeter foundation on sloping ground and crawl space vents that are visible on the rear left side of the foundation.

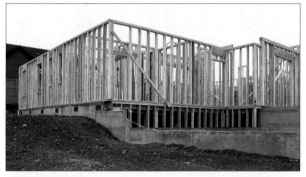

Fig. 7–2. Perimeter foundations can provide level flooring on sloping ground and will result in a crawl space between the ground and floor above.

From a fireground perspective, there are several disadvantages to this system:

- Interior personnel need to be concerned with a floor collapse if a floor is exposed to heat/fire. The base of operations on a floor over this type of foundation is dependent on the type of construction that was utilized for the floor and the distance from the floor to the ground (which can be noteworthy in some cases). This system will typically result in either a combustible (wood) or noncombustible (metal or concrete) floor system, both of which can be hazardous when exposed to sufficient heat and/or fire.

- There may be sufficient area in a crawl space for storage that, if ignited, could either weaken the floor structural members and/or provide a dangerous condition to firefighters who might drop into a fire area with a collapsing floor.

- A foundation that is constructed on sloping ground will result in the area under the grade floor increasing toward the back of the downward portion of a slope. As an example, in figure 7–3 a commercial building is constructed on ground that is sloping left to right. The foundation on the left side has two visible crawl space vents that indicate a perimeter type of foundation with the grade floor level just above the vents and below the windows. However, as the ground slopes away from the front of the building, the perimeter foundation follows the contour of the downward sloping ground until it changes to level ground and becomes a slab foundation that supports a walkout type basement. Therefore, the front portion of this building offers a stronger and safer base of operations for personnel as compared to the back of the structure, which has a significant distance between the upper grade floor and the slab below.

Fig. 7–3. This commercial building has two different types of foundations, which can present different fireground considerations.

When conducting a structural size-up, consider the following items that can add to your size-up perspective:

- Look for the presence of crawl space vents. These vents are designed to minimize the buildup of moisture under a structure and can also give a clear indication of fire within a crawl space under a grade floor.

- If the height of a foundation is visible, it can often indicate the distance between the ground and the grade floor, indicating the volume of space under the grade floor.

- While a perimeter foundation may be covered with stone, masonry, and various types of siding for decorative considerations, the crawl space vents still need to be operational and are usually fairly visible (although you may have to look a little closer to find them, particularly in decorative stone).

Basement/cellar foundations. Basements and cellars are most often found in cold portions of the country where frost lines mandate the use of deep footings to keep the buildings from shifting during the freeze-thaw cycle. They are also common where property is a premium and multiple stories and square footage are desired with a minimal footprint. When considering basements/cellars, there are five definitions to consider as follows:

- **Basements (or cellars)** are habitable spaces that are either completely or partially below the ground floor. For this text, basement and cellar are interchangeable terms. (See the following historical perspective on basement/cellar terminology.) Basements may be unfinished, finished, or partially finished.

- **Daylight basements (or walkout basements)** are found in buildings built on slopes and are under the grade floor (or main entrance), which allows occupants to walk out of the basement on the lower grade level through a doorway to the outside (fig. 7–4).

Fig. 7–4. Daylight basements allow occupants to exit the basement on a lower grade level through a doorway to the outside.

- **Look-out basements** have walls that extend above the grade level so that some of the windows are above grade (fig. 7–5).

Fig. 7–5. Lookout basements have large windows above grade level.

- **Walk-up basements** are characterized by an exterior stairway entrance. The exterior entrance may be unprotected, partially covered, or fully enclosed.

- **Crawl space** is the unfinished space below a ground floor that allows access to under-floor utilities (pipes, ducts, etc.). Crawl spaces are of limited height and typically have a soil surface.

HISTORICAL PERSPECTIVE ON BASEMENTS AND CELLARS

Terminology can be a confusing (and occasionally humorous) topic in the fire service. Basement and cellar are two notable examples. Some fire departments have very distinct definitions for each and others use the terms interchangeably.

As we conducted research for this book, we wanted to acknowledge the various interpretations for various terms, yet provide a historically accurate portrayal of those used.

The term cellar predates basement and was typically used to describe a below-grade space (typically under a building) that was used for the storage of coal, wine, and other items. The depth (or height) of the cellar was usually determined by the frost or freeze line that was common in the geographical area of the building—some were pretty short (duck your head!) and some were tall enough for most people to stand erect.

Originally, cellars were rarely finished—the building foundation and overhead floor beams were visible and the bottom surface was earthen. This environment provided a constant, cool temperature space that was ideal for preserving whatever was stored there—it didn't get too warm and it rarely froze thanks to the earthen insulation.

In rural areas, farmers used cellars for the preservations of roots and seeds (hence the term root cellar) to keep them protected from frost and freeze and allow them to be used as plantings for the next year's crop. This also became a great place to store preserved goods such as canned fruits and vegetables and jerked meats. In some cases, these cellars were dug separately from the farmer's home yet close enough to retrieve the preserved goods. These below-ground cellars became a perfect refuge location from tornadoes and other threats—thus the term storm cellar.

The origin of the term basement is more elusive, although most believe it derived from the fact that builders began digging deeper into the earth for building foundations/footers to help prevent foundation shifting from freeze/thaw cycles—the *base* was deeper. Obviously, deeper (higher) spaces were attractive as spaces for habitation (as opposed to merely storage). If nothing else, the "base space" was perfect for the placement of modern appliances and features such as boilers, furnaces, water heaters, ducting, plumbing, and electrical distribution equipment within the building. These appliances used up some space but the remainder of the basement could easily be finished into dwelling, storage, or other occupancy need spaces.

Several fire departments and noted fire service authors have defined a cellar as those levels that have more than 50% of their height below grade and a basement as those that have more than 50% above ground. For most, those descriptors seem dubious (and not really reflective of the definitions used in their area). The 50%/50% definition is, however, found in laws that have been promulgated in several cities. For example, the New York City Building Code (2008) defines a basement using the 50%/50% rule. The reasons for this are not entirely clear but is likely to do with taxable commercial space (revenue) or for the sake of counting floors (if the basement gets counted) for life-safety or height restriction requirements or inclusions.

Given all of that, we choose to use basement and cellar synonymously.

Basements are customarily constructed from formed concrete walls or concrete block walls (field stones and some other forms of masonry have been selectively used in older buildings) and are normally one story in height for residential structures. Where present, basements typically contain various utilities (electrical, water, etc.) and heating systems, although this can also be geographically specific. The presence of basements can also present an interesting dichotomy of rescue and access/egress considerations. Obviously, these considerations are simplified by the presence of a daylight basement.

Depending on a floor plan, egress from a front basement unit (also referred to as a dungeon unit) may need to be accomplished by an escape window, as illustrated in figure 7–6. This area, which must be checked for the presence of occupants, is often hidden by landscaping.

Fig. 7–6. Escape windows can provide egress from below-grade occupancies.

Basements are renowned for multiple fireground disadvantages, such as the following:

- Occupants use them for storage that can consist of flammable materials of all descriptions.
- In some cases, the grade floor structural members over basements are exposed and are vulnerable to fire. If the exposed joists are lightweight construction such as I-joists, this can present a noteworthy hazard (rapid collapse).
- As basements are below grade, they have minimal ingress/egress avenues and they can be difficult to ventilate for suppression operations.
- A fire in a basement can easily extend upward, and can often be enhanced by the type of construction (e.g., balloon frame, pipe chases, etc.).
- When some levels can only be viewed from the front or rear of a multistory building, confusion can be caused unless proper floor designation is communicated in a timely manner.
- Buildings that have multiple below-grade levels are known to have sublevels (fig. 7–7). For example, if a building has three levels below grade, the first level just under the first floor is labeled *basement*. The next lower level is *sub-basement 1*, and the lowest one is *sub-basement 2*. This approach is merely a guide—local standard operating guidelines/standard operating procedures (SOGs/SOPs) should dictate the labeling of floors.

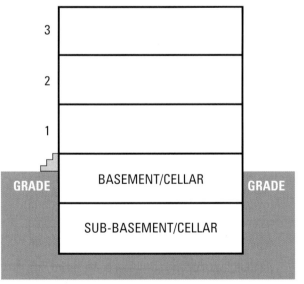

Fig. 7–7. Where more than one basement level exists, the lower levels are labeled sub-basement 1, 2, and so on.

- Although basements and cellars can often be identified by their windows, there may be a lack of crawl space vents. However, it is possible for crawl space vents to be between a basement or sub-basement and the grade floor of a

The Art of Reading Buildings

structure—this denotes an open space between these areas.

- A floor over a basement is in reality a roof over the area. Therefore, when personnel enter a structure with a fire in a basement/cellar, they are in effect standing on a roof over a fire.

- Basement fires can be notoriously difficult to extinguish due to their location, minimal access/egress routes, and challenging ventilation considerations.

- It is imperative to check for vertical extension above a basement, especially in balloon frame structures, pipe chases, and other vertical avenues.

- In some older areas, basements extend under a street to connect buildings on either side of a street (see fig. 5–22). As these basements can be over 100 years old, they will likely contain a noteworthy amount of storage that is flammable. A fire that is encountered in these types of basements will be difficult to access, ventilate, and extinguish. Main Street-type buildings that share a central steam plant are likely to be interconnected by basements or tunnels.

Quick summary

- For this chapter, the four essential components of a structure are foundations, floors, ceilings, and walls.

- The general construction era of a building can often be appraised by the appearance of its foundation (field stones, clay bricks, smooth or rough-cut granite, etc.).

- Field stones, granite, limestone, and brick foundations typically denote an old building that used older building techniques (dimensional lumber, balloon frame construction, knob and tube wiring, etc.).

- The three most common types of foundations are slab-on-grade, perimeter, and basement/cellar.

- Slab foundations are the most beneficial to structural suppression operations, while basements/cellars can be the most dangerous.

- Perimeter foundations are often used on sloping ground to provide level flooring.

- Always include the potential of a walkout/daylight basement and/or escape window for buildings on sloping ground.

- In many cases, the floor construction over a basement/cellar is unprotected and may consist of lightweight construction—a rapid collapse potential.

FLOORS

Floors are that part of a building that fireground personnel depend on for entering a building, conducting interior operations on, and use to exit a building. Floors assist in supporting interior dead and live loads, act as a diaphragm to transfer lateral loads (soil movements, earthquake vibrations, wind, etc.) to the walls of a structure, and separate the area above a floor from the area beneath a floor that includes the ground, utilities, and other similar considerations. Floors consist of four primary parts: supports, joists, subflooring, and floor covering.

Floor supports. Structural elements responsible for carrying the load of a floor. Supports may be foundation walls, structural beams, or stud load-bearing walls.

Joist. A wood or steel beam used to create a floor or roof assembly that supports sheathing or decking. Joists span between primary supporting members such as foundations, load-bearing walls, or structural beams.

Subflooring. The construction industry defines a subfloor as the horizontal platform material that is attached to the top of floor joists and can be made from tongue-and-groove (T&G) planking, plywood, OSB, or even lightweight concrete.

Floor covering. The covering that serves as a durable (and attractive) surface to protect the subfloor.

Although the floors of a building can provide a strong surface for conducting interior operations, they can have numerous fireground disadvantages:

- A primary disadvantage of floors is that they can be constructed from either conventional or lightweight materials. Subfloors are typically made from either 1 × 6 in. sheathing (older construction) or plywood/OSB of ¾ in. thickness or greater (newer construction) for stability and a secure feeling when walking on the floor. However, the same cannot be said for the structural members (floor joists) that support the subfloor.

- Older floor joist construction used dimensional sawn lumber (minimum of 2 × 6 in.) or larger, depending on the age of the building and structural constraints. This provided a degree of resistance to fire before failure. However, in newer construction, the dimensional lumber of yesterday has been replaced by lightweight joists that typically consist of the venerable I-joists that will quickly fail when exposed to heat and/or fire.

- The method used to connect floors to walls has varied widely over the years. Prior to the 1940s, dimensional joists were anchored by inserting their ends into pockets or cavities (see fig. 3–13) in unreinforced masonry construction or resting on ribbon boards in balloon frame construction. After this time frame, joists were normally supported by resting on the top of plates in platform construction. More recently, joists are often supported by resting in metal hangers (fig. 7–8). These metal hangers are (depending on the structural load) typically about 18 gauge (approximately 0.05 in. or 1.27 mm for galvanized steel) and can be inferior to the previous mentioned methods.

Fig. 7–8. Joists can be supported at their ends by thin metal hangers.

- Buildings with multiple floors can present a significant problem, particularly if a building is on sloped ground that results in the appearance of different floor levels depending on which side the building is viewed. Depending on the layout of a building, it is possible to view a building from the front and quickly determine the number of floors, and then view the same building from the side and/or back and see additional floors that were not visible from the front of the building. This is one reason why it is important to view a building from as many sides as possible.

Although this can be easily solved during a 180° or 360° size-up, the important consideration is that all fireground personnel are aware of any variance within a particular structure, the designations(s) that are used for varying floor levels, and that all personnel are aware of the terminology for varying floor levels. If not, then it is possible for interior personnel to be on a different floor than was originally designated for their assignment. As an example, the older apartment building in figure 7–9 is on a corner of sloping ground. From the front of the building there are two stories, and from the back of the building there are three stories. In this case, if an incident commander was in the front of this building and wanted a company that was responding from the rear to enter the building on the second floor (as viewed from the front), the responding resources would

need to know the difference between the number of floors from the front as opposed to the number of floors as viewed from the rear.

Fig. 7–9. This old apartment building has different floor levels depending on the side viewed.

Fig. 7–10. Wooden I-beams offer little resistance to fire and can easily collapse. (Photo by Bill Gustin.)

- When considering the perspective of structural integrity, it is necessary to determine if lightweight construction has been utilized in the building. Chances are, if the roof is lightweight construction (which is often easily discernable from the exterior of a building), so are the floor joists. This is why an early size-up from fireground personnel is imperative in helping to determine the type of construction that resources will be committed to (or have been committed to) and the amount of time available for suppression operations (particularly interior suppression operations).

- Exposed floor joists that are subjected to sufficient heat/fire can be a serious detriment to safe interior operations, and are capable of collapsing in a short time frame. The burned I-beam floor joists in figure 7–10 would not be capable of supporting the weight of firefighters, yet the floor they support could appear normal.

- If personnel need to operate above a fire, they need to know the type of construction that is supporting them and if it is exposed to heat/fire. As an example, let's assume a fire is on the first floor and personnel are directed to conduct a search on the second floor above a fire. Additionally, also assume the second floor joists are I-joists and the walls and ceilings are covered with ½ in. drywall. If a fire is able to expose the I-joists, the floor being supported by these weakened I-joists may not support the weight of firefighters conducting a search. Fires in lightweight construction are a prime reason why personnel should sound their intended path of travel when searching directly above a fire. Additionally, remember that lightweight construction exposed to fire will likely not give any advance warning prior to collapse.

- Most floor coverings are not considered structural—they are merely there to add aesthetics or provide a uniform and protective covering for the subfloor. A notable exception is lightweight concrete used on metal pan subfloors. Flooring such as carpeting can add fire load and flammable/toxic gases to smoke. Tiles, terrazzo, and ½ in. to ¾ in. finished wood floors can add significant dead load to the floor substructure and supports—leading to accelerated collapse (especially in lightweight construction).

Quick summary

- Floors consist of four primary parts: supports, joists, subflooring, and floor covering.

- A primary disadvantage of floor joists is their potential diversity that is not readily apparent. They may be heavy timber, sawn lumber, lightweight truss construction, or engineered wood products.

- Floor joists can be supported by joist pockets, a ledger/ribbon board, a top plate, or metal hangers.

- When buildings are constructed on sloping ground, they can present dissimilar floor level designations from different sides of a building.

- It is imperative to identify buildings on sloping ground that present dissimilar floor levels from different sides of a building and ensure that appropriate resources are aware of the dissimilar levels.

- Most floor coverings are not considered structural, although they can add hazards like additional dead load and flammability/toxicity issues.

CEILINGS

Ceilings are interior overhead surfaces that are not normally structural components of a building, but primarily provide a finished look to a room and also to hide the area underneath a floor or the roof above. From an architectural viewpoint, ceilings are classified by their construction and/or visible attributes such as cove ceilings, coffered ceilings, flat ceilings, and so on. However, from a fire service perspective, ceilings are either *directly fastened* to overhead floor or ceiling joists or are *suspended* and supported by overhead floor or roof structural members.

Directly fastened ceilings

Directly fastened ceilings are most commonly characterized by lath and plaster materials, metal wire mesh, decorative tin, and drywall.

Lath and plaster. Lath and plaster ceilings (which are also known as plastering) were commonly used on older buildings, and drywall ceilings are typically used on more modern buildings. Historically, plaster was used on the interior of a building and stucco was used on the exterior, but both were typically made from lime and sand. In the latter part of the 19th century, Portland cement was added to the lime and sand mixture to improve durability.

To construct a lath and plaster ceiling (which is the same process as used on interior walls), wooden slat laths are nailed to ceiling joists with a slight gap between the laths. These materials are then covered with three layers of plaster: a rough base coat known as a scratch coat, then a brown coat, and finally a finish coat. Although this process results in ceilings that are relatively strong, durable, and have a better fire resistance than many modern materials, the costs associated with installation and labor when compared to modern methods ultimately caused them to be replaced by sheets of drywall.

A lath and plaster ceiling can be considered a relatively strong ceiling as compared to modern drywall ceilings and has a better fire resistive rating than drywall (particularly ½ in.). As an example, the ceiling in figure 7–11 was subjected to heavy fire and is still somewhat intact. Lath and plaster ceilings can absorb more water before collapsing than drywall ceilings and they will typically collapse in smaller sections than drywall ceilings. However, lath and plaster ceilings are a bit more difficult to pull with a pike pole, hook, or other similar tool as compared to drywall ceilings.

The Art of Reading Buildings

Fig. 7–11. Lath and plaster can offer more resistance to fire than modern drywall.

Metal wire mesh and plaster. A variation of lath and plaster that can also be found in older buildings is directly nailing a metal wire mesh (also known as netting or expanded metal mesh) to ceiling joists and then covering the mesh with a rough base coat of plaster, then completing with several coats of smooth plaster for a finished appearance. In some commercial construction, a wire mesh lath with a paper backing is similarly attached to ceiling joists and then finished as previously mentioned. This method results in a strong surface that is also more durable than drywall ceilings, but is significantly more difficult to pull with a pike pole, hook, or other similar tool than drywall as the metal wire mesh is more substantial than lath and plaster and/or drywall ceilings due to its inherent strength. This is the same material that is often used in soffits under facades that are located over public walkways (and is also difficult to pull with pike poles or hooks).

Tin ceilings and decorative wood. In the 19th and 20th centuries, decorative tin ceilings were used as a cost-effective alternative to the attractive and expensive plasterwork that was found in European homes and American mansions. In the 1930s, tin ceilings began to lose their popularity but can still be found in many older commercial buildings (and some expensive residential structures) across this country. Tin ceilings are constructed by first nailing wood furring strips onto the ceiling joists and then nailing interlocking preformed tin panels (a common size is 24 × 24 in.) to the furring strips. This produces a relatively strong surface that can be as ornate as desired and also dictated by the resultant cost of the tin panels. The finished ceiling provides a measure of resistance to fire but conducts heat through the metal to the wood backing. Tin ceilings are not prone to a rapid collapse but are similar to metal wire mesh and plaster in that they can be challenging to pull with pike poles or hooks. (Some departments have developed specialized tools to pull these types of ceilings.) Additionally, tin ceilings do not readily collapse unless the wood furring strips have been weakened by fire.

Decorative wood ceilings are another type of directly fastened ceilings, but due to their diversity of construction, their strengths and weaknesses are dependent on their construction and the materials that have been used. These ceilings range from simple wood paneling to wood paneling finished with solid wood coffers. These ceilings are found in more expensive applications and can be significantly more substantial than drywall ceilings. However, if it is necessary to pull these types of ceilings, remember that large sections of wood of significant weight can suddenly collapse as newer versions are likely held in place by modern adhesives that can soften when exposed to heat.

Drywall. Drywall—also known as plasterboard, wallboard, or gypsum board—is a panel that is usually 4 × 8 ft by ½ in. thick and consists of a gypsum plaster layered between two sheets of paper (other sizes such as 4 × 10 ft panels and 5/8 in. thickness can also be used). When used on ceilings (and walls), the drywall is nailed and/or screwed into studs/joists, and then the adjoining ends are covered with a joint compound, as are the indentations left by the screws or nails. The joint compound is sanded to a smooth finish and the finished drywall is then painted or textured with various finishes. The drywall process is cheaper and significantly faster than the installation of lath and plaster, metal wire mesh and plaster, tin ceilings, and decorative wood ceilings. Drywall will provide a degree of fire resistance but not to

the same degree as lath and plaster, metal wire mesh and plaster, or tin/wood ceilings, and will not retain and radiate heat like the aforementioned ceilings after suppression operations are completed. However, drywall ceilings are prone to collapse in large sections when wet and/or are pulled by pike poles or other similar tools. (This can also be an advantage during overhaul operations.)

Concrete. Occasionally, it is possible to encounter either aggregate or lightweight concrete ceilings (as well as walls and floors). Although strength and rapid collapse of these materials should not be an initial concern, their ability to absorb and radiate heat can be a significant disadvantage to these materials, particularly aggregate concrete. This can be a severe limiting factor in suppression and overhaul operations (see the Central Library fire in chapter 4).

Suspended ceilings

Suspended ceilings—also known as dropped ceilings, false ceilings, and grid ceilings—are commonly utilized in modern construction for their relative ease of installation, cost effectiveness (as compared to older alternatives), and their attractive appearance. A typical suspended ceiling consists of wires that are hung from overhead structural members and used to support a framework of metal channels that form a grid of 2 × 2 ft or 2 × 4 ft cells. Tiles made from various materials are then placed into the cells to complete the ceiling assembly. An older and less common version of the suspended ceiling is the concealed grid system, which utilizes interlocking panels that are secured to the grid system by the use of splines. This system makes the removal of the panels difficult and was used in installations where access to the area above the ceiling was deemed unnecessary.

As these ceilings are hung from overhead structural members, the area above the suspended ceiling can be of various sizes and can vary from 1 ft to over 10 ft, as illustrated in figure 7–12. (Also notice the close proximity of stock and storage in relation to the suspended ceiling.) The area above a suspended ceiling is often known as an **interstitial space** and is typically used to conceal ducting, wires, plumbing, sprinkler supply lines, and other similar considerations. A **plenum space** is an interstitial space used as an air return for HVAC systems. In this configuration, air is drawn through grilles that are mounted in the ceiling and returned to the HVAC system via the plenum space. Additionally, electrical wires within a plenum space (and not within conduit) should use low-toxicity and low-smoke insulation on the exposed wires to minimize these hazards within return plenum airspaces.

Fig. 7–12. Suspended ceilings can result in a noteworthy void above a ceiling.

The design of suspended ceilings and plenum spaces presents several noteworthy benefits and hazards. A benefit of suspended ceilings is they are easily installed, can easily change the ceiling height of a room, tiles are easily replaced to change color and design, and the tiles in a common suspended ceiling can be easily removed by fireground personnel to check the area above the ceiling for extension of fire, heat, and/or smoke.

Disadvantages of a suspended ceiling include the following:

- Firefighting personnel do not know what is above a ceiling until they remove some of the tiles.
- Suspended ceilings can hide other ceiling levels above a suspended ceiling.

The Art of Reading Buildings

- If fire/heat/smoke extends into a plenum area, it will be drawn into the building's HVAC system. It is important to turn off the HVAC system when plenum spaces are known or detected.
- Suspended ceiling spaces can accumulate smoke and create a smoke explosion potential.
- Suspended ceilings can be used to hide substandard construction.
- If the suspended ceiling collapses during fireground operations, firefighters can become entangled in the suspension wires and delay their exit from the building. (This is why some departments issue wire cutters to fireground personnel.)

Quick summary

- Ceilings are often encountered in interior operations and are capable of collapsing and causing death and/or injury to personnel.
- Lath and plaster, decorative tin, metal wire mesh and plaster, and decorative wood ceilings typically present an overhead surface with more permanence than modern drywall materials, but are more difficult to pull and open than drywall materials.
- Drywall is the modern substitute for the older ceiling materials and is capable of failing in fire conditions in large sections and less time than older ceilings.
- When they collapse, the wires used for suspended ceilings can easily entangle firefighters.
- Individual tiles in a typical suspended ceiling can be easily removed if necessary.
- Spaces above a suspended ceiling can accumulate smoke and/or spread fire undetected.
- If the area above a suspended ceiling is used as a plenum, the HVAC system should be turned off as soon as possible.

WALLS

The walls of a structure have numerous important functions such as serving as the vertical structural supports for the floor(s) and roof, a passageway for utilities, a barrier against weather, a division or separation between rooms within a building, and to enhance the aesthetic appearance of the exterior of a building. For the purposes of this chapter, let's first look at interior walls and then exterior walls.

Interior walls

A knowledge of walls is necessary to determine potential areas or pathways of extension of fire. Four types of walls in the interior of a building are typically used in building construction, as shown in figure 7–13.

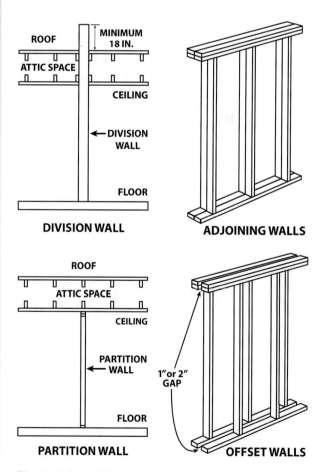

Fig. 7–13. Four different types of walls can be found inside buildings.

Division walls. In the building construction industry, there are two types of division walls. ***Occupancy division walls*** are used to provide major subdivisions within a building for tenant needs, and ***fire division walls*** are used to subdivide a building to restrict the spread of fire. Often, occupancy and fire division walls are served using the same wall. For this chapter, let's focus on the fire division wall. Fire division walls typically travel from the grade floor/foundation through the ceiling and project above a roof at least 18 in. This type of wall is of heavier construction than a party or partition wall and is often made from masonry materials for their ability to provide a barrier to the horizontal extension of fire, particularly in an attic. This remains true as long as the wall has not been breached by incomplete repairs, openings made for utilities and/or cable considerations, an open fire door, or the like. Although division walls are easily identified by viewing the roof of a building (as they project above the roof at least 18 in.), it is important to confirm they have not been breached, particularly if it is important that a wall should be restricting the horizontal spread of fire. This confirmation can be obtained by personnel on a roof by cutting a small opening in the roof on the *uninvolved* side of the wall.

In addition to the aforementioned masonry fire division walls, there are two other types of fire division wall construction that should be briefly mentioned. First, brick noggin (infill brick panels in timber frame buildings), which was used in some older buildings, is often degraded from age and the use of sand mortar. Brick noggin can also present a noteworthy dead load. Second, drywall is a current material that is often used as a type of fire wall in attics and can perform admirably if it has not been breached. As an example, figure 7–14 illustrates drywall that has been used between the ceiling joists, roof rafters, and ridge board.

Fig. 7–14. Division separation walls can be used in an attic to minimize the horizontal extension of fire and smoke.

Partition walls. Partition walls—used to divide areas or rooms into smaller areas or to separate one portion of an area from another—are usually not load bearing. Normally they travel between the *floor and ceiling only*. Common examples are walls that separate interior rooms in single-family dwellings. These walls will not restrict the horizontal extension of fire through an attic or other open space above this type of wall.

Party walls. Walls shared by two buildings or two occupancies within the same building. If the party wall carries beams or structural assemblies, it is a structural element. Party walls are common in townhouses, condominiums, motels, and commercial structures. These walls normally connect the floor and ceiling only. Of these, there are two basic varieties: adjoining and offset.

- Two common walls that are joined together are known as adjoining walls because the adjoining studs are placed together. Although this configuration will tend to confine a fire between the studs, thus limiting horizontal and vertical extension, it will also allow sound to pass through the stud members more easily. Although adjoining walls typically connect the floor and ceiling only, they may project through a roof similar to a division wall.

- Offset walls are those where two common walls are separated by several inches and the studs are also staggered. The resulting air space separation of several inches is for insulation and sound-deadening purposes. However, fire can easily travel horizontally *and* vertically between the offset studs and the gap of several inches between the upper and lower plates.

If personnel open a common wall and see adjoining studs, this configuration should minimize the extension of fire unless it is balloon frame construction. However, if offset studs are found, it is important to place emphasis on checking for horizontal and vertical extension (keep checking until there is a lack of char).

Exterior walls

Exterior walls comprise the outer shells of structures and are constructed of numerous materials. Exterior walls are likely to serve as the principal support (wall column) for roof assemblies (post and beam is a notable exception). The extension of fire through an exterior wall depends on the type of construction, such as wood balloon and platform construction, masonry construction, and so on. Let's consider the various types of construction that are used for exterior walls and how they can affect stability, the potential for collapse, and extension of fire.

Wood framing. *Note: The following information on balloon/platform framing and advanced framing techniques is detailed in chapters 4 and 6, but summarized here for easy reference.*

- **Balloon framing.** After a foundation is completed, the exterior walls (that are comprised of full-length studs) are constructed in a horizontal position and then raised to their vertical configuration. This process results in the exterior wall studs traveling uninterrupted from the foundation to the attic. In addition, a ledger (or ribbon board) is nailed or cut into the exterior wall studs and used to support the floor and ceiling joists either for single or multiple story buildings, and there is a lack of fire blocking in the exterior wall stud cavities (which can be seen in fig. 4–20). This results in open void channels between the stud cavities (foundation to the attic) and also access from the stud channels to the floor joists. The exterior wall is completed with any number of configurations such as brick, wood, stone, stucco, shingles, or other materials. When confronted with this type of construction, it is important to remember that the void spaces in a balloon frame building provide vertical fire spread paths from the foundation to the roof. Additionally, fire can spread from vertical stud spaces into floor joist voids due to ribbon board.

External features that can assist in the identification of balloon frame construction include the following:

- Old wood frame buildings up to about three stories in height
- Wood shiplap siding and asbestos or asphalt-type shingles
- The presence of visible 2×4 in. rafter tails that are spaced wider than 2 ft on center (see fig. 8–3)
- Windows in multistory buildings that line up vertically

- **Platform framing.** After a foundation is completed, 2×6 in. studs are typically used for the exterior walls and travel from the foundation to a double plate that is used to support the ceiling joists for a first floor ceiling and roof rafters for a single-story structure.

If a multistory building is being constructed, then each successive floor is constructed in the same manner. Additionally, fire blocking is used in the exterior walls (as illustrated in figure 4–23). This method provides intrinsic fire stopping from two perspectives: the fire blocking in the exterior walls between the floor and ceiling joists, and the double plate at the top of each floor. Fire blocking minimizes vertical extension of fire, and the double plate eliminates the void spaces of balloon construction that allow fire to extend upward into

the voids between the floor joists of multistory buildings and into an attic. The exterior can be finished in a wide variety of materials such as wood lap siding, vinyl siding, plaster, brick, and other suitable materials.

- **Advanced framing techniques.** This framing method uses techniques that are different from platform framing and are summarized as follows:
  - Walls are framed on 24 in. centers (instead of 16 in. centers).
  - Corners are made from two studs (instead of three or four).
  - Roofs use trusses instead of conventional framing and floors use I-joists instead of dimensional lumber.
  - A single top plate is used instead of two overlapping plates.

Although this framing technique uses 10% to 30% less lumber, it is easy to see that this new version of in-line or *stack* framing results in a structure that has less structural components, potentially resulting in faster collapse times when exposed to fire.

Exterior sidings for wood framing. Once a structure has been rough framed, structural sheathing is attached to the exterior of the framing for two purposes: it provides lateral bracing and strength for the structure, and it also provides a backing for exterior siding materials. In older structures, 1 × 6 in. sheathing was used on the exterior and/or interior. Later on, ½ in. plywood replaced the 1 × 6 in. sheathing and was used on the exterior with ½ in. sheetrock on the interior. More recently, ½ in. OSB is used on the exterior with ½ in. sheetrock/gypsum board on the interior.

Once one of the previously mentioned structural sidings is completed, a finished siding is then applied for not only a finished appearance but also for protection. Additionally, and of great importance, structures in close proximity to each other (fig. 7–15) can present a severe exposure hazard that can be significantly accelerated by the type of siding that is being exposed, such as asphalt and felt, vinyl, and wood.

Fig. 7–15. Buildings in close proximity can present serious exposure considerations.

Let's consider the more common sidings that are used on the exterior of wood framed buildings:

- **Asbestos shingles.** Although asbestos shingles are no longer used due to their significant health considerations, they can still be found on the exterior of older wood frame structures and are normally identifiable by their striated appearance. The fact that they are no longer used is a good reason to consider your involvement with them in fireground operations, particularly in overhaul operations. Asbestos shingles can also be identified by their gray- or silver-tone internal composition. Firefighters who encounter asbestos-containing materials during fires should use SCBA and full protective clothing for suppression and overhaul activities. Minute asbestos particles are a known carcinogen. Exposure to the particles should trigger appropriate decontamination measures. Likewise, incidents where asbestos has been disturbed warrants the notification of local health or building code officials.

- **Wood shiplap or shingles.** Various types of wood (pine, cedar, etc.) can be used for exterior shiplap sidings or shingle sidings. Wood shingles are easily identifiable by their rough

appearance that is somewhat similar to wood roof shingles. However, they will all contribute to the combustible materials in a structure fire (as they are all flammable) and they can ignite from exposure fires.

- **Fiber cement siding.** Fiber cement siding is popular due to its long life, low maintenance, and the fact that it is noncombustible. It is a composite material made from sand, cement, and cellulose fibers, and is made in sheets that can be easily installed. Other types of similar materials can be made in planks that resemble wood shiplap sidings. These sidings are fire resistive and will not add to the burnable materials in a structure fire.

- **Plywood materials.** Plywood materials such as T-11 are popular as they are cost efficient and easily installed. However, they will easily burn and rapidly disintegrate during fire conditions. Part of this problem is the flammable glue that is used to secure their laminations; the other problem is their minimal thickness (likely ½ in.).

- **Vinyl siding.** Vinyl siding has become popular as a replacement for wood siding as it requires less maintenance and lasts longer. However, it is made from polyvinyl chloride (PVC) that will easily melt when exposed to heat and give off noxious fumes that are toxic to fireground personnel. Vinyl siding is easily identified, as it usually has a narrower width than wood/fiber cement sidings and has a hollow, plastic feel and sound when you tap on the surface.

- **Asphalt-felt siding.** Asphalt-felt siding is a highly dangerous type of petroleum-based siding that was popular post-WWII and consists of a fiberboard coated with tar that holds a granular material. It can either look like a plain felt-type siding or the more popular embossed type that imitates the look of brick (fig. 7–16). This siding will rapidly burn and is often referred to as gasoline siding as it has absolutely no fire-retardant properties. Not only is this siding highly flammable from autoexposure from the host building, but also from exposure from buildings that are in close proximity.

Fig. 7–16. Asphalt-felt siding is a highly flammable material that can imitate the look of brick.

- **Masonry veneers.** Masonry veneers are nonstructural and are primarily used for their appearance. Unless a fire is sufficiently capable of weakening the structural wall that the veneer is attached to and causing a collapse of the wall and veneer, there is little danger from masonry veneer materials during a structure fire.

- **Plaster or stucco.** Plaster and stucco materials are normally nonflammable and do not pose a significant hazard during a structure fire.

- **Corrugated.** Corrugated sidings can be comprised of metal (lightweight steel or aluminum) or fiberglass. When exposed to heat and/or fire, fiberglass will quickly fail; aluminum and lightweight steel do not last much longer.

Quick summary

- Balloon frame construction is renowned for allowing fire to rapidly travel up the continuous void channels into the attic of a building and also into the voids between the floor joists.

- Fire walls and separations can be advantageous as long as they are not breached.

- Brick noggin (brick wall infill) can present a noteworthy hazard due to its weight and possible deterioration.

- Lightweight joists and roof systems are commonly used in platform framing.
- Advanced framing techniques have minimized the amount of structural material/members that are used to hold up a structure. This will result in faster collapse times in fire scenarios.
- Exterior sidings vary in the materials that have been used but wood, asbestos shingles, and asphalt siding can be the most hazardous.
- Buildings in close proximity with flammable sidings can be a significant exposure hazard.

Masonry walls

Masonry walls—commonly utilized for their reduced maintenance, heat resistance, and their admirable characteristics of compressive and tensile strengths—are made from common materials such as clay brick, concrete brick, stone, marble, granite, and limestone, with clay brick and concrete block being the most common. Key words that are used in this section are as follows:

Unreinforced masonry. A wall construction method using stacked brick or block and mortar without Portland cement, steel rebar, or strapping. Also, a modern masonry wall that is not designed for load-bearing structural applications.

Concrete infill. This type of construction consists of gaps between parallel courses of masonry units that are filled with concrete and pieces of brick (upper open wall in fig. 7–17), or concrete with vertical and/or horizontal runs of rebar.

Reinforced masonry construction. A wall construction method using stacked brick or block and mortar with steel rebar reinforcement placed in open cells and then filled with concrete, or steel embedded in the mortar joints.

Rebar. Rebar (short for reinforcing bar) is a steel bar that is used as a tensioning material in reinforced concrete and masonry to increase stability and strength. The rough surfaces on the rebar aid in bonding the rebar to the concrete.

Fig. 7–17. Concrete infill is concrete and pieces of brick between parallel courses of masonry.

Wythe. A wythe is a continuous vertical section of masonry, one unit in thickness. A single wythe can be separate from or interconnected with an adjoining wall. In a multiple-wythe wall, the wythes are interconnected for additional strength and stability, and are often used for a structural load-bearing wall. It is common for multiple-wythe exterior walls to use an economical block/brick on the inside of the wall and/or for structural purposes, and a more expensive block/brick on the outer portion that is visible. This was common on older brick masonry buildings.

Veneer. A veneer is a decorative-only wall covering added to help improve the building's appearance. Masonry veneers are usually a single wythe of brick/stone that is nonstructural and is commonly built on the side of a structurally independent, load-bearing wall. Veneers are usually attached to the load-bearing wall using small metal tabs. In figure 7–18, the front wall is a decorative brick veneer on an unreinforced masonry wall (notice the inset windows), and the side wall is wood frame with a plaster exterior (notice the small window is not inset). In this configuration, the veneer wall is predominantly decorative and not structural.

Fig. 7–18. A decorative type of brick is often used as a single wythe veneer for ornamental purposes.

The focus of the balance of this chapter is brick and concrete block. Masonry walls can be constructed in four configurations: unreinforced brick, reinforced brick, weight bearing, and veneer.

Unreinforced brick. Brick buildings that were constructed prior to the mid-1930s are significantly different, both in appearance and structural integrity, from brick buildings built today and there are literally thousands and thousands of these buildings across the country (fig. 7–19). It is safe to assume that the masonry exterior portions of brick buildings constructed before 1935 are an accident looking for a place to happen when exposed to heat/fire, and can present extreme hazards to firefighting personnel under fire or earthquake conditions due to the water-soluble mortar and lack of rebar.

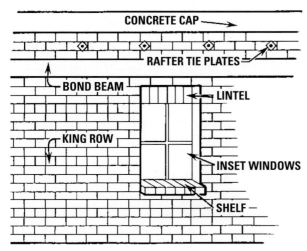

Fig. 7–19. Unreinforced masonry construction can be recognized by inherent characteristics that are typically visible.

Because of the inherent dangers these buildings pose, they have undergone multiple retroactive modifications and can be summarized as pre-1933, post-1933, post-1959, and post-1971.

Pre-1933:

- The mortar consists of lime and sand only, with no Portland cement.
- The masonry lacks steel reinforcing rods (rebar).
- Brick exterior walls are about 13 in. thick.
- Parapet walls are found around the perimeter of a roof. Parapet walls can be 3 ft above the roofline and 5 ft or more if used as a facade on the front of a building.
- The floor and roof joists are let (resting in a cavity) into the inside of the exterior walls.
- The roof and floor joists are often fire cut (the ends are cut with an angle, as illustrated in figure 3–13) so they can pull loose from the exterior walls during a fire and collapse into the interior of the building without pushing the exterior walls outward.

Post-1933:

After the disastrous Long Beach (California) earthquake of 1933, building codes were revised to provide better earthquake safety for new masonry buildings. The following revisions characterize masonry buildings that were built after 1933:

- Portland cement is utilized in the mortar.
- Masonry walls are required to be reinforced with steel rebar.
- Exterior walls are required to be at least 9 in. thick.
- All joists and rafters are required to be anchored to exterior walls. This is usually accomplished by bolting a ledger board to a masonry wall and attaching the joists and rafters to the ledger board with metal hangers.

Post-1959:

After the Tehachapi (California) earthquake of 1959, many building codes (particularly on the West Coast and other earthquake-prone areas) were modified to require the following retroactive corrections on existing buildings of unreinforced masonry construction:

- A 4 in. to 6 in. concrete bond beam cap must be laid on top of lowered parapet walls along public ways and exits.

- A parapet wall should not be higher than 16 in. including the bond beam cap.

- Exterior walls are drilled at the roof rafter level and a steel anchor bar or rod is installed every 4 ft and attached to the existing roof rafter. This modification rendered the fire cut of the roof rafter ineffective. The steel anchor bars or rods are secured to the exterior of the building by a plate and nut, some of which are decorative, that are known as *rafter tie plates*.

- Rafter tie plates are common in virtually every state and can be seen on the front of the old theater shown in figure 7–20. Additionally, the pattern indicates a gable roof is behind the parapet wall and also shows the different heights of the roof behind the parapet wall.

Fig. 7–20. Rafter tie plates are readily visible on the front of this old theater and indicate the location of the roof behind the parapet wall.

Post-1971:

The Sylmar (California) earthquake of 1971 provided the impetus to further modify existing buildings of unreinforced masonry construction. A review by a blue ribbon committee was instrumental in additional retroactive corrections (the earthquake ordinance) that were also partially or completely adopted by states outside of California. The retroactive corrections were designed to prevent exterior walls from collapsing outward by stabilizing a building as follows:

- The walls are anchored to floor and roof systems (with additional tie plates).

- The roof construction is strengthened (plywood on top of roofs with 1 × 6 in. straight sheathing and metal straps across the roof (fig. 7–21).

Fig. 7–21. In some cases, metal straps have been used as a retrofit to strengthen unreinforced masonry buildings and can be detrimental to power saws used for roof ventilation operations.

Depending on a particular area, unreinforced masonry buildings will share all or a portion of the following trademarks and should be familiar to firefighting personnel who are responsible for responding to these buildings:

- Rafter tie plates are on the exterior of a building. (Rafter tie plates can be found on old and remodeled buildings that appear to be new.) Exterior plates on a masonry building indicate that floor joists and rafters of the building are anchored to the exterior walls.

- A bond beam cap of concrete is on top of parapet walls. Concrete bond beams may also have been added for strength over the windows and between the floors of multistory buildings. This is a common technique used for additional strength for these exterior walls.

- Deeply recessed window frames are used. Window frames are set to the inside of a wall, thereby exposing about 8 in. of brick return on the exterior of a building. Remember, these walls can be 13 in. thick and this is a common indicator.

- Windows will have arched or straight lintels.

- The lime mortar between the bricks is white, porous, sandy, and can often be easily rubbed away by a fingernail or knife. In some cases, the bricks have not been uniformly laid and the workmanship appears sloppy.

- In every fourth to seventh row of bricks, one row will have been laid on end. This row of bricks is referred to as the *king row* and is for additional strength. The king row is a visual clue that the wall is a multiple wythe.

- Personnel and apparatus placement in respect to unreinforced masonry buildings should always be considered as exterior walls may suddenly collapse (during fire conditions) outward a distance that is at least equal to the height of the wall (fig. 7–22) and can often be twice the height of a wall. The primary collapse dangers are the front and rear walls. The secondary collapse dangers are the side walls.

The safe areas are as follows:

- The corners of a building are more safe, as buildings normally collapse *outward* (with the exterior walls) and not in the corners.

- Keep a distance at least equal and up to twice the height of the walls away from a building. Placement of personnel and apparatus should be a primary concern when confronted with this type of construction.

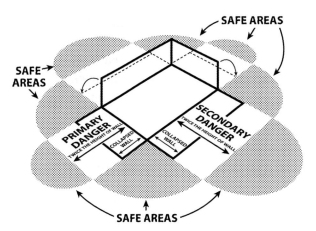

Fig. 7–22. Potential collapse areas associated with unreinforced masonry buildings

Reinforced brick. Brick masonry buildings constructed after the mid-1930s are fundamentally different from unreinforced brick masonry buildings constructed before 1933 and are not willing candidates for collapse. They also have the following characteristics that are *not* shared by unreinforced brick buildings:

- The mortar is comprised of Portland cement.

- Rebar is utilized for vertical and/or lateral strength.

- They use a better quality of brick than unreinforced brick masonry. In figure 7–23, notice the difference in the quality of brick and different characteristics from unreinforced brick.

Fig. 7–23. Reinforced masonry brick has different visible characteristics and strength as compared to unreinforced masonry construction.

Although reinforced brick masonry buildings can be identified by visible characteristics that are not common to unreinforced brick masonry buildings, it is important to remember that some unreinforced characteristics can be incorporated into reinforced brick construction to simulate an old appearance, such as a king row of brick.

Weight bearing. A multiple-wythe brick wall is commonly found in weight-bearing applications and was commonly used in unreinforced masonry construction. Their use in modern applications has virtually been replaced by concrete cinder blocks filled with concrete and strengthened with rebar. The older multiple-wythe walls were constructed by laying two parallel courses of clay brick separated by an infill of concrete and often pieces of brick. Remember that this type of construction typically used a mortar that did not include Portland cement or rebar, so the inherent strength was a result of the wall thickness. Although these walls are classified as weight bearing, they are prone to collapse in fire conditions, particularly when the roof they support collapses.

Modern single wythe walls have replaced older multiple-wythe walls by using concrete masonry units (CMUs) for their ability to support weight, longevity, reduced maintenance, heat resistance, and the ability to provide good protection in tornadoes and hurricanes. CMUs are used alone, for a structural core for brick veneered masonry, or plastered over for decoration. A key structural advantage of this configuration as compared to clay bricks in a multiple wythe configuration is that a single CMU can be used for a weight-bearing/structural wall when the block voids are filled with concrete and tied together with or without rebar (depending on the application).

A common example of this type of construction is the big-box or warehouse-type store such as WalMart, Home Depot, Lowes, and many other commercial buildings. This type of masonry wall provides a strong, fire resistive type of construction that is not a willing candidate for collapse when exposed to fire. Additionally, there are not many instances of these walls collapsing when the roof they support collapses in fire conditions.

Veneer. A single wythe of masonry that is normally constructed on a weight-bearing structural wall of masonry or wood (refer back to fig. 7–18). In this configuration, the veneer wall is predominantly decorative and not structural. The veneer is connected to the structural wall by brick ties that are attached to the structural wall and are placed in the mortar joints between the brick veneer. (Brick ties can be seen on the rough CMU wall in figure 3–9.) This can result in a slight gap between the structural wall and the veneer wall that can increase over time due to soil movement, expansion/contraction, and other circumstances. Although this type of construction is considered a nonstructural type of wall, it is often used for its appearance, longevity, and minimal maintenance requirements.

However, because a veneer wall is basically attached to a structural wall by brick ties, its collapse potential is dependent upon the type of wall it is attached to. Therefore, CMUs provide a superior backing while wood framed walls do not and are subject to collapse along with the exterior brick veneer wall. An additional type of veneer that is similar to the preceding single wythe of masonry is the popular manufactured stone that is bonded to a structure with adhesives as shown in figure 7–24. This presents an attractive surface that can also be fire resistive but can conduct heat to its backing.

Fig. 7–24. Decorative stone is bonded to an exterior surface with adhesives and can be fire resistive.

- A potential collapse hazard from older arched type roofs (bowstring, tied truss, and lamella) on unreinforced masonry construction is the possibility of end hip rafters pushing the corresponding wall outward if the attached truss collapses.
- Earthquake retrofit modifications in unreinforced masonry buildings can result in metal straps that can extend across the width of a roof or 3 ft to 4 ft from exterior walls and strengthened roof decking comprised of sheathing, plywood, and composition.
- Some unreinforced masonry buildings have a higher quality brick on the front of the building to improve the appearance. Look for recessed windows, bond beams over the windows, and rafter tie plates.
- The primary advantages of reinforced brick masonry construction are strength and collapse resistance.
- Although concrete masonry blocks are generally not a problem, the roofs they support can be.
- Brick masonry veneer walls do not often pose a collapse hazard unless a fire is attacking a wood frame wall the veneer is attached to.
- Concrete formed walls are generally considered strong exterior walls that support older conventional full-dimensional lumber inside a building.
- Buttresses and pilasters can be used as an indicator of the location of the primary structural members for a roof unless there are pilasters on all four walls.
- Tilt-up concrete panels that use metal brackets to attach to the slab foundation have been known to collapse when the roof assembly collapses.
- Depending on the type of fire, consider placing resources away from potential collapse zones.

CHAPTER REVIEW EXERCISE

Answer the following:

1. List the three primary types of foundations.
2. List four hazards that are associated with basement fires.
3. List the four parts of a floor.
4. List four ways that floors can be connected to supports.
5. List two ways that ceilings can be fastened to an overhead component.
6. What is a plenum space?
7. What is the difference between a partition wall and a party wall?
8. List five ways that a wood wall can be covered.
9. What is meant by a multiple-wythe wall?
10. List three potential safe zones to be considered for unsupported masonry walls.
11. How are tilt-up wall panels connected wall-to-wall and wall-to-roof?

RESOURCES FOR FURTHER STUDY

- http://Tiltup.com.
- MacDonald, Mary Lee, "Preservation Brief 21: Repairing Historic Flat Plaster Walls and Ceilings," Technical Preservation Services, National Park Service, 1989.
- Thallon, Rob, *Graphic Guide to Frame Construction*, 3rd ed., Newtown, CT: Taunton Press, 2009.

READING ROOFS

OBJECTIVES

- List relevant roof definitions.
- Describe the eight most common roof styles, strengths, and hazards.
- Identify the most common roof coverings.
- List the various types of roof appendages.

THE IMPORTANCE OF READING ROOFS

Fortunately, determining building strengths and weaknesses is often enhanced by reading one of the most important aspects of a building and often one of the most visible—its roof! The importance of roofs in building construction and fireground operations cannot be understated as the roof of a building is normally one-fifth of the primary components of a building (which consists of the foundation, walls, floor, ceiling, and roof). As fireground suppression operations are normally conducted below and/or above a roof, knowing the type of roof and its potential structural integrity can directly affect fireground efficiency and safety. Additionally, the type of roof can frequently help to identify the era of construction, which can be advantageous.

In many cases, identifying a roof can be an easy task as most roofs are visible from the street and can easily be identified by their age and/or characteristic shapes. Obviously, some factors will be dependent on the type of roof construction in a particular area. However, this perspective is enhanced by the fact that East coast, West coast, and no coast roofs have a lot in common in construction methods and styles. The primary difference is the age of a roof. As an example, a gable roof on the East coast is basically the same as a gable roof on the West coast with the possible exception that the East coast roof is likely to have more layers of roofing material (dead load) that have been installed due to its age. In some cases, the edges of a roof (along the fascia boards) can reveal the number

of layers of roofing materials. Additionally, there is also a greater potential degradation of the wood due to termites, wood rot, weather extremes, and other deterioration factors that come with older roofs.

Definitions

Attic. An attic is a large space that is created by a steep pitched roof (arched, gable, etc.) for drainage and/or appearance. Depending on the type of construction, attics are normally large enough for storage, can be modified for additional living space, and can contain HVAC and other mechanical equipment.

Component connections. Also known as framing junctions, where two or more structural members are joined and, for this chapter, how they are joined. As an example, a connection can be secured by nails, bolts, steel plates and bolts, mitered finger joints, gusset plates, and other means. The way structural members are joined can determine how long the connection will last under fire conditions.

Conventional construction. Solid lumber of 2 × 6 in. or larger used in a standard framing configuration.

Cockloft. A cockloft is defined as a small space that is created when a roof is raised above the level of ceiling joists and rafters to provide a pitch for drainage. Cocklofts are common in older buildings (1800s and early 1900s) and are normally of conventional construction.

Dead load. The weight of the building itself and anything *permanently* attached to the building.

Diagonal sheathing. A series of 1 × 6 in. boards that run at a 45° angle from the exterior walls to the primary structural members and provide increased structural stability (compared to straight sheathing) as they cross more roof structural members. This style of sheathing was favored after the 1930s.

Engineered wood. Wood products made from a composite of glue, veneers (layers), and/or wood chips that are pressed together to form components that replace sawn lumber, sheathing, and other structural materials. Common examples are glulams, I-beams, and OSB.

Jack rafter. Roof rafter used in hips or valleys to span between ridge boards or wall plates.

Lightweight construction. Solid or engineered products used to form assembly-built structural elements that are lower in mass than previous construction methods.

Lightweight trusses. Trusses that are comprised of members of less than 2 × 4 in. (or smaller) and are often made from engineered lumber (or metal).

Live load. Any load applied to a building other than dead loads. Live loads are typically transient, moving, impacting, or static (like furniture). Examples include people, snow, and wind. These loads add to the existing dead load of a building but are not a permanent part of a structure.

Roof slope/pitch. This refers to the degree of slope or pitch for a roof and is expressed as a ratio. For example, a 4:12 pitch means that the roof rises 4 in. vertically for every 12 in. of horizontal distance.

Sheathing. All manner of material used to cover or encase walls, ceilings, and roofs of framed structures. For a roof, it is the first layer of covering for joists, trusses, structural beams, or rafters and is found in solid or open sheathing arrangements. Solid sheathing arrangements include OSB, plywood, or other panels as well as solid boards (like tongue and grove or 1 × 6 in. lumber) that are butted tightly together. Open sheathing (often called skip sheathing) is formed when boards (usually 1 × 4 in. or 1 × 6 in.) are laid on the roof with 4 to 6 in. gaps between each. In most cases, sheathing increases the stability of a building/roof.

Straight sheathing. A series of 1 × 6 in. boards that run at a 90° angle to supporting structural members.

Timber truss. Large dimension lumber used to form a truss. Commonly found in older roofs, this type of construction is normally made from

full-dimensional lumber, and is often comprised of multiple members bolted together to form one structural member.

Truss. An engineered structural element that uses groups of rigid triangles to distribute and transfer loads. The triangles create an open web space. Trusses are used in lieu of solid beams in many buildings. A typical truss is comprised of one or more triangular units constructed with straight members whose ends are connected at joints referred to as nodes. A triangle is the simplest geometric figure that will not change shape when the sides are fixed.

Truss loft. An attic space created by the open web nature of trusses.

The definitions presented in the above feature block will help communicate more specific hazards when you're engaged in a roof operation. Additionally, you will see the terms repeatedly as you navigate through the eight types of roof styles, various roof coverings, and the roof appendages that are covered in this chapter.

EIGHT MOST COMMON ROOF STYLES

The eight most common roof styles (in no particular order) in the United States are gable, hip, gambrel, monitor, sawtooth, flat, bridge truss, and arched (fig. 8–1). Below, we'll dissect each of these styles and provide some strength and hazard issues associated with them.

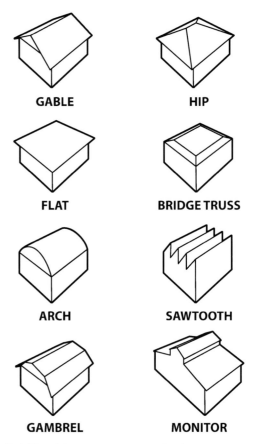

Fig. 8–1. The eight most common roof styles

Gable roof

A gable roof is perhaps the most popular style of roof in the United States as there are literally millions of residential single-family occupancies with this simple to construct and attractive roof. The term gable is derived from an A-frame configuration: two sloped or angled surfaces between two flat ends (gable ends). The two most common methods of construction are conventional (as shown in figure 8–2) and lightweight or truss construction. Conventional construction is often referred to as *stick frame* construction as this name is derived from the fact that each member is literally built stick by stick.

Fig. 8–2. The A-frame configuration of a gable roof

When applied to a gable roof, a ridge board is first erected and then each rafter is attached to the ridge board and the appropriate exterior wall. The ridge board and rafters are typically 2 × 6 in. or larger. Rafters are usually spaced 16 in. to 24 in. on center. A variation of this roof that can be found in older wood frame roofs is **bungalow construction**, which uses rough-sawn 2 × 3 in. or 2 × 4 in. rafters spaced up to 36 in. on center, are butted together at the ridge without a ridge board, and typically use 1 × 4 in. spaced sheathing nailed to the rafters (fig. 8–3).

Fig. 8–3. Bungalow construction uses 2 × 4 in. rafters commonly spaced up to 36 in. on center with no ridge board.

Additional support for gable roof framing is provided by collar beams (a horizontal beam connecting two opposing rafters in the top third of the A shape) and ceiling joists. The gable roof is found in semi-flat to steep pitch configurations and covered by various materials such as tile, composition shingles, shakes, and so on. Metal is being used more often to replace wood ridge beams or rafters in frame construction (see figs. 6–2 and 6–3).

Lightweight gable roof construction uses 2 × 3 in. or 2 × 4 in. wood trusses (sometimes metal), normally held together by metal gusset plate connectors (see fig. 5–13) or glue. Trusses share common features such as top chords, bottom chords, and webbing (supports between the top and bottom chords depending on the style of truss). Metal gusset plate connectors may vary in size, thickness, and depth of penetration; however, 18-gauge steel plates with prongs of 3/8 in. penetration are common. More recently, glue has replaced gusset plates in both residential and commercial truss applications. To achieve structural integrity with glue, the mating ends of structural members are first mitered in a finger-joint configuration, and then glue is applied before pressing the ends together. Random tests (nonscientific) by some fire departments have indicated the glue will soften at approximately 350° to 400°F and allow the mitered ends to separate. Another variation of truss construction is trusses made from metal. This change is gaining popularity due to the increasing cost of wood and the resistance to degradation from weather and termites.

The 2 × 4 in. bottom chord of the truss has replaced the 2 × 6 in. or larger ceiling joists found in conventional construction, and truss systems are also enjoying widespread use in floor and rough window and door openings. The common spacing for trussed rafters is 2 ft (24 in.), but this distance will vary depending on specific applications. Lightweight 2 × 4 lumber trusses with 2 ft spacing typically require OSB sheathing that is glued and nailed to each truss in order to meet strength requirements (assembly-built concept).

Conventional and lightweight gable roof sheathing will then be covered by a variety of materials such as wood shakes, tile, rock, composition,

corrugated metal, and so on. Older roofs are likely to have multiple layers of roof coverings. In some cases, the edges of a roof (near the fascia/trim boards) can reveal the number of layers.

Note: Heavy truss gable roofs are covered in the section titled "Older timber truss roofs," later in this chapter.

Strengths. The inherent strength of gable roofs can be found in conventional construction that uses ridge boards and rafters of 2 × 6 in. or larger and nails instead of gusset plates or glue. This type of construction can last approximately four times longer than 2 × 4 in. trusses when exposed to fire. The strong areas of this roof are the ridge, valleys, and the area where the rafters cross the outside walls.

Hazards. The use of lightweight 2 × 4 in. trusses (wood or metal) with no ridge board is similar in external appearance and size to 2 × 6 in. or larger conventional construction. This similarity can easily mislead unsuspecting fire personnel. The presence of 2 × 3 in. or 2 × 4 in. trusses with metal gusset plate connectors or glue equals a short burning time and early failure rate (one-fourth the time of conventional as a rule of thumb). The trusses are under compression and tension, and when the bottom chord or webbing fails due to fire damage (connector plates that become heated and pull out of the wood, or melting glue), the exposed trusses will fail. Therefore, expect collapse of exposed portions of the roof or total collapse of the entire roof (when involved) in a short period of time. Newer roofs use 7/16 in. or 1/2 in. plywood or 7/16 in. OSB as sheathing instead of 1 × 4 in. or 1 × 6 in. spaced sheathing. Plywood and OSB will burn and fail at a faster rate than spaced sheathing, and offers minimal resistance to fire.

An important additional hazard is possible misidentification of lightweight construction. Identification can be enhanced by an intimate knowledge of a district and prefire planning. Additionally, the presence of 2 × 4 in. rafter tails under the eaves is an excellent indicator of lightweight truss construction. As an example, in figure 8–4, notice the lightweight truss construction in the attic and to the left side of the header beam, while the exposed 2 × 4 in. rafter tails are visible on the exterior of the building. However, if the eaves are covered with a soffit, the rafter tails will not be visible. Be aware that trusses may be modified on commercial buildings to look like 2 × 6 in. rafters on the *front or sides* of a building but in the rear they are normally not modified. In figure 8–5 the I-beams have been altered to look like 2 × 6 in. rafters on the exterior of the building. In many cases, the rear of a building can often allow you to see the real size of rafters used in a building.

Fig. 8–4. Exposed rafter tails can indicate lightweight construction on the interior of a building.

Fig. 8–5. Truss-type structural members can be modified to look different from their original design/size.

Hip roof

The hip roof is similar to the gable roof, but the flat sides (gable ends) of the A-frame are replaced with sloped roofs (fig. 8–6). The hip roof appears more like an irregular pyramid. Similar to the gable, conventional (or stick frame) hip roofs consist of a ridge pole (that is constructed first) and then hip rafters are installed from the ridge pole down to and across the corners at the outside walls.

Fig. 8–6. The hip roof resembles the gable roof but lacks the A-frame end configuration (note that the primary roof is hip-styled whereas the secondary roof on the right side is a gable).

Valley rafters are used where two roof lines join. Jack and common rafters complete the structural members. The ridge pole and rafters are typically 2 × 6 in. or larger with rafters 16 in. to 24 in. on center. Rough-sawn 2 × 3 in. or 2 × 4 in. rafters, 36 in. on center were also used in older wood frame structures (bungalow construction), although not as common as its use in gable roofs. In lightweight construction, the construction materials—both wood and metal—are similar to those used for gable roofs. Various degrees of pitch and roof coverings are characteristic of this style of roof.

Strengths: In conventional construction, the ridges and rafters are 2 × 6 in. or larger. The ridge pole, valley rafters, hip rafters, and the areas where the rafters cross the outside walls are the areas of strength.

Hazards: The hazards of hip roofs are similar to those of gable roofs. The presence of 2 × 3 in. or 2 × 4 in. trusses for rafters will produce results similar to those of lightweight trusses in gable roofs when exposed to fire. Although the spacing of the trusses may be reduced for trussed gable and hip roofs covered with tile or other such heavy materials, the collapse potential will be enhanced due to the increased dead load on the roof. Similar to gable roofs, reading the rafter tails (if exposed) can be an excellent indicator of the size and spacing of the rafters.

Quick summary

- Gable roofs have flat, A-frame style ends (gable ends).
- Hip roofs are similar to gable roofs although there are no gable ends. The roof shape is more like a pyramid.
- When the eaves are not soffitted, 2 × 4 in. rafter tails can indicate truss construction.
- Rough 2 × 4 in. rafter tails spaced farther apart likely indicate bungalow construction.
- Lightweight wood (or metal) trusses will collapse in one-quarter the time that a conventionally constructed ridge board and rafter roof!
- It should be anticipated that older roofs can have additional roofing materials. In some cases, the edges of a roof (near the fascia boards) can reveal the number of layers of roofing materials.

Gambrel roof

A gambrel roof (the name is derived from the Latin word *gamba*, which means a horse's hock or leg) is an attractive design that is very identifiable due to its unique shape. As you can see in figure 8–7, the gambrel roof is two sided, with each side's slope broken by an obtuse angle so that the lower slope is steeper than the upper slope. This design is popular due to its ability to provide the advantage of a sloped roof for rain and snow runoff while also providing maximum headroom in the area below the roof. The most popular use of this roof design

is found on barns on many farms in this country, but it is also used on some residential and commercial buildings. Although a gambrel roof can be constructed from either conventional or lightweight construction methods, most of these roofs have been constructed from conventional construction.

Fig. 8–7. The gambrel roof is a two-sided roof with two slopes on each side of varying angles.

A variation of the gambrel roof is the mansard-style roof. The difference in the two is that all sides of the mansard roof share a hip-style slope. Much like the difference between gable and hip roofs, the gambrel roof has flat walls that rise to the peak on opposite ends, whereas the mansard has hipped slopes on all sides that rise to the secondary (uppermost) slopes that form the ridge top. A mansard roof is often characterized by dormer windows that protrude from the steep slopes of the lower pitches.

Strengths: The strengths of this roof are primarily dependent on the type of construction. Barns are normally of a heavier grade of conventional construction than typical buildings, so these types of buildings should be well built and more resistant to fire due to the size and configuration of their construction. Gambrel roofs on typical residential and commercial buildings that are of conventional construction of 2 × 6 in. or larger can be expected to resist fire for longer periods of time as compared to modern lightweight construction that can be used in some cases.

Hazards: Although this roof is popular for its ability to optimize useable interior space below a roof, it also provides an opportunity for the storage of considerable amounts of content. This content stored in the loft area can create two distinct hazards: the fire load that would be generated by combustible materials, and the weight of large amounts of materials on the floor (or the ceiling over the grade floor). Roof ventilation operations would be extremely difficult due to the design of a gambrel roof (unless attempted from an aerial device). Additionally, some gambrel roofs are constructed of lightweight trusses that will fail in less time than a conventionally constructed roof.

Monitor roof

Monitor roofs have customarily been used on barns and some commercial buildings, but their design affords many advantages for a wide variety of buildings such as garages, dwellings, and some other buildings, hence their developing popularity. A monitor roof (which is sometimes referred to as a lantern roof) is basically comprised of two shed-type roofs that are separated by a center section that is raised above the shed roofs and supports a gable roof. The raised center section is customarily flanked by windows or eave lights along the sides for daylight and/or ventilation (fig. 8–8). The advantages of lighting and ventilation can not only be a benefit for commercial buildings but can also be used for offices, storage, or extra sleeping areas. A side benefit is the center section can provide the necessary height for RV storage.

Fig. 8–8. Monitor roofs can be identified by their characteristic shape that can provide headroom, light, and ventilation to the interior of a structure.

Construction can vary but is typically dependent on the era of construction. Older buildings used conventional construction of 2 × 6 in. (or larger) structural members and were often integrated with pole barn type construction. Older buildings also typically used conventional type construction and materials to finish the interior. While newer monitor buildings can still use conventional construction, lightweight construction is increasingly used as a cost reducing measure to construct these buildings with the same external characteristics. This consideration can include veneer wood-based materials and corrugated-type materials, including corrugated metal or fiberglass exterior panels. In some cases, it is not unusual to find the lower portion (underneath the shed roofs) constructed of substantial structural members and the raised center section constructed of lightweight materials. The shed and gable roofs can be either conventional or lightweight construction.

Strengths: Older monitor buildings are generally well constructed as they are an older design and primarily built with conventional methods and materials. This will afford more time before collapse when exposed to fire as compared to lightweight construction. The strong portions of these buildings are the perimeter of the building and the corners, particularly if pole barn configurations have been employed. An additional strength of this design is the ability to initiate vertical ventilation through the upper windows/eave lights (if the roof construction will support the weight of ventilation personnel).

Hazards: Characteristically, the hazards are dependent on the type of construction, specifically lightweight construction that can readily fail when exposed to fire. This applies to the exterior walls, floors, and roof. Additionally, there will be an additional dead load imposed on the lower structural members from the raised center portion of these buildings that will not only consist of the structural elements but also any interior storage or furniture. An additional hazard is the large open design of this type of building that will enhance the horizontal and vertical extension of fire, heat, and smoke.

Sawtooth roof

Sawtooth roofs date back to the late 19th century and were typically used in manufacturing and industrial buildings as a primary light source (referred to as daylighting) before daylight was replaced with electric light sources. The roof remained popular through the 20th century and began to disappear in the 1940s. The sawtooth roof is a roof system that is unique in that it is comprised of a number of triangular and parallel roof surfaces (hence the sawtooth design as in figure 8–9) with the glass side often facing away from the equator side of the building to capture a diffused light source.

Fig. 8–9. Sawtooth roofs can provide light and ventilation to the interior of a structure.

Additionally, as warm interior air rises, the windows can be used for interior ventilation if they can be opened. This roof is commonly constructed with rafters of 2 × 8 in. or larger, and can use wood and/or metal supports for bracing. The sloping portion is characteristically covered with 1 × 6 in. sheathing and composition roofing material. This type of roof is constructed basically the same today as it was during the 1930s and 1940s, with the exception of the sheathing. Numerous supporting posts are required with the sawtooth design—increasing the strength of this roof as compared with a comparable flat roof with large open areas. If this roof is constructed today with modern materials (which is not common), expect lightweight construction that is covered with ½ in. plywood or OSB.

Strengths: Sawtooth roofs are generally well constructed as they are an old design and primarily built in the 19th and late 20th century. The strong portions are at the perimeter of the building and the areas near the glass panels.

Hazards: Consider the undersides of these roofs to be open or exposed to the interior of the structure. Newer sawtooth roofs may use lightweight construction and can be expected to be covered with ½ in. plywood or OSB sheathing, which have little resistance to fire.

Quick summary

- Older gambrel roofs are likely conventional construction and newer gambrels may use lightweight construction.
- If the presence of lightweight construction is apparent in a roof, it was also likely used in the construction of the floors.
- If there is a noteworthy amount of storage in a gambrel loft area, this could be a significant hazard for personnel working below the loft area.
- Determine the extent of fire (if any) in a gambrel attic. It is easy to visualize what an involved heavy load of storage would do to supporting lightweight structural members.
- Roof ventilation operations would be extremely difficult due to the design of a gambrel roof (unless attempted from an aerial device).
- Corrugated metal/fiberglass exterior sidings will readily fail when exposed to heat and/or fire, as will veneer wood products.
- The large open areas in monitor buildings will enhance the horizontal and vertical extension of fire and smoke.
- Consider the feasibility of using the upper windows/eave lights in monitor buildings for vertical ventilation.
- A diverse interior fire load can be expected in monitor roof buildings and can vary widely from farm animals, hay, and RV storage to offices and multiple commercial occupancies.
- The upper portion of a monitor building may be occupied and needs to be searched.
- If vertical roof ventilation is necessary, the monitor shed roofs will be easier to ventilate than the upper gable roof, which will likely require the use of an aerial device.
- Sawtooth roofs are normally of older design and construction.
- Early collapse of main structural members of conventional construction for sawtooth roofs should not be a primary concern.
- Typically, the complete underside area of a sawtooth roof is exposed to the building contents.
- Numerous supporting posts required with the sawtooth design can increase the strength of this roof as compared to a similar size flat roof with large open areas.
- The sawtooth is very stout and can usually support the weight of ventilation operations. Removing the numerous windows (if possible) can achieve desirable results.

Flat roofs—general

The simplistic flat roof has developed into a very popular style of roof for a wide variety of structures. Instead of constructing a gable, hip, gambrel, monitor, sawtooth, or arched roof, just erect four walls and cover them with a flat roof. This is a cost-effective design when considering labor constraints and even more cost effective when lightweight materials are used. Although simple in appearance, flat roofs can also be deceiving as they can vary in construction methods and can hide modifications to the basic roof components, often making it difficult to determine which type and/or method of construction has been used.

A variation of a flat roof is a shed roof (fig. 8–10), which is nothing more than a flat roof that has been constructed with a pitch that exceeds 10°. (A shed roof may also be considered half of a gabled roof.) A low pitch is defined as a 10° to 20° angle (pitch) and most shed roofs are constructed with a pitch exceeding 20°. Shed roofs are common on sheds of various sizes, but are also constructed on some residential and commercial buildings for a specific architectural appearance that requires the walls to increase in slope from the short end to the high end and the ability to place windows for daylighting on the wall at the high end of the slope.

Conventional flat roof. A true flat roof has a pitch angle between 0° and 10° with a minimum of 6° recommended for drainage. Wood rafters or joists of various sizes (2 × 6 in. and larger) are usually laid across the outside walls (fig. 8–11); however, in some cases, rafters may also be suspended by metal hangers. Rafters/joists in conventional flat roof configurations may or may not have internal horizontal and/or vertical support. The primary method for connecting structural members is metal plates and bolts and/or nails. Some older roofs have bridging or scissor bracing between the joists to increase lateral stability.

Fig. 8–10. A shed roof is a sloped variation of a flat roof.

Fig. 8–11. Conventional flat roof

The focus on the fireground is to try to identify the *type* of flat roof encountered based on a prior knowledge of a particular roof, the style of building, the era of construction, and any additional information from interior/exterior personnel. The wide variety of flat roof construction types can be categorized as follows:

- Conventional
- Wooden I-joists
- Open web construction
- Metal gusset plate construction or glue
- Panelized (all-wood or hybrid)
- Open web bar joist
- Nonstructural lightweight concrete

On-center spacing for rafters is typically 2 ft. The rafters are covered with 1 × 6 in. sheathing, plywood, or OSB (in newer applications) and use composition roofing material, rock, or other similar materials. Older conventional flat roofs on masonry construction commonly used lets (pockets) in the walls to support the rafters/joists. This is an old method (pre-WWII) that is no longer used. Newer buildings use a ledger board that is connected to the load-bearing wall.

Strengths: Susceptibility of these roofs to fire is totally dependent on the size of the rafters, their method of connection to other members, their method of support/suspension, their on-center spacing, and the type of decking. Older roofs use rough-sawn lumber with numerous bridging supports and 1 × 6 in. sheathing as decking, which

offers good resistance to fire. The area of strength for this roof is at the perimeter of a building.

Hazards: The degree of hazard is determined by the span of the rafters, their size and on-center spacing, and the presence of metal hangers used to suspend them. Roofs covered with plywood or OSB instead of board-type sheathing present a significant problem. Plywood and other like materials, typically of ⅜ in. to ½ in. thicknesses, offer minimal structural integrity under fire conditions.

Quick summary

- All of the strengths, hazards, and summary considerations for flat roofs are the same as for shed roofs.
- An older conventional flat roof can be considered substantial compared to modern roofs.
- Unreinforced masonry buildings commonly used lets (pockets) in the walls to support rafters and joists for flat type roofs.
- The older a building, the greater the possibility of multiple layers of roofing material (dead load) that the roof was likely not designed to hold.
- In some cases, roof ventilation operations might be able to be conducted within close proximity to fire on conventional construction.
- Flat and shed roofs can be covered with a wide range of materials such as composition, metal, rock, slate, tile, membrane, and many others.

Flat roof with wooden I-joists. Wooden I-joists, also known as engineered wooden I-beams, were developed over 40 years ago and their dimensional stability assists in helping them to resist warping, shrinking, and twisting. Additionally, they are lightweight and come in long lengths. These traits have made them very popular in both floor (fig. 8–12) and roof construction. I-joists consist of three main components: top chord, bottom chord, and stem or closed web. Most commonly seen are top and bottom chords that measure 2 × 4 in. or 2 × 3 in. Some chords may resemble plywood due to horizontal (longitudinal) laminations. For example, a trade lamination process (referred to as micro-lam or LVL) enables a cheap grade of lumber to be used for structural members. The stem is joined to the top and bottom chords by a continuous glued edge joint and is typically constructed from ⅜ in. OSB.

Fig. 8–12. Wooden I-beam members consist of a top chord and a bottom chord that are separated and joined by a stem.

This construction is very unstable until adequately braced with 2 × 4 in. nailing blocks that run perpendicular to the tops of the chords and 4 × 8 ft sheets of plywood/OSB sheathing (usually ½ in.) nailed to the I-beams. When 4 × 8 ft sheets used for sheathing are nailed to structural members (truss or conventional construction), a method called diaphragm nailing is used. Prior to nailing, the 4 × 8 ft sheets are placed so that the 8 ft dimension crosses the roof structural members and the 4 ft dimension parallels them. The sheets are then staggered every 4 ft like bricks in a masonry wall. Wooden I-beams are either hung from bearing walls by metal hangers (see fig. 7–8) or placed on top of the bearing walls. The common on-center spacing is 2 ft.

Strengths: The strong area of this roof is at the perimeter where the roof ties into the building.

Hazards: The principal hazards are threefold. One, the 3/8 in. stems and the 2 × 3 in. or 2 × 4 in. chords will take little time to burn, weaken,

and cause the rapid collapse of any chords and roof sufficiently undermined by fire. Unfortunately, many buildings can be found with unprotected chords (they are often painted a dark color to minimize their presence). Two, the glue used to form OSB and also used for webs can degrade with convection heat. The glue is hydrated (chemically wet) and once the moisture is steamed off, the glue turns to dust. Three, a common practice is to run heating and air conditioning ducts of various sizes through the web (as in fig. 4–30). This practice removes a significant portion of the stem and gives fire horizontal access to adjacent I-beams, enhancing extension and weakening of exposed materials.

Flat roof with open web construction. Open web construction (OWC) consists of top and bottom parallel wooden chords cross-connected by steel tube web members (fig. 8–13). Another version of this construction is constructed of all metal in the same configuration and is referred to as open web steel (OWS) joists. In both versions, the top chord (supported) that is under load offers a bridging effect, causing it to be in compression and the bottom chord member (unsupported) to be in tension.

Fig. 8–13. The top and bottom chords in open web construction are connected by steel tube web members.

OWC is prefabricated at the factory before installation and is constructed either with parallel chords laid on-edge or flat laid chords. The steel tube web members are prefabricated from 1 in. to 2 in. cold-rolled steel tubing. The ends are pressed flat into a semicircular shape with a hole punched through them. These flattened ends are inserted into slots in the chords and steel pins are then driven through, completing the assembly. Advertised spans of 70 ft are possible using a single 2 × 4 or two 2 × 3s as top and bottom chord members. A single 2 × 4 up to 70 ft is made possible by joining different lengths in glued, mitered finger joints. The normal on-center spacing is 2 ft.

Strengths: The strong area of this roof is the perimeter of the building where the roof ties into the exterior walls.

Hazards: As with most lightweight construction, the hazards of this roof are numerous. It is basically constructed of 2 × 3s or 2 × 4s under compression and tension with sheathing of ½ in. plywood or OSB. These components offer minimal resistance to fire. The chord members are exposed in the interiors of some structures, increasing the exposure hazard to the roof. Unlike conventional construction, the open space between the top and bottom chords will promote the lateral extension of fire, resulting in early collapse of the roof (this is actually horizontal balloon frame construction). Expect rapid failure when exposed to fire due to the size of the lumber and its chord members in compression and tension.

Quick summary

- Wood I-beams will readily fail when exposed to fire or sufficient heat. (See the Las Vegas incident in chapter 1.)

- Wooden I-joists consist of top and bottom chords separated by a stem (or closed web).

- If fire or sufficient heat is exposing I-joists, safe roof ventilation operations are not an option in an *exposed* area of roof.

- If fire has involved I-joists, consider extinguishment from safe areas that are not directly underneath involved areas.

- Regardless of the configuration of wood, metal, or wood and metal, open web construction (OWC) still incorporates all of the attributes of truss construction, including rapid failure rates when exposed to fire and/or sufficient heat.
- OWC consists of wood top and bottom chords connected by steel tube members.
- The openness between OWC chords will readily allow extension in any direction.
- If OWC truss construction is exposed to fire or sufficient heat, safe roof ventilation operations are not an option in an *involved* area.

Flat roof with metal gusset plate construction or glue. This type of truss is the most common type of lightweight truss as it is a favorite method of roof construction for both residential and commercial structures large and small across this country. Trusses for roofs are constructed in a wide variety of styles but all of them share common features. Lightweight wood trusses are predominantly comprised of 2 × 4s held together by metal gusset plate (MGP) connectors that are typically 18-gauge steel plates with prongs of ⅜ in. penetration (fig. 8–14) and used in a wide variety of wood truss applications. One note: While MGPs are often called gang nails, we avoid the term because it grants the connection too much integrity! You may encounter 2 × 4s in spans of up to 80 ft. The sheathing is usually ½ in. plywood or OSB.

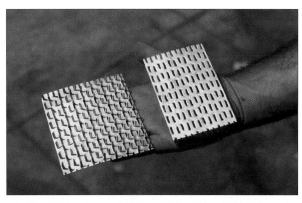

Fig. 8–14. Metal gusset plates are used in a wide variety of wood truss applications with a typical penetration of ⅜ in.

Currently, trusses are being manufactured that utilize glue (fig. 8–15) instead of metal gusset plate connectors! As this is relatively new, there seems to be some disagreement regarding the temperature at which the glue will melt. However, several nonscientific tests have indicated that the glue will melt at around 350°F to 400°F! These types of trusses consist of top chords, bottom chords, and webbing (supports between the top and bottom chords).

Fig. 8–15. Newer trusses may use glue for connection points.

Strengths: The strong part of this roof is the area where the trusses terminate on the outside bearing walls or cross them in cantilever applications.

Hazards: Extensive use of 2 × 4 in. trusses with MGPs equals short burning time and early failure of a roof as the truss members are under compression and tension. When the bottom chord or webbing fails (whether from connector plates that have pulled out or from deep char), one or more trusses can also give way. Rapid collapse is common to lightweight truss members, MGPs, or glued joists when exposed to fire or high heat. In addition to the MGPs, it is easy to visualize that if trusses are encountered that are glued together and the glue melts at a moderate temperature, collapse will occur relatively quickly, depending on the amount and extension of fire. Sheathing comprised of ⅜ in., ⁷⁄₁₆ in., or ½ in. plywood or OSB is common. Plywood and OSB will burn and fail at a fast rate and offer little resistance to fire.

Fat roof—panelized. This roof is only found west of the Mississippi and is most predominant on the West coast—where wood was plentiful—and is commonly found on wood, masonry, or concrete tilt-up slab buildings. However, due to the rising cost of wood, this roof is being replaced with steel open web bar joist roofs. There are two panelized roof systems: all-wood and hybrid wood and steel.

- **Wood panelized roof system.** As the name implies, this system consists of wood and four primary components: beams (normally laminated wood), purlins, 2 × 4 in. joists, and ½ in. plywood or OSB sheathing. After the walls have been erected, the roof is typically begun with laminated beams spanning the length or width of a building. These beams vary in size, but 6 × 36 in. are common. They are supported at their ends by pilasters or saddles and may be bolted together to provide lengths well in excess of 100 ft. Wood or steel posts may provide additional support along the span. These beams are spaced from 12 ft to about 40 ft apart. Supported by these beams, wooden purlins are then installed with metal hangers on 8 ft centers. A common size for a purlin is 4 × 12 in., with the length depending on the spacing of the beams.

  MGP trusses can be substituted for conventional purlins (fig. 8–16), resulting in substantial cost savings. Joists measuring 2 × 4 in. by 8 ft are then installed with metal hangers on 2 ft centers between the purlins, parallel to the beams. Sheets of ½ in. plywood or OSB are nailed over this framework. Composition roofing material covers the plywood or OSB sheathing. An insulation paper is often stapled to the underside of the roof, between the beams and the purlins, and consists of a tar-impregnated kraft paper covered on either side by aluminum foil.

- **Hybrid panelized roof system.** The hybrid roof system combines panelized wood components that are placed on top of open web steel joists. Although similar to the wood system, it consists of girders/primary steel trusses 12 ft to 40 ft apart, smaller steel trusses on 8 ft centers, 2 × 4 in. wood joists (which are part of the panelized sections) spaced 2 ft apart, and roof sheathing. Vertical steel posts provide support for the primary steel trusses. To reduce the time required for construction, preframed panelized roof sections 8 ft wide and up to 72 ft in length are assembled on the ground, lifted into place, and secured to the steel bar joists that are bolted or welded to the steel trusses 8 ft on center. The roof is then covered with composition or other appropriate materials.

Fig. 8–16. Metal gusset plate trusses can be substituted for solid beam purlins.

Strengths: The strengths of the *wood* roof are its beams, the purlins, and the regions along the building's perimeter. The strengths of the *hybrid* roof are the girders/primary steel trusses and the regions along the building's perimeter.

Hazards: Hollow steel pipe of 4 in. diameter can be found supporting the span of the wood beams or girders/primary steel trusses. Expect weakening or collapse of these supports with subsequent failure of large portions of the roof under heavy fire conditions. Moderate to heavy fire intensities will quickly burn through the 2 × 4 in. joists and ½ in. plywood or OSB sheathing, which can result in vertical fire travel and a reduction in horizontal fire spread. Additionally, when the kraft paper insulation is subjected to fire or sufficient heat, the foil covering will peel away from the upper layer of tar-impregnated paper. The paper will give off flammable gases that rise and build up between the insulation paper and the decking. When the

ignition temperature of these gases is reached, they can flash, igniting the insulation paper and charring the surrounding structural members. The burning insulation will fall away, contributing to horizontal extension within a structure. In this case, fire is then able to expose the overhead structural members and ½ in. plywood or OSB sheathing, which offer little resistance. If lightweight trusses are used for the purlins, expect additional and rapid roof failure if exposed to fire.

Quick summary

- Metal gusset plate (MGP) connector trusses have been used in residential and commercial applications for over 50 years, yet firefighters are still injured or killed every year due to the rapid collapse of these trusses.
- MGPs are not gang nails—the plate perforations penetrate the wood by only ⅜ in.
- Identification of MGP construction is simplified if fireground personnel are alert and look for exposed rafter tails and/or pull interior ceilings in a timely manner.
- As MGP trusses can collapse in a short period of time, allowances must be made for glued trusses that will likely fail in a *shorter* period of time.
- If fire or sufficient heat is exposing MGP trusses, safe interior and/or roof ventilation operations are not an option in an *involved* area.
- Incident commanders east of the Mississippi are rarely confronted with panelized roof systems.
- Laminated wood beams will last longer than the girders/primary steel trusses.
- Kraft insulation paper can enhance horizontal extension of fire within a building.

Flat roof—open web bar joist. Open web bar joist roof construction is found in a wide variety of buildings, large and small (fig. 8–17). It is predominately common in the central and eastern states, however it is replacing panelized roofs in western states due to lower costs and an increasing scarcity of wood. The top and bottom chords are usually made from ⅛ in. steel, and the web supports are solid ⅝ in. steel bar. Large buildings may have bar joists for girders spaced up to 45 ft apart. The joists are spaced at 8 ft intervals with corrugated metal decking covered by melted asphalt, then some type of insulation board, and then alternating layers of melted asphalt and composition.

Fig. 8–17. The open web bar joist roof is common in central and eastern states, and is gaining popularity in western states due to the scarcity and price of wood.

Strengths: The strong area of this roof is at the perimeter of the building.

Hazards: Metal exposed to fire or sufficient heat (800°F to 1,000°F for steel) will expand, twist, and likely fail. The time necessary for roof collapse is of major concern when the entire roof is comprised of metal. An additional hazard is travel of fire between the multiple layers of tar paper, insulation materials, and the corrugated metal decking, making extinguishment difficult. This horizontally extending fire is supplied by gases from the hot metal melting the asphalt coating on the deck, giving off flammable vapors that ignite and propagate flame between the metal decking and insulation board (known as a metal deck fire).

A more recent version of this roof uses a layer of ethylene synthetic rubber membrane to seal the roof from the elements. However, this roof performs in a similar manner to the older style roof. In some

cases, an even newer modification uses a noncombustible thermal layer between the metal sheets and the insulation. Although this modification can have satisfactory results, it is unlikely that an IC will know what version is present.

Flat roof—nonstructural lightweight concrete roof. To construct this flat roof, a steel or wood substructure is first covered by corrugated metal (Q or Robertson decking). Then, a specific type of concrete known as nonstructural lightweight concrete is pumped on top of the metal decking to a thickness of about 3 in. to 4 in. To provide additional integrity, a 4 × 4 in. or 6 × 6 in. wire mesh is used in the layer of concrete. Composition roofing material makes up the final layer.

Fig. 8–18. Nonstructural lightweight concrete roofs can provide soundproofing as well as strength.

These roofs are commonly used to provide additional insulation properties for buildings near airports, freeways, expressways, turnpikes, and other high-noise areas (fig. 8–18). This material is also used for floors in newer multistory buildings and condos. Lightweight concrete roofs offer strong, hard surfaces and they are structurally sound and resistant to fire.

Strengths: The strong area of this roof is at the perimeter of the building. The remaining portions can also be strong, depending on the span of the roof.

Hazards: Although the roof is strong, items of consideration are the span of the roof and what holds the roof up. Common vertical structural members are hollow 4 in. steel pipes and wood members. Lightweight concrete roofs are difficult to penetrate without specific equipment. Masonry blades are ineffective, so ventilation may require the use of a rotary saw with a carbide-tipped wood blade or multi-use blade. A chain saw equipped with a carbide chain will also work but can be detrimental to the saw.

Quick summary

- Fires in bar joist roofs have a collapse history of large portions of the roof, or the entire roof.
- A fire between the roof decking and insulation in bar joist roofs can travel horizontally and spread additional fire within the building from dripping/burning asphalt that can also expose interior firefighters.
- Various layers of insulation and membranes can be used to finish a bar joist roof.
- Large buildings commonly have large spans of bar joist construction, which increases the potential of collapse.
- Lightweight concrete roofs will likely not be readily identifiable until viewed after a fire (still standing) or roof ventilation personnel find that it is difficult to penetrate with common ventilation equipment.
- A lightweight concrete roof is capable of resisting fire for an acceptable period of time.
- The best roof ventilation tool for a lightweight concrete roof is a rotary saw with a multi-use blade.

Bridge truss roof

The bridge truss roof is a direct descendant of the bridges used to span rivers and gorges for the railroad industry—namely the Pratt truss and Howe truss. The same engineering principles it took to support a 40-ton locomotive could also be used to span greater distances between walls or columns in buildings. The bridge truss is characterized by

sides that are sloped from the exterior walls to a flat roof portion. The sloped sides are derived from the trapezoidal shape of the truss (unequal parallel chords, with the bottom chord longer than the top). These roofs are found on various types and sizes of commercial buildings constructed during the historical and industrial eras (fig. 8–19). The wooden truss members are built from rough-cut solid mill lumber of 2 × 12 in. sizes or larger, depending on the size of a building. This usually constitutes a heavy grade of construction. Vertical tie-rods between the upper and lower chords may be used for additional support.

Fig. 8–19. Bridge truss roofs can be found on older commercial buildings.

The wood roof joists and rafters used between the trusses are 2 × 6 in. or larger and are covered by 1 × 6 in. sheathing (diagonal or straight) and composition roofing material. Straight sheathing was used prior to the mid-1930s, and diagonal sheathing was favored afterward. Diagonal sheathing provides increased structural stability to a roof assembly compared to straight sheathing. Plywood or replacement OSB sheathing and metal straps can be found on top of straight sheathing in earthquake-prone areas.

The expansive area created between individual bridge trusses and the enclosure provided by the rafters and sheathing, viewed as a tempting space for occupancy needs, led to the addition of dormer windows to the sloped portion of the roof. Viewed from the street, a bridge truss roof with dormers may present as a mansard roof. As mentioned previously, a true mansard roof is really a gambrel style with hips. If a mansard-appearing roof with dormer windows has a large, flat roof section (no center peak), it is likely a bridge truss roof that has had its open web space finished into an occupant area.

Strengths: Bridge truss roofs are well constructed and are easily identified by their characteristic sloping sides and ends. When exposed to fire, early collapse of main structural members should not be an initial concern. In most cases, the bridge truss will sag prior to its collapse (the strength gradually burns away). In fact, there are many examples of bridge trusses still standing even though the rafters and sheathing are totally consumed (see the section in this chapter titled "Unique Roof Construction Considerations"). This type of roof can predictably fail in sections, depending on the type of fire. The strong area is at the perimeter of the building.

Hazards: Its strength is dependent on the size of the lumber and the span of the trusses. The trusses are in compression and tension and can fail under severe fire conditions. The underside of this roof is usually exposed in warehouse-type structures. Due to their age, roofs are likely to have multiple layers of composition materials.

Quick summary

- Bridge truss roofs are easily identifiable from the street due to their sloping sides and flat top.
- Bridge truss chords that have been covered with a flooring and/or ceiling will hide the attic from interior personnel.
- The truss loft space of the bridge truss roof may have been finished into an occupant space with dormer windows. Bridge truss roofs with dormers appear as mansard roofs from the street level.
- Bridge truss roofs can be expected to have multiple layers of roofing materials.

Arched roof

Oversimplified fire service texts categorize all arched roofs as *bowstring truss* roofs. The bowstring, however, is one of many methods used to achieve an arched roof shape. An argument exists that firefighters should treat all arched roofs as bowstring due to the well-documented history of catastrophic collapse involving them. We get that. We can also argue that the tied arch is more dangerous than a timber bowstring. The four arched roof construction methods covered here have unique characteristics that could influence firefighting operations. Let's define each.

Lamella roof. Of all of the older arched roofs, a lamella roof is not considered a true truss roof, but it is definitely one of the most unique due to its method of construction and the amount of lumber that was utilized for its construction. Interestingly, this unique design was developed in Germany in the early 1900s and was first utilized in the Los Feliz area of Hollywood, California in 1929 for the famous Brown Derby restaurant (frequented by Hollywood stars) which gave it its unique "derby hat" look.

A lamella roof (also known as a Summerbell roof) uses an egg-crate, geometric, or diamond-patterned series of structural members that help form a higher (or steeper) arch (fig. 8–20) than the common bowstring roof. It is constructed of 2 × 12 in. wood framing with steel plates and bolts at the framing junctions. The roof uses 1 × 6 in. straight or diagonal sheathing with composition roofing material. This type of arched roof is supported by three methods: (1) the arches terminate on a foundation on the ground, (2) the roof is supported by exterior walls that are supported by exterior buttresses (see fig. 7–26), and (3) in a few instances, there are internal tie-rods with turnbuckles. It is common on gymnasiums, recreational buildings, large supermarkets, and some buildings that may be considered medium-size buildings.

Strengths: Lamella roofs are solidly built, possessing good construction techniques and lumber. The strong area of this roof is at the perimeter of the building.

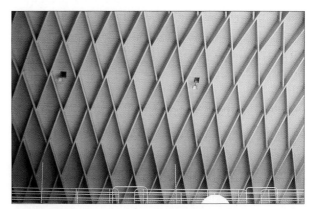

Fig. 8–20. Diamond-patterned lamella roof

Hazards: Although these roofs can offer some protection when exposed to fire, total roof collapse may occur if fire removes more than 20% of the roof structure. In this case, total collapse of the roof can result from the domino effect. These roofs will likely have multiple layers of composition materials and will be difficult to vertically ventilate unless an aerial device is available.

Tied arch roof. *Note: Chapter 3 featured a brief history of the arched truss. Although the tied arch and bowstring roofs (discussed next) are both arched, there has been a difference of opinions regarding the nomenclature for these two roofs. The tied arch roof looks like a bow when viewed from the side, but it is not a true truss as it lacks diagonal webbing.*

Although the tied arch roof is similar in external appearance to bowstring roofs and possibly some lamella roofs, it is significantly different internally in that it uses horizontal metal tie-rods to counter the lateral forces that the arched top beam places on walls (fig. 8–21). Tie-rods (usually ⅝ in.) with turnbuckles are used below each arch member to help achieve the counter-force. The tie-rods may pass through the exterior walls to outside plates, which can facilitate identification of this roof, or they can be attached through the end of each arch end where it meets the wall. The steel used for the tie-rod is ductile, meaning that it can flex and move

as live loads are gained or lost on the rooftop. For this reason, the walls of a tied arch building are typically reinforced with pilasters or buttresses where the arches sit.

> ## Tied arch identification tip
>
> A tied arch is not a true truss. If diagonal web members are attached to the arched chord and the bottom tie chord, the arched roof is a *true bowstring*. Some tied arch roofs have vertical tension rods hanging from the arched top beam to add tension to the bottom tie-rod or to support ceiling/utility needs.
>
>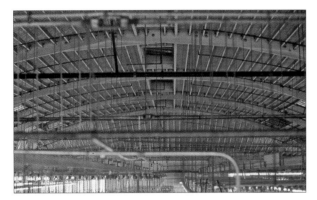
>
> **Fig. 8–21.** The tied arch roof uses metal tie-rods to counter the lateral force of the arch on load-bearing walls.

Proper tie-rod tension is maintained by turnbuckles. The top beams of the arch members may use multiple 2 × 12 in. or larger members that are bolted together, and 2 × 10 in. rafters that are covered by 1 × 6 in. sheathing (straight or diagonal) and likely multiple layers of composition materials. Metal/glass skylights are often placed in the roof to provide light to the interior of the building.

Strengths: This type of roof uses a large size of lumber (2 × 12 in. or larger) and 1 × 6 in. planks as the roof sheathing. The strong area of this roof is at the perimeter of the building.

Hazards: The primary hazard of this roof is early failure of the metal tie-rods and turnbuckles. The tie-rods (which are in tension) provide lateral support for the exterior walls and prevent the arches (which are in compression) from pushing the exterior walls outward, thus thwarting collapse. Because this roof depends on the strength and security of its tie-rods, failure may occur in sections that are exposed to fire. More likely, failure of any part of the roof can precipitate a total roof failure, depending on the type of fire. Compared to other arched roofs, the tied arch roof can offer significantly less structural integrity during fire conditions and should be considered a very dangerous roof. Consider also that the age of these roofs leads to multiple layers of composition roofing materials.

Arched truss. Also known as a rib arch truss or rigid arch truss, the arched truss is a direct descendant of the bridge truss we covered earlier, except for its shape (arched rather than trapezoidal). The first arched truss roofs used for buildings were all-timber with steel plates and bolts used to assembled the wood pieces that formed the chords and web. Usually, large size rough-cut (2 × 12 in. to 2 × 14 in.) wooden members comprise the arch trusses and related members. The bottom chord sits upon the load-bearing walls in an axial fashion, eliminating potential lateral wall forces from the arch. Arched trusses and bowstring trusses look very much alike—adding a level of confusion when conducting preplans. Two clues can help differentiate them: (1) a substantial bottom cord that receives the top chord and web members is indicative of a rigid arch truss, and (2) the abutment where the truss meets the wall. If the bottom chord sits on the wall (the top chord does not touch the wall), it is likely a rigid arch truss if no bottom chord tie rod has been added.

The rigid arch truss roof is covered in the same manner as the bowstring truss roof, which is detailed next.

Strengths: Like the bowstring, rigid arch trusses using large timber members and bolted connections are well constructed. When exposed to fire, early structural collapse should not be of primary concern. Like the bridge truss roof, it usually fails in sections, depending on the type of fire and the structural integrity of the roof. The strong area is at the perimeter of the building.

The Art of Reading Buildings

Hazards: While the arched truss has more inherent rigidity and stability than other arched roofs, it is still susceptible to collapse. All-steel rib arch trusses are much like other steel trusses in that they're prone to rapid heating, especially if the bottom chord is exposed to burning contents. Steel rib arches will try to elongate when heated. Arches that sit in wall pockets will try to push out exterior blocks or bricks where they are attached—a collapse warning sign. Likewise, a rib arch that sits fully upon a wall (no wall pocket) can move outward and form a warning sign that is visible from the exterior.

Timber rigid arch trusses can suffer the effects of aging (like any wood product). The bottom chord of the truss is especially susceptible. Bottom chords that show signs of splits or cracks are a collapse warning sign. Building owners often remedy cracked or split bottom chords by adding a tension rod that spans the bottom chord and is affixed to its ends or to the abutment wall. In these cases, the remedy has converted the rigid arch to a bowstring and should be treated as such. (See bowstring hazards below.)

Bowstring roof. The original bowstring truss is a tied arch with diagonal tension rods added to form a web. All the diagonal web members are in tension, whereas vertical web members are in compression, making the bowstring a true truss. (See the feature on arched trusses in chapter 3.)

The design for this popular type of roof was first used for arched bridges and then made its way into buildings by the thousands from the 1800s to mid-1900s on both small and large commercial-type structures. Although the original bowstring bridge was all-metal (cast and wrought iron), bowstrings used for roof construction can be all-iron, all-wood, wood and iron (fig. 8–22), and, in later years, all-steel. Early bowstring truss roofs used large dimension timbers for chords (2 × 12 or similar) along with iron plates and bolt fasteners—a substantial roof. Likewise, the Hammond all-iron riveted bowstring truss is considered the high-water mark for bowstring truss strength (in the engineering world—not the fire service world!). The bowstring truss made from smaller dimension wood, all-steel, or steel/wood combinations typically require pilasters or buttresses on exterior walls to help absorb lateral forces.

Fig. 8–22. Bowstring truss arches can be constructed from wood, iron, steel, or a combination of wood and metal as shown in this illustration.

Rafters of 2 × 6 in. or larger are covered with 1 × 6 in. sheathing, straight or diagonal, and composition roofing material. Diagonal sheathing provides a stronger bond as each 1 × 6 crosses multiple truss members as opposed to straight sheathing. In some cases, two buildings (each with a bowstring roof) are constructed side by side, with the arches coming together to create a valley in the middle (fig. 8–23). To alleviate the problems of water leakage and snow buildup, a rain roof is often constructed over the valley and joining the tops of the two arches (explained in a following section on rain roofs).

Fig. 8–23. Adjoining buildings with bowstring roofs will create a valley between the roofs that can be covered by a rain roof.

Strengths: Bowstring roofs using large timber members and bolted connections are well constructed. When exposed to fire, *early* structural collapse should not be of primary concern. Like the bridge truss roof, it usually fails in sections, depending on the type of fire and the structural integrity of the roof. The strong area is at the perimeter of the building.

Hazards: Although most heavy wood roofs of this type can be substantial, the age, size of the lumber, and the span of the arches determine the hazards. Some bowstring arches are comprised of iron, steel, or a combination of steel and wood that will not offer the same strength as the heavy wood versions when exposed to fire, especially from the underside. These roofs are often found on commercial/warehouse-type structures, the contents of which may increase the risk of rapid exposure of fire to the underside of the roof structural members. Also, the bottom chord of the arch trusses can be modified (by adding flooring) to allow storage in the attic area and/or ceilings, which can make ventilation, fire control, and overhaul difficult (this can also apply to bridge and gable truss roofs). In addition to the aforementioned hazards, lath and plaster (or other cementitious materials) can be attached to the bottom chords of the trusses, further increasing the weight they must carry (and were not designed for). As bowstring trusses are heated they can move and place additional lateral forces on load-bearing walls to which the top chord is affixed. Often, it is a simultaneous failure of the bowstring truss *and* load-bearing wall—hence the extreme caution and attention given to these types of roofs during fires.

If two bowstring roofs (or other similar roofs) are connected together as in figure 8–23, there is a strong possibility of large open doorways that allow access between the adjoining buildings. This can allow the rapid extension of fire between the buildings. Lastly, remember that the ends of buildings with this roof can present a primary collapse hazard. If jack rafters have been used to slope the roof downward from the last truss to an exterior end wall, the end jack rafters *will* forcibly push the corresponding end wall outward if the end truss fails.

Quick summary

- Four common ways to form an arched roof include lamella, tied arch, rigid arch truss, and bowstring truss. Heavy/large dimension wood lamella, bowstring, and rigid arch type roofs can be quite durable when exposed to fire. The same cannot be said for the tied arch and those that are primarily steel.

- Rigid arch truss and bowstring truss are true trusses, lamella and tied arch are not.

- It is difficult for a lamella roof to have an enclosed attic that can hide fire from interior personnel.

- If a portion of a lamella roof becomes well involved, the entire roof can collapse from the domino effect.

- A lamella roof is normally identified by its larger humped shape than other types of arched roofs. Viewed from underneath, the lamella will have a diamond or honeycomb grid appearance.

- If vertical ventilation is necessary on a lamella roof, it should only be attempted from an aerial device.

- Key to the structural integrity of a tied arch roof are the metal tie-rods, which can easily fail when exposed to sufficient heat/fire. This will cause a rapid collapse of the roof and likely the exterior walls.

- Look for tie plates on the exterior of the long side of an arched roof building. This can denote the presence of tie-rods within the building.

- Due to the presence of the metal tie-rods, do not expect to see an enclosed attic space in a tied arch roof.

- Large timber, rigid arch trusses resist the effects of fire better than other arched roofs.

- An aged timber arched truss may have been repaired using an added tension tie-rod, making it a bowstring truss.

- The bottom chord of a rigid arch truss transfers loads to a wall, whereas the top chord serves that purpose for bowstring trusses.
- If bottom chords in a bowstring roof have been covered to provide an enclosed attic area, expect a higher dead load on the truss members and always verify for extension of fire.
- Bowstring roof arches can be constructed from heavy wood members, iron, steel, or a combination of wood and metal. The metal/wood versions will provide less integrity than the heavy wood versions when exposed to fire.
- Movement of the all-steel or steel/wood bowstring truss during fires can cause a simultaneous collapse of the roof and load-bearing walls.

UNIQUE ROOF CONSTRUCTION CONSIDERATIONS

As you can see, all roofs are not created equal. Each of the eight primary styles have substyles and variations of materials used to construct and finish the roof. We hope to have clarified some of the differences that exist. In addition, we hope that the details we have provided will help correct the oversimplification that can be found in other fire service resources. With that said, there are some unique roof construction considerations that deserve attention if we are to better understand them when making strategic and tactical decisions at building fires.

In this section we discuss some special issues associated with older timber truss, corrugated metal, and inverted roof construction techniques, and roofs that are added to help redirect rainwater.

Older timber truss roofs

The previously discussed bridge truss, wooden bowstring truss, and wooden rigid arch truss are examples of older timber truss roofs. In context here, there are two other older truss examples found in flat and gable roofs: parallel chord and gable truss.

Older parallel chord truss roof and gable truss roof. The characteristic shape of an older parallel chord truss flat roof and the older gable truss roof has not changed since the 1800s. Unfortunately, the current shortage of lumber and the rising cost of labor have rendered the construction of these older roofs obsolete. The parallel chord truss (fig. 8–24) and gable truss (fig. 8–25) are similar with the primary difference being their characteristic shape. Both of these roofs were used on commercial buildings of moderate size and were constructed from the late 1800s to mid-1900s (historic and industrial eras). Both of these roofs were constructed from a heavy grade of lumber (usually rough-cut) and have durable characteristics.

Strengths: These roofs are well constructed and are a good example of older construction that was used when lumber was plentiful. These roofs can predictably fail in sections that are exposed to fire, while uninvolved sections remain intact. The gable truss is easily identified by its A-frame shape, and the parallel chord truss is common on older buildings with flat roofs. The strength of these roofs is the perimeter of the roof.

Hazards: The main hazard of these roofs is twofold. First, although they are constructed of lumber that is significantly larger than modern trusses, they can last longer than modern trusses, thereby allowing interior personnel to overcommit in suppression operations. Secondly, the underside of the gable truss is often exposed and vulnerable to fire. The parallel chord truss is normally not exposed to fire due to ceiling type materials on the underside of the truss; however, it is capable of concealing fire within the trusses. Additionally, these roofs can have numerous layers of composition materials.

Fig. 8–24. The parallel chord truss is a flat appearing variation of an older truss roof.

Fig. 8–25. The gable truss is a variation of older truss roofs and is easily identified by its characteristic A-frame shape.

Older timber truss roof issues. Let's examine the older timber truss roofs from several different viewpoints. In recent years, attention has been appropriately focused on timber truss roofs, particularly as a result of the fire in a building with a bowstring truss roof located in Hackensack, New Jersey that resulted in five firefighter fatalities. This incident focused on three principal hazards that are often ascribed to these types of roofs: truss construction, visual identifiers, and early failure potential. Although most of these hazards can apply to any roof that is subjected to fire, let's consider three perceptions regarding older timber truss roofs:

- **Truss dimensional size and component connection.** Roof configurations that are comprised of a top and bottom chord and separated by webbing are normally referred to as truss construction. However, this viewpoint often does not take into account the size of the truss, which can directly affect the amount of time available before collapse. Compare the size of structural members in older heavy-timber trusses and modern 2 × 4 in. trusses. Although the dead load (roofing materials, HVAC equipment, etc.) on the older trusses can be greater, fireground experience has proven that timber trusses can offer significantly more time before failure than lightweight trusses. Also, the older truss roofs used steel plates and bolts at component connection points instead of gusset plates that only penetrate wood members 3/8 in., or that use glue. These factors can affect the amount of time available to initiate or terminate appropriate operations.

- **Visual identifiers.** Although the perception that "older trussed roofs are easy to identify by their characteristic arched shape" applies to arch truss roofs, it doesn't apply to all timber truss roofs. Although all trusses are constructed according to the same principles, remember that timber trussed roof shapes—bowstring (arch), rigid arch truss (arch), gable truss (triangular), bridge truss (trapezoidal), and parallel chord truss (flat)—are quite different. Therefore, not all timber truss roofs exhibit the characteristic exterior hump or arch shape. Additionally, consider that the timber truss roof is not limited to long, unsupported spans on particular types of buildings. The timber truss roof was one of the most popular roofs constructed until the 1950s and was used on a wide variety of commercial and industrial buildings. These include large warehouses, two-story office buildings, and even simple 20 × 40 ft commercial occupancies. Expect to encounter these roofs on many older commercial and industrial buildings, and know the different types of truss roofs in your area.

- **Early failure potential.** The amount of time for failure cannot be reliably predicted for a roof. As a result, any roof can be dangerous and may collapse during the early stages of a fire. Interestingly, history does not support the perception that timber truss roofs generally fail during the early stages of a fire. Although the definition of early is debatable, let's consider the following case studies of two well-known and noteworthy fires involving timber truss construction.

CASE STUDY: WALDBAUM'S SUPERMARKET

In 1978, resources responded to a fire at a supermarket in Brooklyn at 8:39 a.m. and found a 120 × 120 ft building that had been built in 1952 and was undergoing extensive renovations. The roof construction consisted of seven bowstring roof trusses with a 100 ft span, rain roof alterations, and a tin ceiling attached to the bottom chord of the truss system in addition to a suspended acoustic ceiling system. As resources were battling heavy fire in the two-story mezzanine area with additional resources on the roof conducting ventilation operations, the center portion of the roof (trusses 4, 5, and 6) collapsed, plunging 12 firefighters into the flames with a resultant loss of 6 firefighters. Three factors contributed to the collapse of this bowstring arch roof:

- The double roof (rain roof) alteration
- The extent/severity of fire in concealed spaces (double concealed ceiling system)
- Failure of personnel to recognize the signs of a growing concealed fire that was weakening critical structural assembly connections

The roof collapsed 32 minutes after the initial units arrived.

CASE STUDY: HACKENSACK FORD DEALERSHIP

In July of 1988, resources responded to a fire in a Ford dealership in Hackensack, New Jersey, and found heavy smoke conditions from the attic area of the building. The building had a bowstring truss wood roof. Resources were initially assigned suppression operations on the interior and roof of the building. However, after 20 minutes, operations began to concentrate on a defensive operation. While this change was underway, a portion of the roof collapsed, killing five firefighters. Later it was determined the fire had burned for a significant length of time and was well advanced prior to detection.

Three factors contributed to the collapse of this bowstring arch roof:

- Alterations that consisted of a heavy ceiling of cementitious material on wire lath
- Auto parts storage in the attic
- The burn time before extinguishment

The roof collapsed 35 minutes after the initial units arrived.

Neither of the preceding two incidents supports the view that timber truss roofs fail in the early stages of a fire and fail completely, even when they have been significantly altered from their original design. Fireground experience indicates that timber truss roofs (except the tied arch) can last over 30 minutes under heavy fire conditions. The major hazard attributable to timber truss construction may not be the construction itself but a combination of the following factors: alterations, mass, location, and size.

- **Alterations.** The dead and live loads created by later alterations to a roof often exceed the design criteria of the original roof. Unfortunately, these roofs are easily altered for numerous purposes. As previously mentioned, it is common practice to cover the top of the bottom chords with flooring, which creates a significant storage space within the trusses. Additionally, it is also common to cover the underside of the bottom chords with materials to create a ceiling, particularly cementitious materials in older applications. Consider the following three hazards when confronted by timber trusses with ceiling materials on the bottom chords of the truss members: the added weight of the ceiling itself, the difficulty of access, and the weight of additional materials that may be stored there. Ceilings add additional weight that the trusses may not have been designed to support. They will also hamper access and suppression efforts in the attic area and can hide a well involved fire in the attic that may not be visible to interior personnel located below the ceiling. The ceilings can be indicative of storage, translating into additional weight on the truss assemblies.

- **Mass.** The dimensional mass and inherent strength of this construction can allow a fire to burn for a period of time in the upper portion of the trusses while personnel initiate extended interior attack and/or roof ventilation operations. This hazard and others were present at the Hackensack and Waldbaum's collapses.

- **Location.** There can be a significant difference between timber truss roofs located in the western and eastern states. Although these roofs can be constructed similarly, the timber truss roofs on the East Coast can be significantly older and have been subjected to harsh weather conditions, likely have higher roof dead loads, wood rot, termites, renovations, and other detrimental circumstances for a longer period of time than their west coast counterparts. Not surprisingly, the majority of timber truss roof failures that have injured or killed firefighters have occurred in the eastern states.

- **Size.** Depending on the size of the truss members, timber trusses can possess more strength than they are given credit for. The timber trusses in figure 8–26 span more than 70 ft with a significant portion of the truss members and all of the roof sheathing removed by fire, yet the trusses are still standing.

Fig. 8–26. These timber trusses are still standing after significant portions have burned away.

Using the preceding timber truss roof information as a baseline, let's address the perception of offensive operations as applied to these roofs and how an aggressive attack on fires in timber truss roofs can be successful in some instances. The following factors are suggested considerations for suppression operations in timber truss roofs:

- First, and if possible, determine if a fire is *below* the trusses only and not within the trusses, or in the truss attic area and *involving* the trusses. If a fire is within the building but not involving the trusses, then the collapse potential of the trusses is minimal.

- Second, if the extent of fire is unknown, and/or the underside of the trusses have been covered by materials to form a ceiling and are obscuring the truss attic area, these considerations should require the first company to briefly enter the structure, pull the ceiling, and evaluate the attic/trusses (make an inspection hole just inside the access point). If a fire is in the attic area and well involved, plan for a defensive operation. If a fire is in the attic area and not well advanced or can be extinguished in an acceptable time period, then an aggressive attack should be successful. If a fire is not in the attic area, then an aggressive interior attack should also be successful.

- Third, the presence of flammable stock in close proximity to trusses will dramatically enhance the extension of fire into a truss area, particularly if the trusses are exposed, as in figure 8–27.

Fig. 8–27. Flammable materials in close proximity to trusses will enhance the extension of fire into the trusses and increase the potential of early collapse.

- An incident commander (IC) must carefully evaluate the type of construction, the degree of fire involvement, the approximate time the fire has been burning, and the resources available. As with other types of construction, having a working knowledge of these variables can help in deciding whether to initiate an offensive or defensive attack.

- Avoid conducting an interior attack of long duration. If interior attack operations are initiated, suppression efforts should begin to control a fire. If a fire is not controlled in an acceptable period of time and actually appears to be increasing, discontinue offensive suppression efforts.

- Timely and accurate communications are necessary at any fire, particularly at one involving a timber truss roof and aggressive suppression operations. All personnel must be aware of the type of construction and the intended plan of operation. Interior and roof personnel must continually communicate their progress or problems to the IC.

Corrugated roofs

Corrugated roofs (fig. 8–28) are inexpensive and easy to erect, whether used in large or small applications. They consist of steel, aluminum, or fiberglass over a wood or metal substructure.

Fig. 8-28. Although corrugated roofs aren't a specific style of roof, consider them to be extremely dangerous when exposed to fire.

Corrugated steel is usually of 18- to 20-gauge thickness, and the aluminum and fiberglass is also relatively thin and not generally considered a substantial material, particularly when exposed to fire. These roofs are normally found in flat (with varying pitches), gable, or hip configurations and are easy to identify. These types of roofs can also use fiberglass panels imbedded in the roof to allow light into the structure (daylighting).

Strengths: The strong areas of this roof are the ridge and those portions that cross the outside bearing walls.

Hazards: The corrugated panels may be fiberglass, aluminum, or lightweight steel. Fiberglass light panels are common. Expect rapid failure of these materials, especially of the light panels, when they are exposed to heat or fire. Personnel must consider corrugated materials to be extremely hazardous for above-ground operations.

Rain roofs

A rain roof is a type of roof that has been added onto an existing roof to change the appearance for aesthetic reasons, or to change the pitch of an existing roof so drainage from the roof will be improved. As an example, the two dwellings in figure 8–29 were identical two-story residential structures with flat roofs. A rain roof (hip design) was constructed on the left structure for appearance and/or to improve the drainage from the original flat roof.

Fig. 8–29. Rain roofs are normally a retrofit to an existing roof (shown on the left building) to enhance the drainage of water from an existing roof or to improve the appearance.

Another example of a rain roof addition is used when a double arched roof is constructed. Where the two arches meet results in a significant low point or bowl in the middle of a building (see fig. 8–23). The low point will collect snow and rainwater, which can impose significant loads that can be detrimental. To correct this problem, a rain roof can be constructed over the bowl or valley between the existing roofs. Be aware that the added rain roof will give the building a *new look* that can hide original roof features. The space created by a rain roof can hide HVAC equipment (within the new attic) on the original roof. Also remember that rain roofs are popular on manufactured housing (mobile home) structures.

Personnel may only realize they are on a rain roof while performing suppression operations. For example, after a hole is made they discover a large void and no pressurized smoke emits, or after making the hole they attempt to punch through the ceiling and are met with a resounding thud (the original roof).

Strengths: In reality, this type of roof is a retrofit constructed over an existing roof; therefore, the strengths are dependent on the type of construction that has been used and how it is supported. In any case, the strongest areas are the perimeter of the roof.

Hazards: The hazards of this construction are numerous and dependent on three basic factors. One, a rain roof is likely not detectable, thereby leading fireground personnel to assume there is one roof instead of two. Two, it is obvious a rain roof will significantly increase the weight (dead load) on an existing roof that may not have been designed for the added weight. Third, when roof personnel cut ventilation openings, unless it is obvious there is another roof below the rain roof, they may not observe any smoke/fire/heat from the opening. This can lead to a false opinion of actual conditions below the rain roof—obviously, a dangerous situation.

Inverted roof (raised roof)

On most roof structures, the top structural members carry the load (and roof covering) and lighter elements are found on the bottom to carry a ceiling. The inverted roof is the opposite: The bottom structural elements carry the load and the top elements are lighter to attach the roof covering.

An inverted roof is typically constructed as a complete assembly, not a retrofit like a rain roof. A typical conventional roof is normally comprised (*from bottom to top*) of ceiling materials (plasterboard, etc.), ceiling joists, rafters of 2 × 6 in. or larger, roof sheathing, and roofing materials. An

inverted roof is comprised of (*again, from bottom to top*) ceiling materials, rafters of 2 × 6 in. or larger, 2 × 4s to support the roof sheathing, and roofing materials. This configuration uses the 2 × 4s (or other similar members) to slope the roof for drainage and places the 2 × 4s directly underneath the roof sheathing.

Strengths: As opposed to a rain roof, this roof is constructed as an integral unit, so the dead loads should be accounted for, as opposed to a retrofit type of construction. The strongest area is the perimeter of the roof.

Hazards: Compared to the rain roof, the hazards are somewhat simplified. However, as the roof is inverted, if fire is able to reach the cockloft space, it will not take long to weaken the top members (2 × 4s or similar members) as they are much smaller than the ceiling joists supporting the ceiling. This can be a dangerous situation for roof personnel.

Quick summary

- Older parallel chord flat roofs look like other flat roofs except they are more substantially built.
- The gable truss can be readily identified by its characteristic trapezoidal shape and the fact that it will be found on older buildings.
- The bottom chords of older parallel chord and gable trusses can be covered to give a finished appearance to the interior.
- The older flat roofs typically have an enclosed attic/cockloft and the gable roof does not unless it has been modified.
- Older timber trussed roofs are not always arch shaped. They can also be flat or gabled.
- Timber trussed roofs offer significantly more durability during fires than lightweight trusses.
- Factors such as aging, multiple roofing layers, alterations, higher fire loads, and longer burn times are more responsible for timber truss collapses than the actual truss design.
- Fiberglass, aluminum, and steel roof corrugated panels can look the same from the street, but they will react differently when exposed to heat/fire.
- Use a roof ladder or aerial device to distribute the weight load if it is necessary for personnel to be on a corrugated roof.
- If a corrugated roof has fiberglass daylight panels, they will quickly fail if exposed to heat/fire.
- One of the first indicators of a rain roof is roof ventilation personnel opening a roof and finding an abnormally large attic when one should not be present.
- The presence of a rain roof can give a false indication of a lack of heat/fire in the attic below the rain roof.
- In some cases, the appearance at the eave line can indicate the presence of a rain roof. Look for double converging rooflines.
- A rain roof can add an unwanted dead load to an existing roof that can lead to an earlier collapse of the entire roof assembly.
- Rain roofs are popular on mobile home type structures.
- Inverted roofs will always have a cockloft and necessitates evaluating for the presence of fire.
- Due to the design of an inverted roof, the smaller members (rafters) are underneath the roof sheathing and can be easily weakened by fire, which presents a notable hazard to roof personnel.
- As inverted roofs are generally older, expect multiple layers of roofing materials on the roof.

ROOF COVERINGS

In the preceding portion of this chapter, we discuss the primary styles of roofs, how they are constructed, their inherent strengths and weaknesses, and how they can be readily recognized from the street. Now, let's look at the most common types of roof coverings and how they can also affect a roof size-up, particularly for personnel on a roof.

Hot mop

Hot mop roofs are very common in flat or semi-flat commercial applications. The primary advantage is their low cost of materials (compared to other types of roof coverings), which consists of felt paper that is covered and secured with melted asphalt. Although this roof covering is not known for its attractive appearance, it can be covered with small stone/rocks to improve its look (also known as a tar and gravel roof). The stone covering also helps prevent damage to the asphalt from foot traffic, hail, and ultraviolet rays. This roof is very flammable in fire scenarios, and if stones/rocks have been added, this will significantly increase the difficulty of vertical ventilation operations.

Composition shingle

Composition shingles are the most popular type of roof covering and are used on residential and commercial buildings with pitched roofs. They usually do not have the dimensional look of shakes or tiles, although some high-end, 30- to 50-year varieties have a more structured appearance. Although they are combustible, the primary hazard comes from multiple layers that are built up over the years (composition roofs typically last about 15 to 20 years). The original roofing layer is supposed to be removed before a new layer is installed, but this is often not the case and can significantly increase the dead load on a roof. Therefore, the older a building, the more likely multiple layers of roofing material with an increased dead load can be expected.

Wood shake shingle

Wood shingles and shakes (they differ in thickness) offer a unique look to a roof but will not last as long as most other types of roof coverings (in some cases, less than 15 years). Wood shingles and shakes are not allowed in some areas of the country due to their high flammability, although there are pressure-treated shakes that can offer a degree of fire resistance. An additional significant hazard is their ability to extend fire from flying brands (particularly in wind) and they have been known to cause exposure problems multiple blocks away from the original fire.

Clay tile

Clay tiles are very popular in some areas of the country for their Spanish-style appearance, long lifespan, ability to reduce the interior temperature of a structure by 5°F to 10°F, and their noncombustible nature. The most significant drawback is their weight, which can be significant depending on the type of tile that is used. Obviously this can add a significant dead load to a roofing system, particularly lightweight truss systems even when the trusses have been engineered for the weight of tile. Additionally, their removal for roof ventilation operations can be time consuming and dangerous to personnel on the ground (falling tiles).

Slate

Slate tiles are normally found on more expensive structures and offer many of the benefits of clay tile roofing materials, including their noncombustibility. Similar to clay tile, slate can also be heavy, which increases the dead load on a roof and requires the removal of the slate before roof ventilation operations can be completed.

Concrete tile

Concrete tiles are relatively new and are constantly evolving with various styles and lighter weight materials. Generally, concrete tiles are formed from fiber-reinforced cement products and feature long life, minimal maintenance, lower weight than clay tile or slate, and are noncombustible. Similar to clay tile and slate, they must be removed before roof operations can be completed.

Metal roofing

Metal roofing has become more fashionable in recent years with the most popular style generally referred to as standing-seam steel roofing for its upturned edge at the junction of the panels (fig. 8–30). However, metal roofs can also be made to resemble clay tiles, slate, shingles, shakes, and so on. The advantages of metal roofs are longevity, overall appearance considerations including a wide range of colors, and energy efficiency (they can save up to about 20% on energy bills). The most beneficial advantages to the fire service are their lightweight nature (22- to 26-gauge steel) combined with noncombustibility.

Fig. 8–30. Standing-seam steel roofing can be identified by its upturned edges at the junction of panels.

Although the metal can be cut with a power saw for roof ventilation operations, a significant disadvantage for the fire service is that metal roofing can be installed over existing roofing, which creates a void between the existing roofing and the metal roofing. An example of this is the stone-coated steel roof, which can look like tile or a traditional wood shake roof. This type roof is best identified by being familiar with their use within a particular area and/or by a 4 in. to 6 in. oversized metal trim used on the fascia board.

On newer installations, the metal panels are installed on 2 × 2 in. battens (fig. 8–31). For construction over an existing roof, 1 × 4 in. counter battens are installed to provide a nailing surface for the metal panels. Both of these installation techniques result in a void area between the roof decking/materials and the metal panels. If fire extends into this void area, it is difficult to access from either below or on top. As a result, some departments remove the metal at the ridge area and flow water downward to reach the fire, although with very limited success. As a side note, the edges of these metal panels can be as sharp as razor blades!

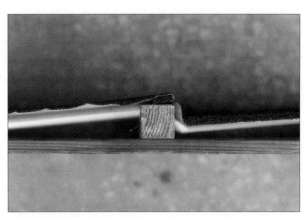

Fig. 8–31. Stone-coated steel roofs can be installed over an existing roof by fastening steel panels to battens that are attached to an existing roof.

Single-ply roofs

Single-ply roofs (or membrane roofs) are basically comprised of a layer (or layers) of insulation on top of a roof decking that is then covered or sealed with a waterproof membrane material. The insulation can be comprised of polystyrene, wood fiber materials, or polyisocyanurate foam. Common materials used to create the waterproof

membrane are polyvinyl chloride, bitumen materials, or vulcanized and nonvulcanized materials. Viewed from the top, this type of roof will appear in different forms, but common appearances are ballasted with a type of loose gravel/small rocks or a white membrane surface that is referred to as an adhered system.

Although there are multiple variations of this roofing system, it should be noted they are capable of releasing toxic gases when exposed to heat/fire, and some are capable of burning underneath the top layer of roofing material. Additionally, not only can roof ventilation operations on these roofs be a challenge, but they are also capable of holding in a higher percentage of heat and smoke within a building due to their superior sealing characteristics compared to conventional tar paper and hot mop roofs. The ability to hold a higher percentage of interior heat can increase the potential of a flashover.

Quick summary

- Hot mop roofs on metal deck, bar joist roofs are dangerous due to rapid failure and fire spreading between the metal deck and top composition covering.
- Older composition roofs can acquire a higher dead load from additional roofing layers and can slow roof ventilation operations.
- Not only are wood shingles/shakes flammable, they can present significant exposure hazards, particularly in windy conditions.
- Tile and slate roofs will considerably increase the dead load on any roof and can also slow roof ventilation operations.
- Although metal roofing can be beneficial, remember to check for a void between the roof sheathing and metal panels.
- Single-ply roofs can be flammable and capable of releasing toxic gases that are highly hazardous to fireground personnel.

ROOF APPENDAGES

Webster's Dictionary defines *appendage* as "something connected or joined to a larger or more important thing." When this general definition is applied to building construction and roofs in particular, we can redefine appendage as "items that are added to a specific roof style for an explicit purpose." From a size-up perspective, let's look at some specific items that can be found on or as a part of various roofs and consider how they can be important clues about the interior environment of an attic and/or interior of a building.

Live green roofs

Live green roofs are covered in detail in chapter 6.

Monitor

A monitor is a roof with a raised extension above the ridge of a roof and is normally found on peaked or arched roofs. Although it can be used for aesthetic purposes, its primarily use is to provide light and ventilation to the area below the monitor. A monitor can also effectively provide a visual confirmation of any smoke and/or fire in the area below the monitor. See figure 8–8 for an example of a monitor roof that shares many of the same features as a monitor that is built on the ridge of a roof.

Division walls

Division walls are designed to partition a building and/or minimize the horizontal extension of fire and generally project at least 18 in. above a roofline. The important consideration to evaluate in fire scenarios is whether a division wall has been breached by renovations, utilities, or for other reasons. Division walls are discussed in chapter 7.

Chimneys/fireplaces

Although the terms chimney and fireplace are often used to denote the same type of construction, their definition can vary depending on their specific use.

Chimney. A structural component used for the venting of hot flue gases or smoke from a stove, boiler, furnace, fireplace, or other appliance.

Fireplace. An architectural structure or appliance designed to contain a fire for heating and/or cooking.

Although the subject of chimneys and fireplaces (for this discussion, let's use the term fireplace to also include chimneys) may seem somewhat simplistic, they can provide some valuable clues during a size-up process as follows:

- A zero-clearance fireplace is not a real masonry-type fireplace but a prefabricated fireplace appliance that can be installed almost directly against combustible surfaces such as walls and floors. The portion that is visible outside a structure is typically constructed from wood framing and can be finished with stucco, wood, or decorative stone and includes a vent pipe and deflector.

    They can be constructed on the exterior (fig. 8–32) or interior of various occupancies and are easily identified as compared to masonry-type fireplaces. The primary hazard is the typical lack of fire-stopping in the vertical channel from the bottom to the top of the fireplace. This channel can be common to other areas of framing adjacent or connected to the fireplace and can be considered a modern version of balloon frame construction. As an example, if fire has extended into the bottom of the zero-clearance fireplace in figure 8–32, it would be necessary to check for extension all the way up the vertical channel space and any additional common structural framing such as the second floor and attic areas!

- Masonry fireplaces are constructed on the exterior or interior of a structure and typically use brick, stone, or concrete block. Exterior fireplaces do not normally present a fire hazard unless they are of unsubstantial construction and/or have cracked and allow fire into the adjoining walls or attic.

    Fireplaces that are constructed in the interior of a structure can offer two substantial size-up considerations. One consideration is that roof ventilation personnel who need to traverse the ridge of a roof for ventilation operations on pitched roofs will find traversing around a fireplace can be challenging. Second, a masonry fireplace of older construction that is of unreinforced masonry (fig. 8–33) and is in the approximate middle of a structure typically denotes an older structure of dimensional lumber and may include balloon frame construction. Older homes typically have their fireplaces in the middle of the home for maximum heat transfer throughout

Fig. 8–32. Zero-clearance fireplaces are often identified by their location and type of construction.

Fig. 8–33. Older masonry fireplaces in the middle of a structure can denote an early-era structure.

the structure. Now that there are vast improvements in heating systems, the central fireplace is no longer necessary and interior fireplaces are constructed in varying locations within a structure.

Attic vents

Attic vents come in a wide variety of shapes and styles, but they all share a common attribute of providing proper attic ventilation for residential and commercial buildings. Sufficient attic ventilation is a necessity for several reasons: to remove or minimize moisture, to reduce the buildup of heat in the attic that could increase the interior temperature of a building, and to increase the longevity of the exposed roofing materials. Attic ventilation is typically achieved by introducing intake air at the soffits and/or exterior walls, and allowing air to escape through vents that are placed in the upper third of the roof.

When viewing a roof, it is important to know what you are looking at and what the particular vent is responsible for as there are numerous types of attic vents, although they all serve the same purpose. This common purpose can also be used to evaluate the interior environment of an attic and building. Attic vents are summarized as follows:

- Turtle vents (look similar to the back of a turtle shell and are low-profile)
- Gable attic vents, normally found directly underneath the ridge of a gable roof
- Wall attic vents, often found in exterior masonry walls
- Ridge vents, located along the peaks of a roof and often barely visible from the street
- Turbine vents (whirlybirds), found on residential or commercial buildings

Cricket

A cricket is the junction of a vertical member (such as a skylight riser, parapet wall, etc.) and a horizontal member (such as a roof) where the intersection junction is covered by a roofing material (fig. 8–34). Its purpose is to keep water from running into and down the vertical member. Although this configuration delivers acceptable sealing, it also provides an excellent avenue for the extension of fire. When this condition is suspected, the cricket must be opened and exposed.

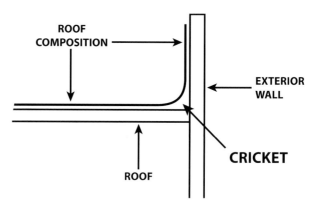

Fig. 8–34. Roof crickets provide can provide an excellent opportunity for extension of fire.

Air shafts/lightwells

Air shafts and lightwells were used in older buildings to provide light and ventilation to interior occupancies. Smoke issuing from an air shaft can indicate the location of a fire (an adjoining occupancy) and the shaft can also assist in the extension of fire to upper occupancies. Also be aware that some air shafts have been covered over with materials (on a roof) that may not support the weight of a firefighter. Look for square or rectangular areas that do not match surrounding roofing materials. Air shafts are covered in more detail in chapter 9.

Skylights

Skylights are good indicators of building floor plans. Skylights in habitations are normally located over hallways. In this instance, if roof personnel can see the hallway/carpet below, the hallway is clear of contaminants. In commercial occupancies, skylights are often placed over manufacturing areas. Again, if the area below a skylight is clear, the area

CHAPTER REVIEW EXERCISE

Answer the following:

1. What is the difference between an attic and a cockloft?
2. Provide the proper name for the following roof styles:

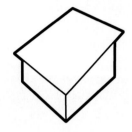

A_____

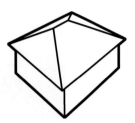

B_____

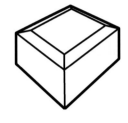

C_____

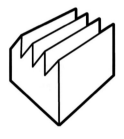

D_____

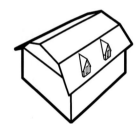

E_____

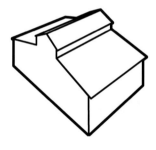

F_____

3. What is a defining feature of "conventional construction"?
4. How do gambrel and true mansard roofs differ?
5. What benefits are provided by a monitor roof?
6. What is a metal deck fire?
7. List the defining features of a bridge truss roof.
8. List four construction methods that can create an arched roof.
9. Which of the four arched roof types is likely to collapse first?
10. What is a rain roof?
11. Name five of the eight common roof coverings.
12. What is a cricket?

RESOURCES FOR FURTHER STUDY

- Ching, Fancis D.K., *Building Construction Illustrated*, 5th ed., Hoboken, NJ: John Wiley & Sons, Inc., 2014.

- Haris, C.M., ed., *Dictionary of Architecture and Construction*, 4th ed., New York: McGraw-Hill, 2006.

- Mittendorf, John, *Truck Company Operations*, 2nd ed., Tulsa, OK: PennWell, 2010.

- Schunk, Eberhard, et al., *Roof Construction Manual: Pitched Roofs*, Berlin, Boston: Birkhäuser, 2003.

- Thallon, Rob, *Graphic Guide to Frame Construction*, 3rd ed., Newtown, CT: Taunton Press, 2008.

BUILDING FEATURES AND CONCERNS 9

OBJECTIVES

- Define the primary types of windows and doors.
- Classify cornices and parapet walls, and their related dangers.
- Describe the term facade and list the inherent hazards of facades.
- Understand the hazards of remodels, renovations, and additions.
- Delineate fireground concerns of fire escapes and roof hazards.
- Describe the hazards of drug labs, pack rat conditions, and abandoned buildings.

THE 800-POUND GORILLA IN THE ROOM

So far, we've introduced the basic building structural elements (foundations, columns, beams, and connections), discussed the ways to classify a building (era/use/type/size), and have detailed various building components (walls, floors, and roofs). Before we can get to the logical conclusion—how to practically size-up a building at your next fire—we need to talk about the 800-pound gorilla that is sitting in the room. Namely, every building has some very obvious features that could pose some threat, obstacle, or challenge for firefighters, yet are often overlooked. Likewise, there are specific features that are hiding and ready to pounce. The obvious and not-so-obvious features we're talking about include:

- Windows
- Doors
- Utility systems
- Overhead hazards
- Renovations and remodels

- Light wells, skylights, and atriums
- Miscellaneous traps

While most buildings contain the first three features, some buildings may host all seven. Regardless, each presents concerns that can impact the safety of firefighters. Therefore, and it warrants repetition, firefighters need to study the information in this chapter and then go out and study the buildings in their district. Look for these features and concerns and denote them in training activities, surveys, and preplans. That way, the 800-pound gorilla won't be so surprising.

WINDOWS

Windows are a readily visible part of building construction, yet they can easily be overlooked during the formulation of a size-up due to their commonality. Nevertheless, windows can provide a host of valuable clues such as:

- Presence of smoke, heat, and/or fire
- Potential access and egress routes (or the lack thereof)
- The likely floor plan of a building
- Probable era of building construction
- Possibility of renovations
- Presence of trapped occupants

Window construction

Although there are numerous types of windows that may be encountered, windows are of simple construction and normally consist of three basic components:

- **Frame:** The structural case or border into which a window is set.
- **Glazing:** The glass and/or thermoplastic (usually transparent) that is set into a window frame.
- **Sash:** The metal, wood, or plastic framework that surrounds and supports the glazing (window glass).

Types of windows

There are four basic types of windows: stationary, sliding, pivoting, and swinging. Stationary windows are firmly mounted and nonopening, sliding windows are made with two overlapping parts (sashes) that move in a horizontal or vertical direction, pivoting windows pivot in the middle of the sash placing the glass both inside and outside the wall plane when opened, and swinging windows are those where the sash and glass swing inward or outward. Examples of swinging windows include casement, awning, hopper, projected, and jalousie types. Because of their simplicity, the most common types of windows are the vertical (single- or double-hung) and horizontal sliding windows, as illustrated in figure 9–1.

Fig. 9–1. The two most common windows that are encountered in fire scenarios are likely to be the vertical and horizontal sliding windows.

Types of glass

Glass is a brittle material that is derived from silica, soda ash, and lime along with other trace materials. These materials are heated to a molten state, formed, and cooled using various methods.

Different types of glass can be created by adjusting the basic heat/cool formula or by the addition of specific chemicals or reinforcing materials. Some of the more common types glass are listed below.

Plate or annealed glass. Annealed glass is the most common type and is formed by a slow, controlled cooling of the silica/soda ash/lime mixture. It is the least expensive type of glass and is used where strength is not required (as in most standard residential windows). In older structures, particularly residential structures, this was the first type of glass that was used and when looked at from the side will appear wavy. Modern glass used in newer structures has a smooth surface. Plate or annealed glass is easily broken and will do so in large, sharp shards. It also has poor resistance to heat and, depending on its thickness, will easily crack with flame contact or ambient smoke heating (usually around 600°F for newer glass, lower temperatures for older/thinner glass).

Heat-strengthened glass. Heat-strengthened glass is used where greater strength is necessary. It is produced by subjecting plate glass to a temperature of 1,150°F and then quickly cooling the glass. This makes for a stronger glass, although not as strong as tempered glass. This type of glass is typically used in modern residential and commercial applications. Heat-strengthened glass takes more effort to break than plate glass and breaks in smaller shards than plate or annealed glass.

Tempered glass. Tempered glass is also heat-treated to increase its strength, and is rapidly cooled and compressed during formation. It can be five times stronger than annealed glass. It is commonly used in glass doors and some windows, and is common in high-rise buildings. Additionally, most building codes require tempered glass on either side of an entry door in a commercial building. Tempered glass can often be identified by the word "tempered" etched in one or both of the lower corners. Tempered glass is not easily broken and has good heat resistance compared to other glass. Although it is tough to break (best to use the pointed end of a pick axe or Halligan-like tool), it crumbles into small pieces and collapses either when gently pushed inward with an appropriate tool or on its own (fig. 9–2).

Fig. 9–2. Tempered glass breaks into small pieces if struck in a lower corner with a sharp or pointed forcible entry tool.

Laminated glass (or safety glass). Safety glass comes in numerous forms, but most contain a layer of polyester or polyvinyl film sandwiched between two layers of standard or tempered glass—adding strength. Common applications are in glass doors, which require an extra measure of strength. In addition to providing strength, the laminated film serves to keep broken glass in the sash. The film can be torn from the sash with common tools and manual effort.

Wired glass. Glass can be embedded with a wire grid that significantly increases its strength for security and fire resistance purposes. Wired glass is among the most difficult to break and remove from the sash.

Thermoplastic compounds. Although the glass-like properties of these types of glazing materials may imitate glass—such as Plexiglas and Lexan—they are not really glass by composition definition. These plastics can be significantly stronger than glass of the same thickness. Such compounds are often used to replace glass when security is a concern. Since thermoplastic type compounds are 250 times stronger than safety glass, they cannot be broken in a traditional manner and often require the use of power tools.

Air blast resistant windows. The General Services Administration (responsible for managing federal office buildings, courthouses, and other governmental buildings) has retrofitted many windows with protection designed to mitigate the danger of flying glass in the event of an explosive terrorist attack. There are different methods utilized to accomplish this degree of protection, but the most popular is a mechanically attached security film. Another method used is a blast-initiated curtain made of a mesh of high-strength synthetic fiber that deploys over a window before the glass is projected into a structure.

Air blast resistant windows can be broken and/or breached with conventional forcible entry tools, although the process will take a little longer than conventional windows. Also, remember that once the glass has been broken in this type of window, the security film or mesh covering the window will have to be removed before access and egress are possible, requiring additional time.

Hurricane-resistant glass. Often called impact-resistant glass, this type of glass has become increasingly popular on the eastern seaboard after hurricane Andrew. Hurricane-resistant glass can take on several forms but is very difficult to remove. For an expanded overview of this type of glass, see the feature titled "Impact-Resistant Windows and Doors."

Ballistic-resistant glass. Mistakenly called *bulletproof*, this glass is formed using various polycarbonate layers clad with glass. The thickness varies according to the firearm projectile size and velocity it is designed to stop. That is why they don't call it bulletproof—the glass can only resist so much, based on its thickness. The thickness also determines the difficulty of forcible entry. For the most part, power saws are necessary to defeat the glass and, even with the proper tools, the progress will be slow. Tests and experience indicate that a timely emergency escape is not an option for occupants or firefighters.

Energy-efficient glass. Energy-efficient glass (e-glass) is one component of an energy-efficient window assembly. The glass itself can be annealed, heat-treated, or tempered type that has had special compounds added to the mixture during production. The finished glass can reflect UV rays and minimize heat conductivity. Taking a step toward improved insulation capabilities and reducing exterior sound, energy-efficient windows are becoming more common and in some states are required on all new construction. A common example of an energy-efficient window is one that has a sash of wood, metal, or vinyl that supports two panes of glass separated by an inert gas such as argon (common terms are double-pane and triple-pane). Another popular style of this window in residential applications uses glazing with heat-strengthened e-glass in vinyl sashes that are set into vinyl frames. Commercial applications can use wood or metal frames that support tempered glass or polycarbonate materials. The prospect of breaking these types of windows is dependent on the type of glass that has been used. Also, this type of glass can allow the interior temperature to rise to a higher level before failing as compared to the common single-pane glass windows, and can increase the chance of a flashover as well as subject interior personnel to higher levels of heat.

Although not a type of glass, there are numerous glass coatings or films that can be added to existing glass for energy conservation, noise reduction, and protection from falling glass in earthquakes. These films are commonly 4.0 mm thick and are applied to windows in new and old buildings before or after the glass has been installed. Breaking this glass is similar to breaking laminated glass.

Window security bars

As a protection against vandalism, security bars are often found on windows and doors (fig. 9–3). Although security is enhanced, they can also present a serious problem since they will prevent occupants and fireground personnel from using a given window as a means of entry or escape. Security bars can be mounted on exterior walls and over the windows of residential or commercial

structures. On some commercial structures, they are mounted within a wall over inset windows.

Fig. 9–3. Security bars are commonly used to deter unwanted invasion but create rescue issues for occupants and firefighters.

Normally, security bars are attached to a masonry structure by lag bolts in lead anchors (creating a compression point), or by embedding their ends directly into the masonry walls. For structures of wood frame and frame stucco construction, lag bolts are used to attach the horizontal bars to the interior studs. The presence of security bars adds a forcible entry and egress challenge for firefighters. Some priority should be given to their removal if firefighters are actively engaged in interior fire suppression and search operations.

Barricaded windows

Barricaded windows have been covered (or boarded over) with various types of materials and are commonly used for security on vacant structures and for protection in states that are subjected to hurricane level winds. Although plywood and OSB are the most common materials used, a wide range of substitute materials can be found, particularly in southern and eastern seaboard states. The obvious concern with this add-on type of construction is that all exterior openings have been securely covered, thereby eliminating timely ingress/egress.

One type of barricaded window deserves special consideration. If four carriage bolt heads are observed on the exterior of a barricade covering, this is an indicator there are two 2 × 4s on the interior of the structure that are used to strengthen and clamp the covering to the structure. This configuration is commonly known as a Housing and Urban Development (HUD) window and will be more difficult to remove than plywood/OSB that is attached only from the exterior of a building.

Other window considerations

When viewing the exterior of a building, the following window characteristics can assist in evaluating appropriate building features that can be helpful or detrimental to fireground operations:

- Windows that are flush with exterior walls typically indicate wood frame construction, as the walls are normally about 4 in. to 6 in. thick. Conversely, windows that are inset in walls normally indicate concrete/masonry walls due to the thickness of the wall, and the greater the inset, the older the masonry wall.

As an example, unreinforced masonry construction can easily have an inset of 8 in. or more, while newer masonry construction will often have an inset of about 4 in. or less. In figure 9–4, the buildings on the left and right rear are unreinforced masonry construction while the building on the front right is frame stucco.

Fig. 9–4. Windows are inset on masonry buildings and flush with the exterior walls on wood frame construction.

The Art of Reading Buildings

- Windows can be an excellent indicator of the type of room or area behind a window. Following are examples:
  - Small frosted windows indicate a bathroom area and are normally above a countertop or shower/bath. These types of windows are poor entry points for personnel and also provide minimal ventilation.
  - Unfrosted windows that are higher on a wall (than normal) indicate a window that is above a countertop and likely a kitchen.
  - Picture type windows indicate family rooms, dens, or other similar types of large rooms. These windows can provide large entry, exit, and ventilation openings.
  - Smaller windows that are between floors indicate a stairway.
  - Rectangular windows on the side or back of a residence may indicate bedrooms. These windows can provide acceptable entry/exit openings and ventilation for sleeping areas.
  - The size of a window near grade level can indicate the presence of a cellar or basement (see fig. 7–5).
  - The presence of air conditioners (fig. 9–5), blinds, shades, or curtains on attic windows indicate the presence of a living area that can have minimal ingress/egress considerations and maximum search implications.

- Large windows (display windows in commercial buildings) may be plate or tempered glass and can provide large openings, but must be carefully broken.
- Older structures (particularly residential structures) used double-hung windows with wooden frames. Typically, more modern residential structures use either stationary windows or sliding windows with lightweight aluminum or vinyl frames. Older structures with aluminum or vinyl frames can indicate window renovations and may be an indicator of additional interior renovations.

The older style wooden, double-hung windows have 1 × 4 in. strips of wood trim around the window on the exterior of a building. These boards (fig. 9–6) cover a cavity on either side of the window that is used for lead weights, typically connected to the window by cotton sash cords. The sash cords (one on each side) are attached to the upper portion of the window and travel over a pulley to the hanging lead weights in the cavities next to the window. These weights minimize the effort necessary to raise or lower the window. On newer buildings, the same type of molding is often used but is purely decorative.

Fig. 9–5. The presence of air conditioners, curtains, or shades in attic windows can indicate the presence of a living area.

Fig. 9–6. On older structures, wood trim around the windows indicates covered cavities on either side of the window that contains a lead weight that helps in raising or lowering a window.

- Windows on a wall that are a noteworthy distance above the grade level (typical for concrete tilt-ups, etc.) will not be usable for access and egress and also may not be viable ventilation openings. Quite often, these types of windows were primarily installed for daylighting and are not operable.

- Ballistic-resistant glass should immediately raise two noteworthy concerns: other security features in exterior doors and increased time needed for fireground operations. As this glass is a viable option for protection of occupants (commercial and high-end residential homes), it should be expected that additional money would be spent to guarantee the same level of protection for exterior doors. Secondly, if this type of protection is encountered at a structure fire, the incident commander should be made aware that the windows (and likely exterior doors) cannot be opened in an acceptable period of time.

IMPACT-RESISTANT WINDOWS AND DOORS

The history of impact-resistant products can be traced back to 1974, but it was not until 1992, when Hurricane Andrew ripped through South Florida, that impact-resistant products began gaining widespread popularity. This single event became the catalyst for sweeping changes in impact-resistant building requirements. The state of Florida and individual Florida cities have promulgated numerous changes to building codes that require impact-resistant features for new buildings as well as those that are being significantly repaired or rehabbed. Public schools in Florida are now built with impact-resistant coverings on all of the openings (windows and doors). Many hurricane-prone coastal states have followed suit.

The scope of impact-resistant construction isn't limited to hurricane-prone areas, though. The influence of the September 11, 2001 attacks has also put many government and public buildings (including public schools) in the scope of impact-resistance standards. The apparent increase in tornado severity throughout the central and southern states will certainly lead to increased impact-resistance requirements for buildings.

Constructing an impact-resistant building includes attention to more details than doors and windows. The impact-resistant building has numerous features:

- Wind uplift protection for roofs
- Increased strapping and anchoring for walls and floors
- Wall and roof sheathing penetration resistance
- Garage door strengthening
- Improved protection from horizontal, wind-driven rain
- Blast resistance (for explosive devices) in certain buildings

Most of these features help add strength and collapse resistance during structure fires, although they may limit traditional ventilation options. Forcible entry and emergency escape challenges create the greatest concern for firefighters when handling an incident in an impact-resistant building. For this reason, we focus on windows and doors. To better understand these challenges, we first discuss some of the testing processes that impact-resistant products must withstand. Following that, we discuss some of the design characteristics that can be found in impact-resistant windows and doors. Lastly, we provide some firefighter considerations that you may find useful.

IMPACT-RESISTANT WINDOW TESTING

There are many types of tests that are used by third-party entities that are tasked with certifying that a given window assembly can withstand an impact event. While varied, the most common form includes a progressive series of performance tests such as the following:

- Large missile (up to three stories). This test mimics a 2 × 4 in. wood stud projectile in a storm with 110 mph winds. To achieve this, a 9 lb 2 × 4 in. 6 ft Douglas fir or white pine stud is shot into the window from an air cannon at 35 mph.

- Small missile (greater than three stories). This test mimics rooftop gravel that is being propelled by a 110 mph wind. Here, a 2 gram steel ball bearing, 5/16 in. in diameter, is fired at a window at 55 mph (130 fps).

In both tests, the window glass may shatter but the impact membrane must not rip or tear. The missile-damaged window then moves to the next test:

- Rain test (also known as a water infiltration test). The damaged window is now mounted in a wall assembly as recommended by the manufacturer. A 2,800 hp radial engine, with propeller mounted in place, is then positioned in front of the window. This propeller creates winds of 110 to 120 mph. Water is injected into the prop wash, which mimics 8 in. of water an hour. There is a provision for some water leakage, but it must stay in the weep track of the window frame.

If the window passes the test above, it moves to the next test:

- Air leak test. The window is then clamped into a machine that creates pressure changes similar to those caused by wind during severe storms. The applied pressure pushes and pulls the membrane of the fractured window over a period of 5 hours.

The window that passes the above tests will likely receive approval to be marketed as impact-resistant for hurricane zones. Door assemblies follow similar testing patterns, although the criteria are adjusted accordingly. There are also specialized tests that may be required for other applications, such as the fire, bomb, and hurricane (FBH) test and military specification testing that may be done to assess the protection against different caliber projectiles.

WINDOW CHARACTERISTICS

The design of the window is determined by the types of testing that it must undergo. For example, if a window is to withstand the impact/rain/pressure test demanded by the State of Florida, it cannot be a jalousie or louver style window. These windows will not withstand the testing. Most impact-resistant windows are of the fixed, single- or double-hung, horizontal rolling, and/or casement types. The glazing part of the impact-resistant window usually consists of an impact-absorbing material (membrane) that is sandwiched between and laminated to two sheets of common glass (fig. 9–7). The membrane is typically polyvinyl butyral (PVB)

Fig. 9–7. Impact-resistant glass includes a membrane sandwiched between two sheets of common glass.

or polycarbonate. The PVB membrane is flexible and similar to that used in car windshields (although thicker and tougher than the car application) and seems to be the most common type of membrane found.

Polycarbonate is an extremely rigid and dense plastic that is very difficult to breach with hand tools (imagine Lexan between two sheets of glass). Polycarbonate membranes vary in thickness and typically provide greater impact resistance than PVB.

The window sash holding the glass to the frame is well bonded and not removable like a typical window. Window locking hardware has also been strengthened (fig. 9–8).

Fig. 9–8. A window sash that holds impact-resistant glass is well bonded and not removable like a typical window. Locking hardware is also stronger—usually secured at multiple points.

DOOR CHARACTERISTICS

Most impact-resistant doors resemble typical standard doors when closed. Open the door and you will see the difference. The doors are typically hinged, but the hinges are considerably more formidable in size and strength and they do not have removable pins. You will also see reinforced steel on the doors and door bucks (frames). The screws attaching the door to the jamb are more stout, beginning at a length of ¾ in. and diameter of ¼ in. and going up from there. The technique of using the Halligan claw to remove the hinge will not work with these doors.

The locking mechanism on these doors is also beefed up. While it is difficult to keep up with the changes in this technology as it occurs at a rapid rate, as of this writing the impact-rated doors have an internal wind bracing mechanism similar to a dead bolt lock. By turning this lock, a 1 in. stainless steel pin is thrown at the top and bottom of these doors. There are other doors that throw these pins into the jamb in six to eight different locations on the door (similar to a bank vault door).

Door and window mounting frames are secured to walls using high-performance screws that are 3 in. long and ⅜ in. diameter. These screws have roughly 1,000 lb of sheer strength, and they can be found spaced at 8 in. to 10 in. intervals along the frames of all windows and doors.

The only real way to determine if you have these products in your area is by preplanning. There is no tip or trick to determining if they are impact-resistant without up-close inspection. Some of the window glass will appear to have a light green hue, some will have stamps/stickers or information etched on the

glass identifying it as impact tested, some will have nothing. The easiest way to tell is to simply walk up to it and tap on it. Impact glass has a distinct "thunk" sound to it—you can tell it is not your average glass when you tap on it. That would be a good time to notify the IC.

FIREFIGHTER CONSIDERATIONS

Impact-resistant building features create a significant forcible entry challenge for firefighters because at some point it becomes an act of futility to breach their windows and doors. Some buildings to keep in mind when thinking of difficult entry would be:

- Military facilities
- Courthouses
- Detention centers (jails)
- CDC (Centers for Disease Control) test labs
- Embassies
- Public buildings (libraries, schools, city halls)

Preplanning forcible entry options for these buildings is essential, and should not be just one option—come up with two or three! Forcible entry is but one challenge. Forcible exiting (evacuation or firefighter emergency) will also be challenged. Additionally, the impact features will affect fire behavior within the building: the box is tighter, flow paths (air in, heat out) are more restrictive, and tactical ventilation options are limited. The potential for flashover and explosive smoke ignition increases dramatically as the building retains more heat.

TACTICAL TIPS

Through actual training using donated impact-resistant products, we've tried numerous techniques to not only enter impact-resistant windows, but egress as well. Some ideas worked, others didn't. We encourage you to do the same and share your findings in trade magazines, blogs, and firefighter social networks. The following is a list of tips I would like to share.

Saws: The quickest way into these windows is with a powered rotary saw (K-Saw) using the maximum size blade it will hold. As there are numerous types of blades that can be used, we recommend that you test and determine which ones work for you. I have found that the diamond blade, the composite blade, and the wood blade (aggressive teeth) all work well depending on the technique you are using. Keep in mind the sash is no longer removable with hand tools, so you will need to use the saw to clean the window.

The chain saw is also another option that works well. There are some considerations to keep in mind when using a chain saw:

- Thoroughly shatter the glass in the path the saw will travel. Glass is a notoriously hard surface—it will dull a chain quickly if you do not first shatter the glass.
- Be careful how deep you plunge the bar into a window. If the chain gets bound up on a window treatment (drapes or blinds), the saw will be rendered useless.
- Don't use the chain saw to cut through the sashes.

Hand tools: There are a number of tools that can be used for breaching the glass from the inside. They include the axe, Halligan bar, sledgehammer, and splitting maul. These are the more common tools that can normally be found on a fire apparatus. There are numerous other tools, such as the GlasMaster

and collapsible pruning shears, that would work great if you get a purchase point and have shattered the path of travel. The reality is, breaching impact-resistant glass with a hand tool entails effort and technique. If you need to make an exit through an impact-resistant window, you had better be acquainted with proper technique using the tools you would commonly have with you. Training will help.

The technique for breaching these windows is relatively universal: You start chopping near the edge of the frame where the glass inserts into the frame. This technique causes the glass to begin to pull away from inside the frame. As you continue on, the rest of the glass follows suit. (In figure 9–9, notice the glass is pulling out of the frame on the left side of the door.) We like to create a three-sided cut, a shape like a doggy door. This does two things: First, it eliminates one cut, making for a quicker exit. Secondly, it leaves us some control of the opening for ventilation purposes.

We have found in large storefronts that to cut this pattern is not realistic and would make it difficult to get in and out of a building. We generally cut an upside-down doggy door shape and flap the window to the ground if ventilation permits.

Fig. 9–9. Force by hand tools allows you to pull impact-resistant glass away from a door frame.

FINAL THOUGHTS

Material technologies and impact-resistant engineering are rapidly changing—they have changed dramatically in the last 10 years that I have been teaching. By the time this book is released, there will undoubtedly be more changes. Site visits, web searching, preplanning, and hands-on training with donated materials is your best strategy to stay abreast of the rapid changes.

The impact-resistive building is here to stay and will continue to gain popularity due to the number of storms that have recently ravaged the Northeast. The rest of the country will experience a surge in these products as communities prepare for and recover from tornadoes and security threats.

Contributed by Ric Jorge, firefighter, Palm Beach County Fire-Rescue, FL.

DOORS

Similar to windows, doors are also easy to overlook in conducting an initial size-up as all buildings have them. However, doors can offer some advantages over windows. Doors are usually the largest openings, which enhances deploying resources through them. Doors are the normal point of access into a structure (and they open to the normally traveled routes), and in some commercial structures (e.g., most concrete tilt-ups), access can only be gained through doors due to a lack of windows. Lastly, doors allow personnel to enter a structure at a lower level (closer to the floor) than will a window, and that is important if heat and smoke are present. Let's briefly summarize common construction considerations when considering doors.

Door assembly

The major portions of a door assembly are best summarized by the following four components:

- **Door.** A moving panel or other cover used to close an opening in a wall. Doors can be comprised of a single material (e.g., wood, metal, glass), or a frame that supports various materials (as in a metal frame that surrounds glass).
- **Jamb.** The structural case, border, or track into which a door is set. A jamb supports and may contain the stop for a door. Jambs are normally constructed of wood or metal. The jamb is also the mounting location for some sort of hinge on swinging doors.
- **Strike.** A receptacle that receives a dead bolt or latch from a locking mechanism. A protrusion that stops a door and keeps it from swinging past the jamb.
- **Lock.** Various types of locking devices used to provide security. Common examples are key-and-knob locksets, tubular dead bolts, rim locks, mortise locks, exit locks, and auxiliary locks.

Types of doors

There are four basic types of doors: swinging, sliding, overhead, and rotating.

- Swinging doors are hinged on one side and either swing to the left or right and inward or outward.
- Sliding doors move horizontally (either left or right) to open or close an opening.
- Overhead doors are mounted above an opening and travel vertically or swing upward.
- Revolving doors consist of multiple panels that revolve around a center hub. They are almost exclusively glass with a metal frame and are found in high-traffic applications (like hotels, airports, and department stores) to help conserve energy by minimizing the transfer of heat and cold air.

Door construction

Basic door construction involves three basic materials: wood, metal, and/or glass.

Wooden doors. May be hollow core, panel, or solid core as follows:

- A hollow-core door consists of a solid wood frame surrounding a core of wood strips or cardboard covered with a wood veneer. This is a cheap, lightweight door.
- Wood panel doors use a solid wood frame inset with panels. The panels are the weakest portion of a door.
- A solid core door is a strong type of door, compared to the previous three types of doors, and made of solid lumber or an engineered wood product (such as glulam).

Metal doors. May be Kalamine (solid wood covered in sheet steel) or hollow, and are normally stronger than wood doors.

Overhead doors. Include rolling, sheet curtain, sectional, and tilt-up. Descriptions of each are as follows:

- Rolling steel doors are characterized by interlocking steel slats that are manually or mechanically raised or lowered by cables winding on or off a drum at the top of the door. This is a strong type of door, but it can be cut with power tools.
- Sheet curtain doors are solid doors made of thin corrugated metal, are manually operated by cables winding on or off a drum at the top of the door, and are similar in appearance to rolling steel doors. These doors are not as substantial as rolling steel doors and can be easily cut with power tools.
- Sectional doors have sections made of thin metal interconnected by steel hinges, or wood sections made from a frame surrounding solid wood panels. The sections are interconnected by metal hinges. These types of doors can be easily forced/cut with power tools.
- Tilt-up doors are solid doors made of thin metal or wood that tilt with the use of spring-loaded hinges. Power tools can be used to cut these doors but must be able to cut the interior strengthening members.

Glass doors. There are two basic types of glass doors: frameless and aluminum framed.

- Frameless glass doors use tempered glass of about ¾ in. thickness, with an aluminum cross member (stile) at the top and bottom to support appropriate locking mechanisms. This door can be categorized as a substantial door.
- Aluminum frame doors employ an aluminum frame around heat-treated, tempered, or laminated glass. This door can be found on commercial buildings of all types and can also be a substantial door.

Bulkhead doors. Although bulkhead doors (i.e., "Bilco" doors) cover stairs that allow access to or egress from a basement or cellar, they are not often considered true doors. However, they are a type of door that opens outward and is normally securely locked. From an intruder's viewpoint, access to a cellar/basement is a potential opportunity to be able to gain easy access to the interior of a structure. Therefore, bulkhead doors are normally made of metal and have substantial locking mechanisms, and can be a formidable opponent against being forced in a timely manner.

Door security bars

Doors equipped with security bars fall into two categories: interior and exterior.

- **Interior.** Wood or metal outward-swinging doors may have a bar of wood or metal for additional security. The bar is placed horizontally across the interior side and is supported by two brackets attached to the door. Since the bar extends past the door jamb on both sides, the door cannot be opened outward unless the bar is removed. The presence of a security bar can often be recognized by a pair of bolt heads on the exterior of the door (very common on the rear doors of strip malls).
- **Exterior.** Security bars over the exterior of doors are somewhat different from security bars over windows in that they are not solid to the structure but must swing outward (normally) to allow passage through the door opening. It is usually necessary to remove or open them to allow initial access to the standard door into a structure. These types of security bars usually consist of two parts: a metal jamb that is secured to the doorway opening, and a metal frame that swings outward and is connected to the jamb by hinges. The frame supports the horizontal and vertical security bars to complete the assembly. These types of security bars can be removed by manual means or power tools.

Other door considerations

- Does a door swing inward, outward, move vertically, move horizontally, or rotate? For example, a swinging door swings *outward* (toward exterior personnel) if the door is flush with the exterior of a wall and the hinges are visible (fig. 9–10, right), and swings *inward* (toward interior personnel) if it is recessed into the wall and the hinges are not visible (fig. 9–10, left).

Fig. 9–13. The main shutoff, multiple meters, transformer, and cable/phone boxes are colocated on this multifamily dwelling.

- **Power distribution circuit protection.** The breaker box may be colocated with the main shutoff. Firefighters can expect multiple breaker boxes in larger buildings.

- **Transformer(s).** Buildings that require higher voltage (220 V, 480 V, or higher) and/or multiphase electric supplies will also have transformers to help regulate the voltage needs of various equipment, appliances, and general use (120 V). Transformer boxes are often located externally (pad mount, underground, or rooftop), although they can be found internally in older buildings. *Hazard note: Older, internally-located transformers present a significant smoke, fire, and shock potential for firefighters. Transformers manufactured from 1920 to 1977 may contain polychlorinated biphenyl (PCB), a very toxic carcinogen. Significant efforts have been made to replace all PCB transformers, although they remain in many buildings. Suppression efforts involving PCB transformers should trigger appropriate hazmat decontamination procedures.*

- **Distribution wires.** Knob and tube wiring (K&T) was a common, ungrounded wiring system used from the late 1800s through the 1930s (historical and industrial eras). The K&T system is easy to spot as it includes single, minimally-insulated wires supported by ceramic knobs used as spacers/supports between the wire runs. The hot wire and neutral typically run parallel to each other, separated by the ceramic knobs. A grounding wire is not present. Porcelain or cloth tubes were used wherever a wire passed through a wall stud, floor, or junction box. A cloth tube or *loom* was used where two wires crossed or where a wire entered a junction or outlet box. From the exterior, K&T wiring can be identified by a two-wire service feed to the building. The K&T method presents several hazards—mainly heat, age deterioration, and lack of grounding.

  Most modern electrical runs within a building are accomplished using wires that are dual-insulted, multiwire (hot and neutral), and ground wire, all wrapped in a single, nonmetallic sheath (often called Romex, although that is a trademarked name). Wire runs are typically located behind interior walls, below floors, and above ceiling coverings. Permanent wire runs that are exposed are required to be within a protective conduit.

- **Outlets/switches/junction boxes.** Individual equipment, appliances, devices, switches, and light fixtures access electricity by plugging into outlets or are hard-wired into junction boxes. Common outlets include the familiar duplex outlet (120 V) and larger 220 V type (for electric clothes dryers, some air conditioners, etc.). Modern electrical codes require a ground fault interruption (GFI) for an outlet circuit located near a water source (faucet, bathtub, pool, etc.). A GFI outlet typically includes a reset button as part of the outlet. The purpose of the GFI is to disconnect the electrical current and help prevent electrocution when the current finds an open ground (water and electricity don't mix!).

Outlets and switches located on the exterior of a building or potentially wet areas (such as a wash bay) should have some kind of cover or cap to protect them from elements. Wires and appliances in these areas should be utilizing marine-type plugs that form a water seal when connected to the outlet.

Hazards associated with outlets are typically tied to the device that is plugged in. Wires going from the outlet to the device create trip hazards and those that are damaged from trauma or fire present an electrocution hazard. Outlets, wires, and devices that become wet from fire suppression operations present similar hazards—the best policy is to shut off electrical service at the breaker.

The most common form of electricity used in buildings is alternating current (AC). AC is a form of electricity in which electrons move back and forth within a material (wire). The other form that can be used is direct current (DC), in which electrons move steadily in a single direction. Batteries and solar cells supply DC power. Buildings that have incorporated batteries and solar cells for power augmentation require an AC/DC inverter to make the DC useful in an otherwise AC system. Important to note is that DC power is always trying move forward from its source. Simply stated, you can't shut off a battery. Large battery backup rooms may have a disconnect switch between the batteries and the inverter, but the batteries remain live.

Buildings with automatic backup generators typically use batteries for uninterruptible power supply (UPS) purposes. When power from the grid is lost (or when firefighters shut off the feed), a series of relays trip, which starts battery draw until the generator can start, warm up, and accept the load. Once generator power comes online, the batteries get recharged through other relays. Shutting down this backup power system is often beyond the technical abilities of firefighters. Closing individual circuit breakers can help isolate a damaged portion of the building if interior operations are underway. Otherwise, it's best to consult the building engineer or mechanical technician who represents the property.

The main hazard of electricity is obvious—a shock. Electrons are constantly looking for a return to zero potential (seeking ground). Damaged electrical systems create the potential for electrons to seek ground through unintended conductive materials. Most metals and water (that is, the minerals that are suspended in water) are excellent conductors of electricity. Firefighters need to be ever mindful of the shock hazard—awareness is the key. Short of an obvious electrical arc (electrons flashing through air to a ground), there are few warning signs. Some warnings of electrical shock potential include:

- Burning odors from electrical equipment
- Wires, connections, and outlets that buzz
- Tingling sensations through boots and gloves
- Metallic materials that are discolored, glowing, or have paint that is blistering

Ground gradient is a term used to describe electricity that is returning to zero potential through nonconductive surfaces like soil, concrete, and masonry. Downed wires can create a ground gradient in concentric waves from their contact on those materials and travel for several yards. Firefighters who feel tingling through their boots should shuffle-step away from the gradient and notify others.

Electrical utilities that are visible on the exterior of a building, and particularly in the area of a roof, can present a formidable hazard to personnel using ladders for roof access and egress, especially at night. Transformers that are mounted on or in close proximity can present a significant hazard to personnel considering roof access. Specifically, transformers that are mounted on a roof (fig. 9–14) will not only increase the dead load on that portion of a roof, but will also place a notable electrical hazard in close proximity to personnel on that roof. Transformers mounted on racks or poles close to a roof can also result in charged wires that can be hazardous to ground ladders/aerial

device placement. These hazards can be magnified during nighttime hours when visibility is reduced or minimal.

Fig. 9–14. Electrical utilities can pose significant hazards on roofs.

An additional electrical hazard to roof personnel is visible wires that run along the inside of parapet walls and mounted on ceramic insulators for security. This condition can occasionally be found on older buildings of unreinforced masonry construction, where it was necessary to bring the wires from a pole to one side of a building and run the wires to another portion of a building where they then enter the building to an electrical panel. This hazard is worsened by brittle and deteriorated or missing insulation on the wires.

Solar electrical generation equipment can present unique hazards to firefighters also. We detail those later in this chapter in the section titled "Alternative Energy Systems."

Quick summary

- Electrical system components for buildings include a supply feed, main shutoff, circuit protection, transformers, distribution wires, and outlets.
- Older, interior located transformers can include PCBs. Leaking PCBs and smoke from PCB-laden transformers are extremely toxic.
- Knob and tube (K&T) wiring is an outdated wiring method, although many historic- and industrial-era buildings may still use it. This wiring is easy to spot because of the single-wire parallel runs and ceramic or porcelain insulators and tubes, and a two-wire service feed on the exterior of the building.
- Modern wiring, often referred to as Romex, combines three or more wires that are insulated in a single nonmetallic sheath.
- AC is the most common form of electrical current. DC current, created by batteries and solar cells, needs to inverted to AC.
- DC power sources cannot be shut off, although they may be isolated.
- The key to safety from electrical shock is awareness. Warning signs include burning smells, arcing, ground gradient (tingling), and metallic distortion (glowing, discoloration, peeling paint).
- Roof-mounted and overhead transformers and electrical distribution equipment present multiple firefighter hazards, especially at night.

Utility gas systems

Propane and natural gas (methane) are the two most common utility gases used in buildings for the firing of various appliances. Both can be stored in a liquid form (propane as liquefied petroleum gas [LPG] and liquefied natural gas [LNG]). Propane is heavier than air, while natural gas is lighter. Both gases have low flashpoints, narrow flammable ranges, and high Btu potential. Both gases are used as fuel for flame-producing appliances that include:

- Heating furnaces
- Radiant space heaters
- Cooking appliances
- Dryers
- Manufacturing equipment

Most developed towns and cities rely on a natural gas supply and distribution system that is piped underground to individual buildings. The gas can be delivered at various pressures that are reduced at individual buildings. The individual buildings will have an outside gas meter and shutoff that is either attached to an exterior wall or in very close proximity to a wall. Multifamily dwellings and multiple-tenant commercial buildings can have numerous meters and shutoffs feeding a single building. Shutting off a simple residential system is accomplished using a wrench or spanner and turning the inline valve 90 degrees (perpendicular to the pipe). Commercial gas shutoffs may be of the Nordstrom type that requires a simple quarter-turn. It is important to not attempt to turn the Nordstrom valve a full half-turn, as damage may result. Likewise, commercial facilities that use large quantities of gas will likely have a rotary vane meter. These can be spotted by the single or twin circular casting of the meter. Firefighters can use this clue to indicate that the occupancy contains larger gas-fired equipment, requiring higher gas pressures and larger diameter supply pipe.

Buildings using propane are most likely to have an external propane tank on site or in the complex. Shutoff is usually accomplished at the storage tank by opening the top cap and spinning a hand valve until it is fully closed. In both cases, the shutoff valve or cover may be locked (especially on commercial buildings or in areas prone to vandalism). Propane (and, rarely, natural gas) stored in tanks outside of a building contain product in liquid form (LPG/LNG). When exposed to flame, they present a significant boiling liquid expanding vapor explosion (BLEVE) hazard. Cooling with judicious gpm is essential to help prevent the BLEVE.

Tank storage of LNG is not common but increasing. Where found, it may indicate that the building contains motorized equipment (forklift, pallet tractors, etc.) that are powered by the cleaner-burning natural gas.

Within a building, the gas distribution is achieved by various types of pipes—black iron pipe is the most common. More recently, plastic pipe is being used for natural gas. Piping that is run within a wall or floor will likely have an appliance shutoff valve where it leaves the wall or floor and feeds the desired appliance. Most appliances use low-pressure gas, although some industrial-sized equipment can require high-pressure feeds.

HVAC systems

Heating, ventilation, and air conditioning systems (HVAC) can take on many forms that have evolved over time, creating a potpourri of possible combinations. In all forms, HVAC systems contain the following components:

- Heating or cooling appliance(s)
- Energy generation source (oil, steam, utility gas, and/or electricity)
- Conveyance system (ducting, fans, pipes, etc.)
- Temperature regulation system (thermostat, relays, switches, etc.)

Table 9–1 provides a brief overview of the more common HVAC systems and appliances that can be found in buildings. Noteworthy is the reality that buildings can rely on a single central system to provide all the HVAC needs (like a single-family dwelling), or may include dozens of HVAC systems to maintain climate control in large spaces or highly-compartmentalized buildings. Hotels and multifamily dwellings are interesting in that they are likely to have a central HVAC system for common areas and individual HVAC appliances for each room or unit. HVAC systems that use ductwork to distribute heated and cooled air add challenges for firefighters—those ducts are also fire and smoke conveyance systems! While high-rise, large, and mega-box buildings are likely to include automatic dampers or shutoffs to minimize smoke spread in ducting, most buildings don't benefit from that protection.

HVAC systems can be mounted on the ground, in the interior, and/or on the roof of buildings. Roof-mounted HVAC units are a concern to firefighters for several reasons:

- Dependent on the size of a system, a significant dead load can be imposed on the roof. Although the supporting structural members are normally engineered to carry the weight (some retrofitted modifications may not have been properly engineered), supporting structural members attacked by fire will collapse faster due to the imposed additional weight. This can be a dangerous condition to roof and interior personnel as evidenced by two firefighter deaths on February 14, 2000, in a fast-food restaurant in Houston, Texas. A fire that originated in the kitchen spread into the structural elements of the open-span roof, causing the HVAC unit to drop. A similar event took place in 2007 in Boston. Two firefighter LODDs occurred when an HVAC unit dropped during a restaurant fire.

- Associated with roof HVAC units are various ducts that can be mounted on and travel across a roof from the unit(s) to respective areas within a building. Depending on the amount and size of ducting, they can hamper or negate vertical ventilation operations. In all cases, HVAC ducting can provide channels for the rapid extension of heat/fire/smoke and must be checked if this is a possibility.

- Rooftop HVACs typically have utility gas *and* higher voltage (220 V) power supplied to them—adding more hazards for the firefighter.

- Roof mounted HVAC units are also capable of providing a source of fire from overheated electric motors and/or drive belts.

Table 9-1. Common HVAC systems and appliances

Appliance Type	Use	Power Source	Conveyance	Notes
Boiler	Heating	Oil, steam, or utility gas	Rigid piping to radiator-like devices	The boiler heats water used in radiators
Forced hot air	Heating	Utility gas	Air ducting and fan	Most common form of heating. Is often a central unit in homes and multiple roof-top units in large+ commercials
Electric baseboard	Heating	Electric	Metallic heat exchanger (and perhaps a small fan)	Usually individual units in each room
Radiant space	Heating	Utility gas	Metallic radiant tubes	Ceiling-mounted with a reflector to radiate heat downward
Radiant in-floor	Heating	Utility gas	Tubing grid embedded within or below a floor covering like tile, wood, or poured concrete	A small boiler heats water pumped through the tubing
Water heater	Hot water for faucets, etc.	Utility gas or electric	Copper or plastic piping	Heats fresh water to 100–130°F for most applications
Central air conditioner	Cooling	Electric	Air ducting	Consists of a condenser, fluid, and heat exchanger. Can be a separate unit or combined with a forced hot-air system
Evaporative chiller	Cooling	Electric	Air ducting and a fan	Water is circulated over heat exchanging fins that cools ambient air
Individual room combination units	Heating and Cooling	Electric	Small fan	Typical for individual hotel rooms—wall or window mounted
Ventilators	Air exchange, heat removal	Electric, mechanical	Large fan	Typically roof or wall mounted, used to remove accumulated heat or to exchange air within a space

Other building utilities

As mentioned earlier, buildings contain many utility systems that may not be easily recognized by firefighters. They are highlighted here.

Communication/data systems. The digital era has created more hazards for firefighters. In a simpler time, a building had a few low-voltage phone lines and perhaps a cable TV feed that presented very minimal hazards. Now, the communication systems within buildings can include morphed phone lines, digital cabling, routers, wireless receivers/transmitters, cellular repeaters, data storage computers, cameras/displays, and high-tech entry/security features. We highlight cellular repeaters and antennas in the last section of this chapter.

The features themselves add fire load to the building. Additionally, many buildings had to be retrofitted for these features, leading to exposed cable trays, wall/floor/ceiling poke-throughs, and higher electrical demands.

Freshwater systems. Many fire departments respond to water problems in buildings—and those firefighters can attest to the damage that can occur. Buildings have been condemned from the structural damage incurred by a failed water system. For this reason, firefighters are encouraged to denote water shutoff options when conducting building surveys and prefire plans. *Important clarification: Firefighters need to differentiate freshwater and fire sprinkler system shutoffs.* Most commercial fire protection systems require a separate supply feed for fire protection. Newer residential sprinkler systems may combine freshwater and fire sprinkler supplies. The rapidly-expanding use of plastic water pipes increases the chances that a building fire will also become a water problem (flooded basements, electrical equipment saturation, degradation of engineered lightweight wood products, etc.).

Wastewater (sewer) systems. Unlike freshwater systems, the sewer system can't be shut off. Wastewater systems are considered open systems in that they are vented to the outside, which assists in gravitational draining. The vent stacks (pipes) used for this venting provide a ready conduit for fire extension (especially plastic types). Smoke emitting from these stacks, as viewed from the building exterior, are cause to investigate fire spread potential in walls and the attic/truss loft.

Buildings not served by a wastewater treatment plant or connection to a municipal sewer system will have a septic system. The septic system usually includes a holding tank and/or leaching field that are outside, but in close proximity to the building. The primary hazard of these is the potential of sinkholes and collapse from apparatus impact loads.

Quick summary

- Propane and natural gas are the two most common utility gases. Propane is heavier than air, and natural gas is lighter.
- Utility gases stored in tanks outside a building are usually in liquid form (LPG and LNG). Fire impingement on these tanks can lead to a BLEVE.
- Gas shutoffs are typically located outside. Natural gas is shut off at the meter with a half turn of an inline valve. Nordstrom valves are quarter-turn.
- Propane is shut off with a twist valve located under a protective cover atop the tank.
- HVAC systems include a heating/cooling appliance, power or fuel source, conveyance system, and temperature regulation system.
- Rooftop HVAC systems create significant firefighter risks—namely a concentrated dead load that can cause rapid collapse during fires.
- HVAC air ducting can also be considered smoke and fire ducting.
- Communication/data systems have expanded in modern times and now present more hazards to firefighters.
- Flooding from failed freshwater systems can cause structural and electrical system damage. Knowing shutoff locations is imperative.

- Firefighters should differentiate freshwater shutoffs and fire protection water shutoffs.
- Vent stacks used for wastewater (sewer systems) are often plastic and can spread fire and smoke to upper portions of a building

ALTERNATIVE ENERGY SYSTEMS

The previous section on utilities briefly mentions solar panels in the discussion of electrical systems. The evolving (and rapidly expanding) use of alternative energy systems warrants a separate section. Alternative energy sources are not limited to electric solar panels. Solar energy is also being used for hot water and heating. Other alternative energy sources include the use of wind, geothermic, and fuel-cell energy for building systems. While wind and geothermic are evolving, their use in individual buildings is quite limited and not covered here.

Solar energy

Solar energy can be harnessed in multiple ways, although most associate the rooftop solar panels as the primary method. Before we discuss those panels, some mention is warranted for passive solar energy. Buildings can harness the sun's energy for heating purposes through the use of large collector windows that heat an interior air space. The heated air is then circulated through the building using a series of ducts and shutters (a passive system). Some passive solar systems have fans and high-tech dampers that help distribute heat. During building fires, the passive solar system can accelerate fire and smoke spread throughout the building given the open-flow nature of the system. Buildings with passive solar features are easy to spot as they will typically have an entire exterior wall (usually multiple stories) that is glass (or high-density acrylic). Often, the passive solar collection room appears as an addition to the building.

Solar energy that is collected by roof-mounted panels can be used for heating water or for creating electricity. Each is described in more detail below.

Solar water heating panels. Solar water heating (SWH) panels or solar hot water (SHW) systems are familiar to most people and are easily identified by their common placement on the roof of a structure to collect the most solar radiation during daylight hours and by their typical characteristic black rectangular shape, with PVC pipes to transfer water (or other fluids) from the panels to lower levels of a structure. These panels can use collected energy to heat domestic fresh water, heat water used in radiators for general space heating, and/or circulate hot water through swimming pools to increase their ambient temperature.

In a close-coupled system, a storage tank is mounted above the panels on a roof (fig. 9–15) and heated water naturally rises into the tank.

Fig. 9–15. In a close-coupled system, a storage tank is mounted above the panels on a roof.

In a pump-circulated system, the tank is mounted below the level of the panels and a pump moves the water between the tank and the panels. Although both of these systems are relatively simple, they can add additional complexity to a roof and possibly negate roof ventilation in some

cases. Most solar water heating systems don't circulate the actual freshwater through the panels. They use a more viscous fluid (e.g., mineral-, silica-, or glycol-based) that absorbs and releases heat more efficiently than water. This fluid passes through a marine or immersion type heat exchanger located within the building to heat the freshwater. In all cases, the panels, piping, and fluids add a significant amount of weight (depending on the size of the system) to a roof that may or may not be designed for the additional weight. Additionally, solar fluid panels are hot to the touch during daylight hours. Solar water heating panels are typically larger and weigh much more than the solar photovoltaic (PV) panels.

Solar PV panels. Solar cells (or solar panels, solar modules, photovoltaic modules, or photovoltaic panels) are normally a collection or array of interconnected photovoltaic cells (commonly referred to as PVs). The name photovoltaic comes from *photo*, which means light, and *voltaic*, which means electricity. Unlike solar water heating panels that are used to heat fluids, these panels are used to generate and supply electricity in residential and commercial applications. PVs are visually different from fluid panels—they are thinner and have a grid of cells (silicon wafers and contact grids) below the glass, whereas fluid panels are thicker and have tubes below the glass.

The PV cells convert sunlight into electricity as the light strikes semiconductors within the cells. Although various materials can be used for the semiconductors, the most common is silicon and micro-thin reflective metal compounds. Solar cells are normally comprised of an array of various sizes of solar panels (depending on the desired output), an inverter (DC to AC), the necessary wiring, and quite often a battery complex for storage and backup.

The electronics (voltage regulation and relays) that are necessary to operate PV cells can be mounted externally away from the cells, or more recently, some companies have begun to embed the necessary electronics into the PV modules.

Solar cells can be found either mounted on fixed racks on the ground (tilted to correspond to a location's latitude) or, of more interest to fireground personnel, mounted by frames to supports on a roof (fig. 9–16). Rooftop PVs can also be integrated into the roof covering in the form of shingles—they appear as tile shingles that are flush with the other shingles but have a glass-like finish. Electrical power is always being generated by the PV cells if light is present (even on a cloudy day). Therefore, they should be treated as a live electrical utility. Some PVs may have a disconnect on the PV unit, although most feed wires (within conduit) that lead to an inverter junction box near the main electrical panel (exterior wall). Power can be shut off at the main electrical panel or the inverter. *Be warned that DC power will remain in the PV panels and wires up to the shutoff.*

Fig. 9–16. Solar cells can be mounted on frames and supports on a roof.

There are two basic types of solar cell systems:

- Battery backup system. In this system, electricity created by the PV panels is fed into batteries. When lights and/or appliances are turned on, power is drawn from the batteries (and inverter). The battery backup system also provides electricity when the PVs are not generating (at night or when snow covered). Some call the battery backup setup an *off-grid* system. Most batteries used in solar panels are the lead-acid type. When these batteries are

charging, they emit highly flammable hydrogen gas. Additionally, if these batteries are exposed to fire they can spill sulfuric acid, which is extremely hazardous and toxic.

- Grid-tied system. This system is tied to the local electric utility's meter and distribution grid. When the PV panel generates more power than the individual building is using, the excess is diverted to the grid system and back to the appropriate power company which then buys it (or credits the generator). Conversely, when more power is used in a building than the PV system creates, the appropriate power company supplies the difference. Grid-tied systems may also include a battery backup system for grid power outages.

Of major importance to fireground personnel are the questions of how much do PVs weigh and how much power/voltage/amperage do they generate. As there are a multitude of PV panels that can be used in various configurations, the following is a general overview based on current technology:

- A range of from 12 to 24 volts DC is commonly employed for operation (output), which can be considered a low voltage hazard. Depending on the system, expect the amperage to be in the range of 5 to 7 amps (relatively low). Wattage ranges from a low of 10 to highs between 175 and 210 watts.

- Weight can vary significantly due to the size and type of a cell array, whether it is mounted in aluminum or steel frames, and so on. A common 3 × 4 ft residential PV panel weighs about 40 lb. There are 32 panels distributed on the roof in figure 9–17. That is an additional dead load of 1,280 lb plus the mounting hardware and conduit feeds.

In all cases, solar fluid heating panels and PV panels can negate and/or significantly increase the difficulty and time required for roof ventilation operations.

Fig. 9–17. This residential solar array adds approximately 1,280 lb of dead load to the roof.

Fuel cells

Fuel cells are not new, but the emphasis on them has accelerated as a result of the hydrogen fuel initiative that is a part of the Energy Policy Act of 2005. A fuel cell is an electrochemical energy conversion device that converts oxygen and hydrogen into water, with direct current (DC) electricity being a by-product of the conversion process. Although most people are aware of their use in automobiles, fuel cells are also growing more popular as a replacement for UPS systems, primarily in commercial buildings. Therefore, the fire service can sooner or later expect to encounter fuel cells in fireground operations. Fuel cells are capable of generating electrical power with the sole by-products of heat and water, hence their adaptability to green buildings. As compared to a lead-acid type battery, fuel cells do not stop working as long as there is a flow of fuel into the cell. However, there are two obvious disadvantages with fuel cells. Because they produce DC voltage, there must be an inverter (for DC to AC) and all the associated wiring/relays, both of which can present an electrical hazard. The most concerning hazard for firefighters, however, is the storage and use of pure hydrogen—an extremely flammable gas with the widest flammable range of any gas.

Chapter 9 — BUILDING FEATURES AND CONCERNS

> **Quick summary**
> - Solar energy can be harnessed for passive heating, fluid heating, and electrical generation.
> - Passive solar systems are identified by an entire wall of windows.
> - Solar fluid heating panels are thicker and include internal tubes for heat collection.
> - Photovoltaic (PV) solar panels are thinner and include an internal silicon wafer/metallic grid.
> - PV panels create DC power when light is present. The actual cells cannot be shut off, although they may be disconnected at the inverter or main shutoff box.
> - Most batteries used in solar panels are the lead-acid type. When these batteries are charging, they emit hydrogen, which is a highly flammable gas. Additionally, if these batteries are exposed to fire they can spill sulfuric acid, which is extremely hazardous and toxic.
> - All solar panels can add a significant dead load to roofs that may or may not have been designed into the load limits of the roof.
> - Solar water heating panels and solar panels can negate and/or significantly increase the difficulty and time required for roof ventilation operations.
> - Fuel cells are becoming viable replacements for lead-acid batteries. However, they introduce the presence of pure hydrogen, which is highly flammable.

OVERHEAD HAZARDS

Due to their inherent danger and historical hazard to firefighters in particular, a special section on cornices, parapets, and facades is appropriate. Although we mention these overhead hazards in several places throughout the book, some further detail and emphasis is warranted. Dozens of firefighters (if not hundreds) have been killed by unstable and falling overhead hazards. Further, modern varietals present interesting challenges that could surprise working firefighters. In almost all cases, cornices, parapets, and facades should not be used to support ladders (ground or aerial) and they should be viewed as a collapse hazard at any fire that impacts the roof or upper exterior walls.

Cornices

A *cornice* (which comes from the Latin word meaning ledge) is a horizontal molding that is a common construction feature on older commercial buildings and was originally used to divert rainwater from the building's walls. (In residential structures this is accomplished by roof eaves and gutters.) However, as the classical and Greek style of architecture became more popular, cornices were also used to alter the architecture of the front of a building, often by using a sizeable overhang supported by corbels and/or decorative horizontal supports. To further enhance the appearance of a cornice, the underside was often finished with decorative tin panels that covered the upper wood planking used to construct the cornice. Interestingly, cornices are still used on modern buildings, although they are dramatically different in their appearance, construction, and stability. Let's look at older and newer cornices.

Older cornices. Cornices were commonly built on multistory residential and commercial buildings until the mid-1900s for the purpose of capping a wall and enhancing the architectural appearance of a building. Unfortunately, this style of construction can appear as a substantial type of construction, when in reality it may not be an integral part of the exterior wall of a building and is often significantly degraded by time, weather, and other similar considerations. Additionally, the older cornices should not be depended upon to support the weight of personnel, ground ladders, or aerial devices. Stone and wood are common materials that were used to construct the older type of cornice.

- **Stone:** Cornices made from stone are supported by a ledge constructed in a brick facade or parapet wall for support and commonly use mortar that acts as an adhesive to the parapet/facade wall. It should be mentioned that if mortar was used as an adhesive prior to the mid-1930s, it likely lacked Portland cement and was basically comprised of lime and sand. Although this type of construction can appear to be an integral part of the exterior of a building and relatively strong, the use of a substandard mortar can result in a weak bond that has further weakened over time.

- **Wood:** Wood, tin, and some plaster type components were commonly used to construct these types of cornices and are more common than stone cornices. The construction of a wood cornice often began with leaving lets in a masonry exterior wall and then inserting short, visible wood supports (that look similar to the ends of beams) into the lets or decorative corbels. In figure 9–18, notice the wood cornice is collapsing due to age, and the two corbels with the wood supports are still visible on the left and right sides of the photo. However, although the cornice is still intact on the right side, it obviously does not possess any inherent strength and would not support the weight of a ladder or personnel.

The top portion of wood cornices are covered by extending the roof sheathing over the cornice, and the underside is covered by tin to give a finished appearance. This type of cornice presents two noteworthy hazards: fire extension and collapse potential. A concealed hollow space in the cornice can allow horizontal extension of fire and the potential of extending fire into areas behind the cornice that are common to the cornice. A partial collapse of this cornice can result in the collapse of the entire cornice, creating a significant danger to fireground personnel.

Newer cornices. Cornices on modern buildings are not quite as ornate as older cornices but they serve the same primary purpose: to alter or enhance the exterior architectural appearance of a building. Although they can vary in size and complexity, they all share some common characteristics. Modern cornices are most commonly made from synthetic materials such as closed cell polystyrene that is then glued to the exterior wall(s) of a building and often finished with stucco and paint. Although this procedure results in an attractive look to a building (fig. 9–19), it can also present a dangerously flammable and weak element for firefighters engaged in aboveground operations.

Fig. 9–18. Decorative corbels and wood supports were commonly used to support wood cornices.

Fig. 9–19. Modern cornices are typically made of foam and glued to the exterior of a building to alter the external appearance.

In order to access a roof, firefighters will normally place a ground ladder or aerial device to the junction of a roof and exterior wall, which is also the location of foam cornices. Because the structural connection of the cornice to the building is

dependent on a bonding agent (adhesive), this connection will likely not provide sufficient strength to support a ground ladder and will definitely not support the weight of an aerial device. Additionally, it is also likely that a foam cornice will not support the weight of a firefighter stepping from an aerial device to the cornice before continuing onto the exterior wall and/or roof. Unfortunately, this style of construction is currently very popular, particularly in strip malls and many commercial buildings.

Parapet walls

A parapet is a continuation of an exterior wall (normally the front wall) above the roofline of a building. The term parapet comes from the Italian word *parapetto*, which means to cover and defend. Today, parapets can be used to prevent the spread of fires and improve wind-uplift resistance, but more often they are similar to cornices in that they can alter the architectural appearance of the front of a building. Architecturally, they provide an aesthetic function by increasing the apparent height of a building and hiding the roofline and HVAC equipment on a roof. Parapet walls can extend from 1 ft to well over 8 ft, depending on the occupancy and type of wall. Several examples of parapet walls and their hazards are as follows:

- Unreinforced masonry construction is markedly different from newer types of masonry construction. The lack of Portland cement and rebar plus the addition of old age and deteriorating mortar (lime and sand) combine to result in a weak, freestanding wall above the roofline that is either an extension of the wall it is on or is supported by a steel I-beam (lintel) that is embedded in the wall below the parapet wall. Recall from chapter 2 that masonry walls achieve strength and stability by the compressive load that is imposed upon them. A parapet lacks compressive loading and can be considered like a vertical cantilever. Depending on the age of a building, this type of wall can be capped by coping stones, tiles, or wall caps (fig. 9–20) that have lost their adhesive connection through age and weather and are only attached by gravity. These conditions can present an extension of an exterior wall (normally the front and side walls) that can readily collapse during fire conditions and/or when struck by heavy streams.

Fig. 9–20. Coping stones on parapet walls can present a significant hazard to fireground personnel.

- Concrete-tilt up slabs normally extend above the roofline. Although the stability of this type of parapet wall is not a concern, these parapet walls normally encircle an entire building and are capable of hiding the roofline below the parapet wall (which is a danger if personnel are going to the roof) and hiding the type of roof (flat, arched, etc.). In this case, the presence of scuppers can indicate a hidden roofline. In figure 9–21, the scuppers are well below the signage on the building.

Fig. 9–21. Modern masonry parapet walls can be substantial in height.

- Modern masonry parapet walls can be constructed from concrete cinder block (fig. 9–21) or modern brick, which has similar concerns to concrete tilt-up slabs. Stability of this type of construction (particularly concrete cinder block) is normally not a concern due to the added strength of concrete and reinforcing rebar.

Quick summary

- Neither the older style masonry or wood cornices or newer foam style cornices should be considered structurally sufficient to support the weight of personnel and/or ground ladders/aerial devices.
- Newer foam cornices can be flammable, as graphically demonstrated in the Monte Carlo Hotel fire in Las Vegas in 2008 (see chapter 4).
- Unreinforced masonry parapet walls and older cornices are potential collapse hazards to fireground personnel. In both cases, suitable collapse zones should be identified when appropriate.
- In many cases, an older unreinforced masonry parapet wall is supported by a metal I-beam that will lose its strength between 800°F and 1,000°F. Although the brick wall on the front of this type of building can look substantial, it can easily collapse outward if fire/heat is able to weaken the metal I-beam.
- Master streams are easily capable of dislodging brick or portions of unreinforced masonry parapets, particularly when applying water with pressure/volume in a horizontal direction. These considerations also apply to older conventional wood type cornices.
- The height of a parapet can hide the distance from the top of the wall to the roof below. The difference can range from several feet to a full story.

Facades

An important construction feature that has become popular in virtually all areas of this country is the facade (or fascia). The term can be defined in a few ways:

- As a figure of speech, the word *facade* is used to describe "the face that people show other people, as opposed to what they really think or do."
- Webster's Dictionary defines *facade* as "an imposing appearance concealing something inferior."

Applied to buildings, the facade can be defined as "an exterior construction feature that is used on the walls of a building to alter its visual appearance." For this discussion, let's use this definition and the term facade to encompass fascias, false mansards, cantilevers, eyebrows, and overhangs.

The venerable facade has recently become a popular construction feature for the simple reason that it can make the plain, economically-constructed, boxy shape of a building look more appealing. Many local ordinances actually require a building to be finished in a décor that matches surrounding buildings or styles. A good example is that of Spanish architecture. The creative use of stone, stucco, and tile for towers, window and wall construction, and roof finish is very attractive—and cost-prohibitive for most buildings. Adding a facade can give the look of a classic Spanish building, even though the construction method is simple wood frame or lightweight Type III.

Facades are typically built using a lightweight form of construction, yet can be finished with heavy tiles and other decorative materials that add tremendous dead load. While this dead load should be engineered into the supporting structure, it will likely have lower safety tolerances, as it is not considered a part of the building that people occupy.

Facades also serve to provide shade and protection from weather to the front of a building and to hide the presence of equipment and machinery (such as HVAC systems) on a roof. Like cornices

and parapets, the facade feature can be extraordinarily dangerous to firefighters. The good news is that facades are usually easy to identify.

Specifically, the facade feature presents six considerations for firefighters. As a reference, we'll use figure 9–22 to discuss the first five considerations (labeled A through E in the figure).

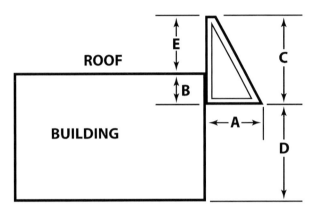

Fig. 9–22. Facade size-up considerations: (A) overhang, (B) attachment, (C) height and shape, (D) support, (E) height from the roofline.

and plaster over a public walkway, that adds to the difficulty of opening the void space above the soffit for examination. It should be mentioned that some facades do not have a soffit. Although these facades can still collapse when exposed to fire, they are normally not common to an attic. Additionally, they are normally only exposed to fire from auto exposure from the front windows, and fire can be extinguished without having to remove a soffit.

Fig. 9–23. When completed, facades often create hidden voids.

Overhang. Regarding the overhang of a facade, there are three primary concerns that should be evaluated:

- The overhang is the distance a facade extends outward from a building. As the distance increases, so should your concern about structural integrity whenever a facade is exposed to fire. As the size of a facade increases, so does its complexity and the materials used in its construction, which can often be of lightweight construction. The size of a facade can have a direct affect on the area, path, and travel of fire. Unless proven otherwise, expect any facade to lack fire-stopping and sprinklers and that it will be common to an attic.

- The underside of most facades is covered with soffit paneling. A soffit creates two problems of note: hidden voids and difficulty in breaching. The facade creates a hidden void above the soffit (notice the void that will be behind the facade in figure 9–23 when it is completed). As most building codes require compressed metal lath

- If a facade on the front of a building collapses, it can block the primary access and egress opening of a building, and it can easily compromise any hose lines that may be stretched through the front door. Most importantly, if personnel are standing within the collapse zone, they are in immediate trouble.

Attachment. Facades come in two distinct varieties: those that are an integral part of the construction, and what we call an *add-on* or *retro* method of construction.

- The integral (or structural) variety is simply a part of the roof construction that is normally cantilevered over the front exterior wall and extends upward and downward in relation to the bearing wall. The amount of cantilever is based on at least a ratio of 3:1, which means a roof that has a span of 30 ft can be cantilevered 10 ft over the front of a bearing wall. Although this type of facade can have a stronger method of support than an add-on/retro facade, it can also be common to the attic space of a building

and not protected by some type of partition or fire wall. Interestingly, if a fire wall is present, it is primarily designed to partition the attic but normally will not extend into the void space of the facade. Therefore, a horizontally extending fire in an attic can easily go around a fire wall and back into the facade and continue to extend on the other side of a fire wall!

- The add-on/retro varieties are attached by various methods (depending on the type of exterior walls) such as ledger boards and concrete lag bolts to the exterior of a building, either during the construction phase or as a retrofit to an existing building. This may result in a partition between the attic space and facade (which would be the exterior wall). However, the attachment of a facade to an existing building can result in minimal strength in the attachment method, which likely will be unknown to fireground personnel.

There is some debate regarding the structural integrity during fire conditions of a structural facade as compared to the add-on/retro facade. As most facades are of lightweight construction, it is probable there is more of a difference in the relative size of a facade, the distance of its overhang, and the type of covering than there is in the lightweight structural vs. lightweight add-on/retro construction. However, both types are capable of collapsing at least the same distance outward from a structure as the size of the facade, which could be considered a noteworthy distance (fig. 9–24).

Fig. 9–24. Facades are capable of collapsing outward a noteworthy distance.

Height and shape. The height and shape of a facade affect its structural stability, the quantity of building materials present, and the potential path of fire. Remember, a facade is an external common attic that can easily conceal and spread the travel of fire around the exterior of a building. Keep such factors in mind during ladder operations. On buildings without facades, the roof type (arched, sawtooth, flat, etc.) can often be identified from the ground. This is helpful (and often necessary) when laddering a roof. The style (curved or flat) of a facade and/or distance out from the building can hide the type of roof and/or also keep a ground ladder from reaching its top (fig. 9–25), further hindering access to a roof. Thankfully, on most buildings, the facade does not extend to the back of the building—a good place for ground ladders.

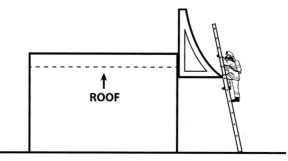

Fig. 9–25. A facade can hinder the placement of a ground ladder to a roof.

Support. Some facades are equipped with vertical supports such as pilasters and/or columns that can serve as a decorative addition but can also serve as an external support (fig. 9–26). In this case, the facade will be supported by its attachment method to the building and the column/pilaster. Obviously, these supports can enhance the strength and safety of a facade and reduce the leverage effect of a cantilever that is only supported by its attachment to a building. Unfortunately, supported facades cost additional money and are thus less frequently used.

Chapter 9   BUILDING FEATURES AND CONCERNS

Fig. 9–26. Some facades are supported by columns, which can increase their strength.

Height from the roofline. As facades can conceal a roofline, ladder operations can present an additional challenge. When laddering a facade and the roofline cannot be seen, the height of the facade above the roof should be determined. If this is not done, an additional trip back to the ground for another ladder may prove necessary. Additionally, personnel on a roof without a safe and easy means of egress to the top or back side of a facade are likely in trouble if the roof begins to collapse. If the distance from a roof to the top of a facade (or parapet wall) exceeds about 5 ft, a ladder from the roof to the top of the facade/parapet wall is necessary. The following considerations can assist in determining the roofline of a building:

- Rafter tie plates. These indicate the location of the roof rafters, which will designate the roofline.

- Windows. Rooflines run between the top floor windows and the top of a facade/parapet.

- Equipment on a roof. Air conditioners and heating units that can be seen above a facade indicate that a roof is in close proximity.

- Attic vents. Rooflines are between the attic vent and top of the facade/parapet.

- Scuppers. A scupper is the actual level of a roof. The scuppers in figure 9–21 are well below the signage on the building.

- Perimeter. Facades are normally constructed only on the front and sides of a building. Normally, they are not on the back of a building (as the public does not normally view this portion of a building). If present on the back, the facade will have an opening for roof access/egress.

Fire spread investigation. An additional consideration (not diagramed above) is that of fire spread potential and the need for investigation. A facade that has been exposed to fire or hot smoke *must* be investigated for fire (and/or overhauled). There are three basic ways to open a facade: from the front, bottom, or back (fig. 9–27).

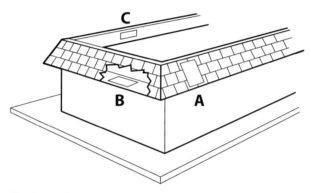

Fig. 9–27. Access to the interior of a facade is normally accomplished from three perspectives: (A) front, (B) bottom, (C) back.

- If a facade has a slope (or pitch) on the front portion, it is possible to work from the exterior (ground ladder or aerial device) down into the facade (A). However, this is difficult and dangerous from a ground ladder, as the facade must support the ladder. If the facade is vertical on the front portion, it is possible to pull the finish material from the front with a tool (pike pole, etc.), but personnel on the ground will be located in the potential collapse zone! In these cases, working from an aerial device is superior to working from the ground.

- Personnel can stand underneath a facade and pull the bottom portion to access the interior (B). As compressed metal lath and plaster is normally required over a public walkway, this can be a difficult operation at best. Additionally, personnel will be standing in the potential

This is not an uncommon problem and firefighters should be acutely aware of its possibility.

Covered windows. Windows that have been covered over or sealed are an indicator that a renovation of note has changed the basic floor plan of a building. This type of modification is often found on older concrete and unreinforced masonry buildings, where brick has been used to seal a window opening. Although windows may be covered over for security purposes, the more likely change is a result of changing a floor plan from smaller individual rooms to larger rooms/areas. Depending on the type of modification, the increase in the size of a particular area and/or removing partition walls will allow a fire to increase without the resistive affects of walls, and can also increase the difficulty of search operations.

Soffits. Soffits seem to be more common in newer homes than in older homes and are likely a result of modern styling (coved ceilings) and space that is necessary to run HVAC ducting. These styles of renovated construction are easily recognizable, particularly in older homes that did not use them, and they are also ideal avenues for the extension of fire. As an example, the soffit under construction in figure 9–29 is being used to hide exhaust ductwork for clothes dryers. If appropriate, soffits must be opened to check for extension of fire/heat/smoke. Also remember that soffits are often found above kitchen cabinets and present voids that can conceal fire.

Fig. 9–29. Soffits can be used to hide plumbing and ducting, resulting in voids that can conceal the spread of fire.

Suspended ceilings. Suspended ceilings are a common method to lower the ceiling of an existing structure, change the appearance of a room, and/or create an interstitial space between the ceiling and the floor/roof above that can hide HVAC ducting, sprinkler supply lines, electrical conduits, communication cables, and other similar components. Although suspended ceilings are easily identified and the individual tiles can be removed with little effort to check the area above the ceiling, they can present several noteworthy hazards:

- They are suspended by thin wire and can readily collapse when exposed to fire.

- Large interstitial spaces can be created above these ceilings (see fig. 7–12) that can conceal multiple avenues for extension of fire/heat/smoke.

- Firefighters can become easily entangled in the suspending wires during or after a collapse, which can affect a timely egress from a building.

Tie-rods and joist/rafter tie plates. Tie-rods and joist/rafter tie plates are a common retrofit renovation and are primarily found on unreinforced masonry buildings in an attempt to strengthen the building. Sagging, unstable lime/sand mortar, soft brick erosion, and wood shrinkage are aging factors that can lead to the need to install reinforcements. Joist and rafter tie plates indicate that screws or L-shaped bars have been added to help secure gravity pocket joist/rafter ends to the wall. The ends of the joist/rafter plates can be found on the building exterior and are recognized by anchor plates (small flat washers or small squares) with a lag screw head or an adjustable nut. Additionally, the anchors will have an orderly arrangement and even spacing of approximately 16 in. to 24 in. on center (see fig. 7–20).

An additional method that is used to strengthen older masonry buildings and connect opposing walls to resist lateral forces is the use of tie-rods that run horizontally through a building. Tension on the rods is supplied by turnbuckles within the building and spreaders located on the outside of the building. Spreaders are typically larger than joist

tie plates and may have an ornate shape like a star, diamond, "S," or coat of arms shield. Unlike anchor tie plates, the tie-rod and spreader arrangement may not be orderly but random, as their location is dictated by weak areas of a building. The presence of spreaders/tie rods and/or joist/rafter tie rods denotes an older unreinforced masonry building that is a candidate for collapse in a fire.

Electrical utilities. Electrical utilities are normally visible on the exterior of residential and commercial buildings and are often a practical indicator of an interior remodel from two viewpoints: the type of wiring used and the number of electric meters.

- Knob and tube wiring was normally used in older buildings and is easily identified by a two-wire lead into a structure from a pole. In addition, this wiring usually leads to a fuse box (with removable glass fuses), often on the porch of residential structures. Because this was an ungrounded system, it has been replaced with the familiar three-wire grounded system on residential and commercial structures. When older residential structures have a newer three-wire system, expect that the electrical system has been remodeled with the possible addition of other changes to the interior.

- Commercial buildings will often have multiple electrical inlets/meters on the exterior of a building, and in many cases these utilities look old (outdated, rusted, etc.). If the older utilities have been abandoned with newer looking (and possibly additional utilities) replacing the older utilities, expect that the interior of the building has changed as well. For example, one likely reason for the change would be that the older electrical utility was not sufficient to supply a change in the current occupancy. In figure 9–30, notice a quadruplex system has been installed above the lower abandoned three-wire insulators.

Fig. 9–30. The original three-wire service has been replaced with a quadruplex service, indicating a noteworthy change within this building.

Residentials. Renovations in residential structures such as single-family dwellings (SFDs) are common and can transform these structures into a stealthy hazard that can be categorized as noteworthy. The key is to look for visible changes and how they can affect fireground operations. Several examples follow:

- In some cases, single-family dwellings have been converted into commercial offices. These are often structures located in the perimeter of the residential neighborhood. In these conversions, the exterior is usually unchanged (except for an exterior business sign) and the interior will be more spartan than a SFD; however, the need for a search at night will be reduced.

- Depending on a particular area, some single-family dwellings have been converted into commercial rest care homes. Although the exterior remains unchanged to blend into the surrounding area, the interior may or may not still contain the approximate original floor plan. In any case, these types of conversions will definitely have a higher search potential at all hours of a day than a typical SFD.

- The multistory residential dwelling in figure 9–31 was originally a single-family dwelling but is now a commercial residential that is common in many areas of this country. Although the exterior remains similar to the original, notice the fire escape on the left side of the building.

Depending on the area, it is likely that the interior has been divided into smaller rooms to increase the occupant load (often illegally), and it has multiple narrow hallways, increased fire load (clothes, bedding, etc.), and an attic that has likely been converted into a living area (and is probably difficult to access). In short, this type of structure can present a significant hazard from the perspective of what there is to burn and life safety of potential occupants.

Fig. 9–31. A fire escape on the exterior of a multistory single-family dwelling is an indicator of a dwelling that has been converted into a commercial residential building.

Commercials. Commercial buildings commonly undergo many conversions during their lifetimes. Common examples are mini-malls, older taxpayers, freestanding commercials (markets, auto repair shops, etc.), and so on. Although the types of buildings vary widely, they share a common characteristic of leaving the exterior of the building unchanged (other than the name of the business) but periodically changing the interior of the building as needed for the type of business.

Generally, the interior does not significantly change other than erecting a partition wall or two and likely changing the level of a ceiling to correspond with the business. In most cases the primary hazard will come from the contents within the building (e.g., electrical shop, clothing outlet, shoe store, etc.), which can usually be recognized by the signage (name of the business) on the exterior and easily readable from the street. Depending on the type of occupancy, remember that many adjoining buildings have a common attic over multiple occupancies, and that older taxpayers often have a residential occupancy over the commercial occupancy.

Remodels

A remodeled occupancy can present a multitude of hazards that normally accompany a change to a building. A familiarity with these types of buildings begins with prefire planning that allows fireground personnel to walk through and evaluate changes that can affect potential fireground operations (should they be necessary) in a particular building. Three classic examples that illustrate this viewpoint are as follows:

- The remodeled four-story Ford building in Portland, Oregon (see fig. 4–15). This old unreinforced masonry building of mill construction sat vacant for a number of years until it was completely remodeled into a modern office building with numerous internal features of modern buildings. However, this remodel resulted in a building that now features suspended ceilings, numerous voids formed by additional compartmentalization, some lightweight construction that is mated to old mill construction, and the list goes on. It is easy to envision a fire in this building burning faster and extending to other areas of the building that would not have been possible in the original building.

- An example of a remodel that has been a benefit to an original building is the commercial building in figure 9–32. This row of commercial occupancies was originally built in the 1930s and featured unreinforced masonry (URM) construction with a conventional flat roof. During the remodel, metal I-beams were used to support the interior of the URM construction, a metal roof with steel supporting members was installed, and the front of the building was modernized. Even though the added steel will weaken with heat (between 800°F and 1,000°F), it is significantly stronger than the original

building. Prefire planning is the key to understanding the hazards in these occupancies.

Fig. 9–32. When remodeled, older buildings are often strengthened to meet modern building codes and/or earthquake resistance, which can be beneficial to building stability in fire conditions.

- For an example of a remodeled building that has been completely changed on the exterior but has retained the original look of the interior, go back to chapter 1 and look at figures 1–1 and 1–2. In this case, interior personnel would certainly notice the inside does not match the outside, yet the focus of this dichotomy of "appearance vs. reality" is twofold. First, can the remodel affect fireground operations from the perspective of mitigation and safety? Second, is the incident commander aware of the building's construction dichotomy?

Closely related to renovations and remodels are additions—that is, expanding the footprint of the building by adding a new wing or space. Likewise, additional floors can be added to a building. Throughout this book we call attention to the need to visually recognize additions and treat them as a separate building classification even though they are attached to the original building (see fig. 5–15).

Recent remodels and renovations are likely to include green construction methods, materials, and technologies. Chapter 6 includes details of these green considerations.

Quick summary

- Windows, doors, and roofs can often give clues to changes made to a building, particularly to an older building.

- Attic rooms can pose a significant danger due to their location, limited access, and in some cases substandard construction.

- Anticipate the potential of rapidly changing conditions when opening knee walls.

- Suspended ceilings pose two significant hazards: the thin wires that hold the assembly in place, and the open space above the ceiling.

- Tie-rods/spreaders and joist/rafter tie plates are excellent indicators of an unreinforced masonry building that has been strengthened to prevent an early collapse.

- Exterior electrical utilities that are new or that appear to be added can be an indicator of a renovation/remodel.

- The rear of commercial buildings (particularly older buildings) can often yield valuable clues to changes such as a modified electrical service, type of construction, blocked openings, and other issues.

- Depending on the extent of a renovation/remodel (particularly in older buildings), a basic rule is that interior remodeling can remove original barriers to the spread of fire, can result in voids that were not present in the original construction, and often substitute lightweight construction for conventional construction. Additionally, it can be anticipated that fire will burn differently as it has new avenues for extension due to the remodeling.

- Renovated/remodeled occupancies are an indicator that a building may have a higher occupant load and a different floor plan than the original building.

LIGHT WELLS, SKYLIGHTS, AND ATRIUMS

Light wells

A light well or air shaft was originally an unroofed vertical shaft within a building and primarily used to provide sufficient light and ventilation to rooms within the interior of a building. Their use dates back to the Egyptians and Romans! Light wells travel from the bottom of a grade floor to the roof or, in buildings that have a commercial occupancy on the first floor and residential units above the first floor, travel from the second floor to the roof. Both types are open to the outside at the top of a building (unless they have been covered over). Because of advances in HVAC systems, atriums, skylights, and multistory building construction techniques, a traditional air shaft/light well is no longer necessary. Modern energy-efficient features that allow natural light and air into a building are categorized here as skylights and will be discussed later.

Traditional light wells were commonly used in older multistory residential buildings (some of which have been converted to commercial applications), and in both attached and freestanding buildings.

A variation of the light well design is the multifamily residential building constructed with a footprint like an H, U, E, or W shape, which allows more rooms to benefit from exterior wall windows. (The upward extension of heat, smoke, and fire is still a problem that needs to be addressed in these types of buildings.) Another variation of the light well design is the half well. Half well is a term that is given to a light well that is between two interconnected buildings. This results in half of a light well serving one building and the other half serving the adjoining building. This condition can result in a fire in one of the interconnected buildings easily spreading to the other building via the light well shaft.

Practically speaking, there are open and enclosed light wells. Open light wells have at least one side open to the exterior of a building (fig. 9–33). Enclosed light wells are completely surrounded by a building but open to the exterior at the top of the light well. In addition to providing light, both types provide a path for warmer air to rise (like the transom over doorways in older buildings). Unfortunately, this same principle also results in light wells creating a vertical channel for the upward extension of heat and smoke, thereby creating a significant exposure problem to windows above a fire within the well.

Fig. 9–33. Open light wells have at least one side open to the exterior of a building.

Two types of wall construction can be encountered in light wells. The most common wall is combustible siding over a wood frame. Wood frame and wood sidings will present an additional combustible hazard as compared to a noncombustible light well. Noncombustible light wells are constructed from concrete, masonry products, or brick, which will resist the upward extension of fire. Regardless of the type of construction, there will always be a junction of the exterior walls and the roof at the top of a light well that can allow the upward extension of fire to expose combustible roofing materials. Smaller light wells present a greater hazard than large light wells as their smaller size tends to allow a higher concentration of heat, smoke, and fire within the light well. Larger light

wells tend to dissipate and/or reduce the concentrations of heat, smoke, and fire.

Additionally, open light wells will also tend to dissipate concentrations of heat, smoke and fire as compared to enclosed light wells, particularly those that have been covered over at a later date. If a light well has been covered over, this modification can not only increase concentrations of heat, smoke, and fire within the light well due to inadequate ventilation, but also increase the possibility of a backdraft if oxygen is suddenly introduced within the light well by opening windows. Likewise, the light well cover may not be constructed in a manner that can hold a firefighter's weight—creating a fall hazard.

Skylights

Skylights allow daylight into interior building spaces and can range from a simple glass panel to a highly functional energy-conserving device. More common skylights include those made of glass, plastic, and reflective tubing.

Glass panel skylights. These are normally panels of wired or strengthened glass that are held in a metal frame that is either flush with or raised slightly from the plane of a roof. Found mostly in older buildings, the traditional glass panel skylight was prone to leakage as it aged and seals deteriorated. It is common to find that tar has been used as a sealant to minimize leakage issues. Likewise, additional flashing and synthetic sealants may be present, making the panels difficult to remove. Remember, if the panels are broken, the resulting shards can fall into the building and become a hazard for interior personnel.

Glass panel skylights can be fixed or include hardware that makes the skylight openable. Energy-efficient skylights may include automated mechanisms that open or close the panels or operate sun shades or shutters. Electrical power for the automation can be solar, battery, or direct-wire AC.

Plastic skylights. The telltale 4 × 8 ft plastic bubble is a popular skylight found in legacy and newer buildings. Although the size may vary, most include a plastic bubble-like panel in a metal frame that is attached to a wood or metal riser attached to the roof. Smaller residential-type plastic skylights in an aluminum frame can be easily removed with hand tools. Larger, commercial-grade types with thicker metal retaining bands will likely require power tools for removal. Plastic skylights can also be flush with the roof plane and appear as colored or frosted plastics or as fiberglass panels. Corrugated metal kit buildings use plastic skylights as part of the roof covering—these panels can easily blend in with the other roof covering, but they will not support any significant weight.

Tubular daylight device. Part of the energy conservation movement includes building features or improvements that maximize natural daylight. The tubular daylight device (TDD) is a relatively new skylight product that can be incorporated in new construction or added to existing buildings. The TDD consists of a light-collecting optic, small diameter reflective conduit, and diffusing fixture. The optic is visible above the roofline and appears as a small glass bubble or parabolic lens that focuses light into a reflective tube. The tube channels the light down through the attic space to a ceiling-mounted light diffusion fixture. Diameters for the TDD can range from 10 to 28 inches, although larger diameters are likely to be utilized in the future. The reflective conduit can be found as a straight or bent tube. Some are even flexible so they can bend around existing trusses or structural members. Small diameter TDDs should not be considered structural.

Skylights can be good indicators of building floor plans. Those in residential occupancies are normally located over hallways and bathrooms. TDDs can add light to any room or space where there are no windows. In commercial occupancies, skylights are often placed over open manufacturing or storage areas. Skylights can also be installed over stairwells, open light wells, or elevator penthouses. Skylights offer an attractive ventilation option for firefighters assigned to the roof. If the building has an attic space, the area below a skylight is normally

boxed off from the attic. Therefore, if the skylight is opened by firefighters, the interior of the building will be ventilated but the attic will not.

Atriums

Atriums are used frequently in residential and commercial applications due to their ability to provide a noteworthy amount of daylighting and because of the feeling of openness and space that they provide to a structure (fig. 9–34). Atriums of moderate size can be found in residential occupancies, but can be of a significant size in commercial buildings, often traversing from the bottom floor to the roof in multistory buildings. The open atrium in the middle of a hotel or mall is a good example of this design configuration.

Fig. 9–34. Atriums can provide daylighting and openness to a structure. (Photo by Michael Gagliano.)

Although the basic design of atriums is relatively simple and straightforward, their openness also results in the following disadvantages:

- Reduce compartmentalization. Walls, ceilings, and floors can provide barriers to the extension of fire/heat/smoke, and when they are removed or minimized, extension will be enhanced. As an example, a fire on the lower floor of the two-story building in figure 9–34 will quickly expose the second floor and any occupants.

- Increased air movement. As warm air naturally rises, the air currents within an atrium will be amplified, resulting in an increase in the upward extension of heat/smoke/fire.

- Atriums in larger commercial buildings, malls, and hotels are often used for displays or exhibits of materials that are flammable (e.g., consumer products, books and magazines, artwork, and even automobiles). Although many atriums have engineered ventilation systems, it is possible they may have not been designed for the additional flammable materials that exhibits and displays can provide.

- Depending on the size and complexity of an atrium, it may be necessary for an incident commander to confer with a building engineer in order to fully utilize the building's ventilation system.

Quick summary

- Fires in lower floors of multistory buildings that access a light well are potentially more dangerous than fires in upper floors.

- It should be anticipated that more windows in a light well will be open during summer months than winter months.

- Due to the proximity of windows to light wells, it is common for flammable trash to collect at the bottom of a light well shaft. This presents a sizeable hazard to windows within the shaft that are adjacent and above a trash fire.

- In areas where light wells are common, roof personnel should quickly recon the roof to determine the presence and status of light wells.
- Skylights come in various forms including glass panel, plastics, and tubular daylighting devices.
- Plastic and fiberglass skylights that are flush with the roofing covering may not hold the weight of firefighter.
- Atriums provide an appealing look, but they can allow an increased vertical and horizontal extension of fire and can be more difficult to ventilate, particularly in commercial applications.
- If roof daylighting is integrated into a design, the skylight can often be used to enhance ventilation operations.
- In buildings with large or complex atriums, an incident commander may need to confer with a building engineer in order to fully utilize the building's ventilation system.

MISCELLANEOUS HAZARDS

Fire escapes

Fire escapes were originally designed as exterior emergency exits for building occupants and can be traced all the way back to the 18th century in England. They were also commonly used in America until they were essentially replaced by interior stairways in multistory buildings. Although fire escapes can provide a primary means of egress for building occupants during an emergency, they can also be used by fireground personnel and can create a substantial dilemma due to their age. Because fire escapes have been in use since the mid-1800s and continued to be built until the mid-1950s, they are at least over 60 years old and pose a danger due to deterioration. This problem is compounded by their location (outside a building) that has allowed them to be exposed to all types of weather, particularly moisture. The deterioration over the years has caused many fire escapes to be dangerously inadequate from a strength and safety perspective.

A basic fire escape is nothing more than a horizontal platform—one at each story of a multistory building and with stairs or ladders connecting each platform—and may include a ladder that can be lowered from the lowest platform to the ground. In some cases, there is a fixed ladder that goes from the top platform to the roof that is often called a *gooseneck*. Additionally, fire escapes are commonly constructed from steel or iron and affixed to the exterior of a masonry building by lag bolts that are held in place by lead anchors (which is a compression fitting) or that pass through a wall and are secured on the interior, or by lag bolts that are screwed into siding and wall studs on wood framed buildings. Obviously, a bolt that is secured on the interior of a wall is far superior to a compression-type connection.

Note: Fire escapes can also be constructed entirely of wood (both the platform and stairs). These types of fire escapes are commonly found on the back or side of low-rise multistory buildings and do not have stairs to the roof or a moveable escape ladder at the lowest platform. Hazards of wooden fire escapes include the obvious combustibility, age/rot deterioration, and, historically, the convenience of use as a storage area. For this discussion, we concentrate on the metal fire escapes.

Several factors should be considered when evaluating fire escapes. They include type, ladders, age, and location.

Type. Although there are some slight differences (depending on the area of the country), there are three types of fire escapes that are commonly encountered:

- Party escapes (or balconies) do not have a stairway or connecting ladder between the balconies as they are for the most part a balcony that is common to two adjoining occupancies. The emergency exit from one occupancy is afforded through the adjacent occupancy via the balcony. Today, these types of balconies are not

made from metal but wood and/or lightweight concrete. As a result, strength is not a concern for the more modern concrete versions.

- Screened stairways are different from standard fire escapes in that they are firmly affixed to a structure with no movable escape or gooseneck ladder, and consist of a permanent metal stairway that is enclosed by a metal screen for protection and security of exiting occupants. They are typically installed on the top floors of public assembly buildings such as theaters, schools, and other similar structures. This type of fire escape is a safer and sturdier fire escape than the standard version.

- Standard fire escapes are the most common and are easily recognizable by their all-metal construction features that include a metal balcony at each floor of a building, a ladder between each balcony, a ladder that often goes to the roof from the top balcony, and a horizontally pivoting or vertically sliding ladder from the bottom balcony to the ground.

Ladders. There are four kinds of ladders that are used on fire escapes:

- The permanent or stationary ladder is normally associated with screened fire escapes and also travels between each balcony on standard fire escapes. Their age is a determining factor in their overall condition and stability. Newer installations should be adequately constructed.

- A gooseneck ladder travels between the uppermost balcony and the roof. If the ladder is of older construction, it can be of questionable stability.

- Both drop ladders and counterbalanced stairs provide access from the bottom balcony to the ground. The drop ladder (also known as a guillotine ladder) is made from steel or iron, is positioned on one of the sides of the lowest balcony, and travels downward when a hook is disengaged. Counterbalanced stairs (fig. 9–35) are not as common as the drop ladder as they are more expensive and complicated to operate.

The stairs are held in the horizontal position by a counterweight, and when a person walks out on the stairs and exceeds the weight of the counterweight, the stairs will gradually travel downward. These stairs are heavy and can be dangerous, particularly the older ones.

Fig. 9–35. Counterbalanced stairs are activated by a person walking toward the end of the stairs.

Age. As the fire escapes age and deteriorate they can become subject to failure, particularly fire escapes that are on the East coast and in other areas that are subjected to weather extremes. Additionally, fire escapes that are installed on unreinforced masonry construction can present a notable hazard due to substandard construction and crumbling mortar.

Location. Depending on a particular area, the location/position of fire escapes on a building can vary, so it is important to know what can be expected. Some general guidelines follow:

- A fire escape on a building such as a multistory residence that would not normally have one is an indicator the building has been modified into a multifamily residential structure.

- Balconies that are not interconnected by a ladder likely serve two units that are divided by some type of separation wall, and if the balcony serves more than one window, it likely serves more than one room and probably more than one unit.

- If there is a fire escape without a gooseneck on the front of a building, it is likely there is a fire

escape on the back of the building with a gooseneck ladder.

- If there is a fire escape with a gooseneck on the front of a building, it is likely there is no fire escape on the rear of the building.

- A fire escape in the middle of the front of a multistory building (such as a residential) indicates a center hallway type of building.

Firefighters have learned many lessons from incidents involving buildings with fire escapes. Several of these lessons are listed here:

- Are occupants visible on a fire escape? If so, these people need to be rescued and there is a high probability there are more people inside the building.

- In some cases, the higher priority may be trapped people inside a building as opposed to the people on the exterior fire escape.

- It is normally advisable to lower a fire escape ladder to keep it from inadvertently lowering and causing an injury to civilians and/or firefighters during incidents.

- Remember that objects that have been placed on fire escapes (flower pots, etc.) can fall during an incident and injure personnel below.

- In most cases, it is advisable to ladder the lower balcony with fire department ladders rather than use fire escape drop ladders.

- Fire escapes on unreinforced masonry buildings can be considered substandard as compared to their newer versions on post-1935 masonry construction.

- Fire escapes that show signs of neglect or severe age deterioration can collapse from the weight of escaping occupants. Likewise, these fire escapes should not be used for fire suppression access.

- Wood fire escapes (which can also be categorized as secondary access/egress routes) are normally found on the back side of buildings (in some cases on the sides) and must be evaluated as to their age, storage considerations, and their ability to burn.

Quick summary

- Fire escapes are common for multistory residential and commercial occupancies that were built between the mid-1800s and approximately 1950.

- Factors that should be evaluated by firefighters include the fire escape type, ladder, age, and location.

- Types include the party escape (balcony), screened stair, and standard escape.

- Ladder arrangements can include the standard or permanent type, gooseneck, drop style, and counter-balanced stair.

- Fire escapes on unreinforced masonry buildings can be considered sub-standard as compared to their newer versions on post-1935 masonry buildings.

- The presence of a gooseneck roof access ladder on the front fire escape of a building can indicate that there is no rear fire escape.

- Fleeing occupants on a fire escape is a visual clue that indicates a high priority for rescue of victims still within the building.

Razor wire

Razor wire was developed to provide additional security to property and structures, and in some cases this also applies to the perimeter of roofs. Razor wire can be constructed of steel, aluminum, or aluminum strengthened with high-carbon steel. For the most part, razor wire is inexpensive, comes in a variety of styles, and is easily installed. It may or may not be supported or entwined with barbed wire. Usually it is under varying degrees of tension and is extremely effective in accomplishing its intended purpose. Regardless of the style encountered, razor wire can present a challenge to personnel considering roof access and egress, and poses a serious hazard of severe lacerations and entanglement.

Hands-on operations with razor wire should be employed only when necessary or as a last resort. Razor wire made from aluminum is easily cut, and can then be separated (with great caution) if absolutely necessary. Remember that razor wire may be under tension and can spring backward when cut. If razor wire is entwined with barbed wire, the barbed wire should also be cut, enabling the removal of the entire assembly. Razor wire reinforced with steel wire must be cut with bolt cutters. Note that this type of wire can also be under significant tension and can result in the wire springing back and unraveling when cut. Cutting this wire is *not* recommended.

Cellular antennas and other communications equipment

An emerging hazard that can confront fireground personnel is communications equipment on rooftops, specifically cellular antennas that are attached to the roofs of commercial buildings (although they can be found other types of buildings as well). Although standalone cellular towers are commonplace and present a minimal hazard, firefighters may not be familiar with the hazards associated with rooftop cellular antennas. A cellular base station and all of its components provides a circuitous connection between cell phones and the traditional telephones in residential and commercial occupancies by using radio frequency (RF) transmission between the cellular antennas on rooftops and standalone towers. Because there are more components to this system than just the visible antennas, let's look at a basic system and the types of hazards it can present to fireground personnel.

Base station. A base station consists of the receiving and transmitting antennas, the room that contains all of the necessary system electronics, coaxial cable for connecting the equipment room to the antennas, and a power supply with a backup such as an uninterruptible power supply (UPS).

Antennas. Rooftop antennas, commonly referred to as flat-panel or sector antennas, are approximately 1 ft wide by 4 ft high and can be found in single and multiple configurations (fig. 9–36). Customarily, one is used for transmitting and others are used for receiving, but a single antenna can be used for both. A cellular repeater is just a cellular antenna that accepts and strengthens a signal to relay to another antenna. Rooftop antennas are either mounted on brackets/supports that are attached to the side of a building, or are secured by a ballast and can be moved. This allows the antenna to be adjusted for a better location or moved for roof repair. Because the look of cellular antennas is not appreciated by some municipalities, they can be disguised or camouflaged to be more visually appealing. Examples are standalone towers that resemble trees, and plastic/fiberglass panels that look like the exterior of a building but enclose the antennas. These panels are often referred to as *roofmask* panels.

Fig. 9–36. Cellular antennas are commonly found on roofs of buildings and can be a significant danger to personnel on a roof. (Photo by Mark McLees.)

Equipment hut/room. The equipment hut (or equipment room) can be located in various locations within a building or on a roof in an appropriate enclosure. Equipment rooms contain some of the key elements of the overall system:

- It has the hard-wired power for the electronic equipment that consists of 240 V.

- An uninterruptable power supply (UPS) system for backup of the hard-wired system is located

in the equipment room. This backup system can consist of lead-acid batteries or diesel-, natural gas-, or fuel cell-powered generators.

- It houses a rectifier (inverter) that converts AC voltage to DC voltage for system operation.

- A fire suppression system may be present in some equipment rooms and capable of discharging an nonconductive extinguishing agent such as a waterless fire system, halon, or other material.

- An air conditioning system that is capable of minimizing the heat within the equipment room is often needed.

- A connection between the cellular network and a link to the public service telephone network (link to traditional telephones) would be found in the equipment room.

Coaxial cable. Coaxial cable connects the antenna(s) to the equipment room and is similar but considerably larger than the coaxial cable that is used for computers, audio visual systems, and other household items. Although power runs through this cable, the power level is fairly low and normally less than 5 watts. The cable has a plastic outer jacket (polyethylene, etc.) that is flammable and will support flame along the jacket covering. The number of cables that will be found in an installation is determined by the number of antennas that are used. The cables are usually enclosed in trays that are designed for multiple cables or metal conduit, both of which offer security and protection from the elements.

Hazards. The hazards to fireground personnel are summarized by the following primary considerations:

- The electrical hazard by the system operating voltage of 240 V.

- Depending on the type of backup power supply used, firefighters may be exposed to sulfuric acid and hydrogen gas, diesel fuel, cylinders of natural gas, and/or pure hydrogen for fuel cells.

- The plastic outer covering on the coaxial cables is flammable and will give off noxious gases if exposed to fire.

- Being exposed to radio frequency waves, either from visible or camouflaged antennas, is a significant danger. Exposure to radio frequency (RF) waves can result in a *thermal effect*, which is the inability of the human body to dissipate excessive heat generated by the RF waves (also known as RF burns). Personnel should maintain a minimum of 10 ft in front of antennas, 3 ft from the sides and back of antennas, and never touch an antenna.

Other communications equipment. Other communications equipment can be found on rooftops (and attic spaces) that have similar hazards as cellular features. Namely, firefighters may find other antennas or satellite dishes that can result in a significant dead load on a roof. Additionally, depending on the type of antenna, the emitted transmissions can also present a hazard to roof personnel who may stand in front of or too close to one of these antennas. In most cases, the size and type of these antennas can be identified from the ground (fig. 9–37). However, remember that nighttime conditions and smoke can obscure these installations, and personnel going to a roof may inadvertently place themselves in unnecessary danger. These devices are likely mounted as a retrofit and can include ballast blocks, guy wires, cables, and other trip hazards.

Fig. 9–37. Roof mounted antennas are becoming commonplace and can often be identified from the ground. Some can add a sizeable dead load to a roof.

Signs

Advertising signs that were often mounted on the roofs of older buildings (refer back to fig. 8–37) are a less common consideration on modern buildings (due in part to lightweight construction). Although some signs are quite large and capable of imposing a significant load on a roof, most of these roofs were engineered for the additional stress in combination with conventional construction. However, similar to HVAC units, supporting structural members will collapse faster when exposed to fire. In all cases, roof personnel must report any abnormal roof conditions that could negatively affect fireground operations. Remember, the collapse of a roof is a danger not only to roof personnel, but equally to interior personnel.

Signage on the exterior of buildings comes in various sizes and styles and can also present a hazard at fires if it is dislodged and falls. Signs are often attached to facades, adding dead load and electrical concerns. Signs that are oriented perpendicular to the building face are often strengthened against wind loads by using a diagonal tension wire that attaches the outermost portion of the sign to an anchor point above the sign (wall, parapet, or roof edge). In many ways, the perpendicular sign arrangement is a cantilevered beam and places an eccentric load on the wall. As unreinforced masonry walls and parapets are not designed for eccentric loads, a fire can cause the sign—and thus the wall—to collapse early.

Clandestine drug labs

A clandestine drug lab is any laboratory that is used for the manufacture of illegal drugs/substances. These labs can be found in numerous areas, including residential occupancies, and can present a significant hazard to first responding personnel. Unfortunately, these labs are designed to be hidden from the exterior but can present a multitude of significant hazards within the *interior* of the building. Although it would be difficult to cover all of the issues that are associated with this particular hazard, the following is a brief summary of considerations:

- From the exterior, common indicators are chemical odors, security bars on doors/windows, fans in unusual locations, propane tanks with unusual valves/attachments, odd looking pipes/ducts coming from windows and/or walls, and comments from neighbors about questionable conditions. Remember, when a structure is suspected of being an illegal lab, it may also be booby-trapped, which should place a different emphasis on entering the building.

- Once inside a building, some common indicators are propane tanks, containers of muratic or sulfuric-hydrochloric acid, bottles or jars with rubber tubing attached, multiple containers of lye, abnormal amounts of common household products (aluminum foil, paint thinner, lithium camera batteries, mineral spirits, coffee filters), and other materials.

These buildings/conditions can present numerous hazards such as a flammable/explosive environment, toxic chemicals, electrical hazards, incomplete chemical reactions, pressurized LPG containers, water reactive materials, and so on. Obviously, a full set of personal protective equipment (PPE) with self-contained breathing apparatus (SCBA) is mandatory, and defensive operations may be more practical and safe than offensive operations. If unsuspecting firefighters have engaged in interior operations prior to the lab discovery, full decontamination efforts are warranted and the incident becomes a technician-level hazmat one. With drug labs, attention to detail can save your life!

Pack rat conditions

Building occupants who exhibit hoarding behaviors can easily fill spaces with tons of added fire load. Common titles given to buildings that are packed with abundant fire load are pack rat, hoarder, and Collyer's mansion. Any structure that is encountered with pack rat or hoarder type

conditions can present a significant dilemma for fireground personnel as the interior of a building will have an inordinate fire load, and access throughout the building can be severely restricted. In some cases, this abnormal condition can be recognized by exterior conditions, as it is not uncommon for the exterior to be used for additional storage. In most cases, however, this condition will not be identified until personnel enter a structure. When this is the case, the amount of storage will quickly dictate the additional difficulty of access, a possible search, and the additional fire load. Additional hazards can include the following:

- Extended burn times
- Blocked windows and doors (no secondary escape path)
- Elevated collapse risks
- Firefighter entanglement
- Hidden fire or deep smoldering pockets
- Difficult hose advancement and high potential for kinks or rupture

In all cases, the IC must be quickly notified of a pack rat condition if one is encountered. As a final note on this subject, if a pack rat condition is exposed to fire, an intensive overhaul should be anticipated.

- If a cellular equipment room is equipped with an extinguishing system, remember that halon can result in toxic gases and lower oxygen levels, while the waterless system removes the heat, not the oxygen.
- Remember to follow cellular coaxial cables to check for extension, particularly if the cables have been routed through the interior of a building to the roof (unsealed openings).
- Advertising signs and communication antenna/discs will add to the dead load on a roof, and there may also be the presence of electrical utilities that are associated with these hazards.
- Clandestine drug labs often look like a typical residential occupancy but can contain booby-traps and a variety of extremely hazardous conditions/materials. Early recognition and appropriate caution is the key.
- Pack rat (hoarder) buildings should immediately raise a red flag that should underscore the viability of interior operations.
- Pack rat conditions have the potential to minimize interior access/egress, timely searches, and effective operations.

Quick summary

- Razor wire can be extremely effective at restricting access to a roof. Handling razor wire is normally not recommended.
- The primary hazard from cellular base stations is from personnel who are too close to the antenna, particularly roof personnel and roof-mounted antennas. If they are present, the IC should be immediately notified.
- An air conditioning system in an unlikely location can assist in the early identification of a cellular equipment room.

The Art of Reading Buildings

CHAPTER REVIEW EXERCISE

Using the following photographs, examine the visible building features and then list the hazards that they can present to firefighters.

Photo exercise 1

Feature/hazard

1.

2.

3.

4.

Photo exercise 2

Feature/hazard

1.

2.

3.

4.

Photo exercise 3

Feature/hazard

1.

2.

3.

4.

Photo exercise 4

Feature/hazard

1.

2.

3.

4.

The Art of Reading Buildings

Photo exercise 5

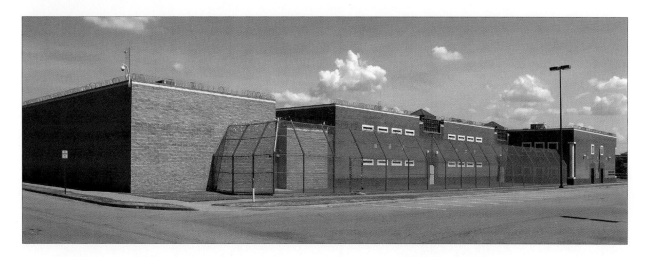

Feature/hazard

1.

2.

3.

4.

Chapter 9   BUILDING FEATURES AND CONCERNS

Photo exercise 6

Feature/hazard

1.

2.

3.

4.

RESOURCES FOR FURTHER STUDY

- A presentation on cell towers is available at http://faculty.sunydutchess.edu/walsh/powerpoint/misc.%20stuff/misc.htm

- Avillo, Anthony, "Fireground Strategies: Fighting Shaft Fires," *Fire Engineering*, December 2007.

- Ching, Francis D. K., *A Visual Dictionary of Architecture*, New York: Van Nostrand Reinhold, 1995.

- Fenion, Wesley, *10 Technologies Used in Green Construction*, HowStuffWorks.com, a Discovery Company, March 2011.

- Flynn, John, "Operating Safely on Fire Escapes," *Fire Engineering*, March 2009.

- Gustin, Bill, "Avoiding the Hazards of Overhangs," *Fire Engineering*, April 2010.

- Mittendorf, John, *Truck Company Operations*, 2nd ed., Tulsa, OK: PennWell Corp., 2011.

- Murphy, Jack J. and Jim Tidwell, "Green Building Challenges for the Fire Service," *Fire Engineering*, January 2011.

- Spadafora, Ronald R., "Green Building Construction and Daylighting: A Chief Officer's Perspective," *Fire Engineering*, October 2010.

- Viscuso, Joseph, "Safe Operations Near Roof Cellular Base Stations," *Fire Engineering*, March 2008.

- Yago, Jeffrey R., "The Power to Roam," *Home Power*, June & July 2012.

READING BUILDINGS: HOW TO SIZE UP A BUILDING

10

OBJECTIVES

- Describe the two kinds of building size-ups for first-due decision makers.
- List the three building size-up considerations that should be reported as part of an initial arrival at structure fires.
- Define the eight primary building groupings that can be used to communicate a first impression for an initial radio report.
- Describe the five basic construction considerations that should be analyzed as part of a 180/360° size-up.
- Define the four categories that can be used to describe building status.
- Define the fireground clock and its relevance during building fires.
- List the six tactical challenges that must be considered for buildings.
- List five hazardous game changers that can impact tactical challenges.
- Name the four analytical inputs for predicting the collapse of a burning building.

TIME TO PUT IT ALL TOGETHER

The preceding nine chapters have set the stage for this chapter. Here, we want to show how your investment in understanding building construction equates to rapid street reads and tactical decision-making. Before we make that transition, let's review the journey.

The ability to truly *read* a building starts with an understanding of basic building engineering concepts like loads, imposition of loads, and the resulting forces of imposed loads. From there, we strive to understand the characteristics of various building materials (wood, steel concrete, etc.) and how those materials are impacted by shape, mass, and the effects of heat and fire. Using the aforementioned

essential concepts, we can explore the various methods used to construct a building. All buildings are built using a structural element hierarchy:

- Foundations and footers transfer imposed loads to earth.
- Columns transfer loads to the foundation through compression.
- Beams transfer roof and floor loads to columns through a combination of compression, tension, and shear.
- Connections help attach the structural elements to one another.

The history of humankind is one of constant change and, over time, humans have developed many methods to make the basic building hierarchy perform. That history has been influenced by myriad factors like occupancy needs, material engineering, wars, ravages of fire and earthquakes, and the resulting constant change of building codes. These influences present the firefighter a significant challenge when trying to classify a given building. This book argues that the five classic types of buildings (NFPA 220) is overly simple and can lead to traps. Alternatively, firefighters need to classify buildings using a formula of *era/use/type/size*.

The balance of our journey examined some specific building features (floors, roofs, etc.) and the associated issues that firefighters have experienced or can expect to experience with them. We hope the journey has helped to improve your understanding of building construction and that you have been able to use some of the information to go out and examine the buildings in your response area. If you have, you may have already improved your ability to read a building. The balance of this chapter (and book) can help you refine your ability to street read. To help achieve that refinement, we present the following:

- A mental approach to building size-up
- The six tactical challenges that need to be considered with most buildings
- Perspectives for triaging a building and predicting collapse

BUILDING SIZE-UP

The term *size-up* has been used extensively in the fire service to describe the constant mental evaluation of fireground factors. We evaluate those factors and make judgments about how each can be used to predict upcoming events. Obviously, those judgments help us make better strategic choices that guide our tactics. Increasingly, we are trying to make tactical assignments that achieve a desired outcome while making fireground safety a primary consideration. If fireground safety is a primary consideration, then conducting a size-up that enhances an ability to mitigate an incident in a safe and timely manner should be a cornerstone of efficient and safe fireground operations. Fire service teachings emphasize that any size-up actually begins before an incident is dispatched (in prefire planning, training programs, etc.). Retired Chief Alan Brunacini has called pre-incident size-up efforts *front-loading*. Throughout this book, we encourage you to go visit buildings—an essential front-loaded size-up. Upon arrival at a building fire incident, the first-due decision maker actually performs two kinds of size-ups:

- The initial size-up made when first approaching the building—we'll call it the 180
- The 360° expanded size-up

The purpose of the initial size-up is to gain a starting impression for the sake of an on-scene radio report. As the first-arriving officer approaches the building, he or she can usually pick up one or two sides of the incident building before even leaving the vehicle (hence, the 180). Some standard operating guidelines (SOGs) suggest that the first-due apparatus actually tries to get a three-sided view of the fire building by pulling past it. In either case, an initial size-up of observable factors (the building and the smoke) can be made, which can lead to an initial judgment (regarding risk taking) and the on-scene radio report. Although you may not see all of the visible features of a particular building, a "first or initial" size-up can often yield important clues that can begin to formulate

your impression/familiarity with the building and provide indicators that dictate additional time that should be devoted to a more comprehensive examination—such the 360° expanded size-up. It is important to acknowledge that a structure fire size-up can be a challenging dilemma for a first-arriving officer who can be easily pressurized into spending a minimal amount of time reading a building as opposed to first addressing the needs of an escalating incident—not to mention the supervision of an anxious crew who want to immediately "put the wet stuff on the red stuff." Therefore, it is imperative that every first-arriving officer take the necessary time to read a building in order to plan for the effective and safe mitigation of the incident.

After arrival, the first-due decision maker should expand and fine tune the 180 size-up by conducting a 360° evaluation.[1] We acknowledge that a complete 360° look at the building may not be possible or practical due to the building footprint, attached buildings, fences, or the time constraint of an immediate rescue need. However, we buy into the recommendation that a 360° is the rule unless an exception is presented. When conducting the 360° building size-up, undress the building in your mind. Look past the exterior and visualize what is inside the building (its strengths and hazards) as *what you initially see may not be what you get!* Remember, when first arriving on scene, every building tells a story (fig. 10–1). Most often, the short amount of time required to quickly analyze a building that is under demolition by fire normally pays huge dividends in timely and safe operations during suppression operations.

Specific to the building, size-ups should include the following:

- The general use and size classification of the building (initial size-up)
- Basic construction considerations like the prevailing methods, era considerations, exterior features, and interior features (360° size-up)
- Building status (360° size-up)
- What is burning (180/360° size-up)
- The fireground clock (360° size-up)

Let's dig into each of these size-up needs and give some specific guidance to help you improve your rapid read.

Fig. 10–1. When conducting a structural size-up, undress the building in your mind.

General use and size classification (initial size-up)

In previous chapters we make the argument that classifying a building by a single dimension is full of traps. When preplanning buildings (front-loading), it is best to classify buildings by using the *era/use/type/size* method. The era/use/type/size classification system helps you develop better preplans and helps you teach building construction to future generations of firefighters. Even with a significant investment in building front-loading, there are likely to be responses to buildings that you have never had the opportunity to walk through and study. Given that, the first building size-up clue you'll receive when dispatched to a reported building fire is the address. That address will likely correspond with the general use grouping of the kinds of buildings you find on that street or in that area. This can provide a jump-start to your initial size-up.

The Art of Reading Buildings

As towns and cities grew, planners found that it was important to group buildings based on community desires, traffic flow, utility (infrastructure) needs, and convenience. The planners developed building use envelopes and zoned those areas accordingly. Responding firefighters can use the building address as a way to start their size-up efforts by grouping the building according to the use zoning found in that area. For example, you're dispatched to a report of a building fire on 1234 E. Main Street. You consult the maps (or your memory) and know that the address falls in a single-family residential subdivision that was built in the 1950s with one- and two-story, legacy platform, wood frame buildings that are 800 to 2,000 sq ft. We can make a soft or tentative size-up use grouping based on that address. We say *soft* because obviously you have to confirm that grouping as you roll up to the address. It is realized that not all buildings can be grouped based on address alone, but most buildings you respond to can be easily grouped based on your first visual impression when you arrive.

On arrival, firefighters can look at a building and tell what it is likely being used for, such as a restaurant, home, retail store, church, or some other structure. While not absolute, the first impression for building use provides a good and mostly reliable starting place for size-up (and initial radio report).

In chapter 5, we list the official (building code) occupancy categories when explaining building use. Here, we create some broad groupings that don't necessarily align with the building code—they are a bit more firefighter related. While broad, remember they are just a starting place for a size-up. We further subdivide the groupings in the rapid street-read guides found in the last section of this book. These broad use groupings are

- Single-family dwelling
- Multifamily dwelling
- Main Street commercial
- Manufacturing/warehouse
- Office/hotel
- Institutional
- Public assembly
- Miscellaneous buildings

Single-family dwelling (SFD). Single-family residential buildings and single-family dwellings are the same thing—dwelling is just a quicker way of saying residential building. SFDs are those that were built for the purpose of providing living accommodations for one family. One can expect rooms for sleeping, bathing, cooking/eating, and relaxing as well as storage areas and perhaps an attached or detached garage. SFD rooms are typically free-flowing, meaning they have open stairways and hallways to connect the various rooms and simple doors (non-fire-rated) to add privacy from room to room. While a building may have been originally built as an SFD, occupants may have subdivided the home to provide separate living accommodations for others. Some visual clues that can help you recognize that an SFD has been altered are multiple mailboxes/doorbells, divided porches or patios, add-on exterior stairs, structural add-on areas (see fig. 5–15), and separate entry pathways.

Multifamily dwelling (MFD). Multifamily dwellings are those built to provide distinct separations between the accommodations of more than one family or tenant. While terminology varies, the MFD includes multiple living *units*. Each unit has its own set of sleeping, bathing, living, and cooking rooms for each tenant. MFDs typically share one or more building features such as a common yard, parking area, utility infrastructure, elevator(s), primary access stairway, or recreational amenity. Apartments, condominiums, tenements, townhomes, and row homes (row frames) are all examples of MFDs.

Main Street commercial. This broad grouping is named for "Main Street, USA"—the part of most towns and cities where residents go to purchase the goods they need. The Main Street concept grew from the early settlers who designated a "commons," which was a central area where everyone brought their harvest (crops and animals)

to share, trade, or sell (like a farmer's market). As towns grew, they did so using the commons as a central town square—an open, park-like gathering area. Permanent retail buildings were built around the square as products and services became commercialized. The delivery of bulk goods by railroad further advanced the Main Street concept.

Most retail buildings were built right near a town's rail stop. All manner of commerce took place there; dry goods dealers and grocers, clothiers, barbers, blacksmiths, and bankers built their buildings up and down Main Street. Fast forward to suburban sprawl. We now have strip malls, megamalls, and big-box store shopping districts that harken to the Main Street concept. In essence, the grouping of Main Street commercial includes those buildings that sell goods and services to the general public. For an initial on-scene radio report, you're probably best off by saying the actual business type (strip mall, retail store, small restaurant, etc.) rather than "Main Street commercial."

Manufacturing/warehouse. In early times, warehousing shared Main Street (right next to the railroad tracks), and small-scale manufacturing could be found mixed in with the retail stores where people shopped (retail in the front, manufacturing in the back). As populations increased, the manufacturing and warehousing sectors increased their building sizes and became industrialized. They moved off Main Street and got their own spur rail tracks and delivery roads. As grouped here, manufacturing and warehouse buildings are those that are typically large, house unique processes or operations, and are generally not open to the public. Buildings that deal with the processing and storage of hazardous materials are also grouped here.

Office/hotel. By and large, hotels and office buildings are very similar in nature: They have individual rooms or areas used for sleep or work and communal areas for reception, meetings, provisions, and light recreation. Most share common parking areas, utilities, and outdoor areas. While building codes vary, they also share very similar requirements.

Institutional. Grouped here are those buildings that are used for civic or societal needs. Schools, jails, courts, hospitals, and other human service facilities are included. For brief, initial on-scene radio reports, the word "institutional" may throw off other responders. It's best to just say school, hospital, etc.

Public assemblies. Buildings used for the purpose of human gathering present the risk of large loss of life when fire strikes. Some of the most stringent (fire resistive) building codes have been developed for public assembly occupancies to help prevent the reoccurrence of large loss-of-life tragedies that have scarred our history. Some public assembly buildings such as stadiums, theatres, and churches are quite large and can include vast parking areas, morphed utility systems, and a web of service and supply tunnels. However, other public assembly buildings can be small to medium in size, such as the restaurants and churches that are common to most areas of this country.

Miscellaneous buildings. In virtually every town or city, there are buildings that don't fit any of the aforementioned groupings. A classic example is the telephone central exchange or switch building (fig. 10–2). These block and brick buildings are typically windowless, have few entry/exit doors, and minimal signage. Grain silos, kit buildings, utility substations, and toll plazas are all examples that can be grouped as miscellaneous.

Fig. 10–2. Some buildings, such as this telephone central switching center, do not fall within common building groupings.

Once you have grouped the building on arrival, you form an impression of the group subset (see the index at the beginning of chapter 11) based on visual clues about the building. Likewise as you arrive, you can form an impression on the approximate size of the building. In chapter 5 we offer some general guidelines and the associated hazards that building size can present.

At this point, the arriving fire officer should have enough information to give an initial on-scene radio report. Most best practice or standard arrival reports include a *brief* description of the building as well as visible smoke and/or fire conditions. The size and use grouping can satisfy the need to be brief for the building description (e.g., "Dispatch from Engine 5, on scene of a large two-story, single-family dwelling. Smoke showing from Side B, second floor . . .").

With the initial radio report delivered, the first-due officer needs to fine tune the building size-up as part of a 360° or three-side walk around the building. We realize that it may not be physically possible to get a complete 360° around a building—especially in the urban environment. Regardless, the first-due officer should invest some time to further evaluate the building (and smoke/fire conditions) where possible. The good fire officer will visually scan the building exterior with attention to three things: building construction considerations (covered next), smoke and fire conditions (including rate of change), and the status of the building (to help determine if people are likely in the building).

At night, the fire officer should use a good flood-style hand light or lantern rather than a narrow pencil-beam style flashlight when conducting the 360°. The reason for this is twofold: the floodlight will allow you to see more features with each sweep, and the broad light beam will refract on smoke and allow you to read it faster. A pencil or focused light beam minimizes what you can see with each sweep.

Quick summary

- The initial building size-up can be jump-started with the dispatched address of the incident. Most communities have building zones that indicate the use classifications of buildings in that area.

- There are two building size-ups: the initial 180° size-up for the on-scene radio report and the 360° focused size-up.

- The initial size-up includes a brief judgment on the building size, apparent use of the building, and obvious fire/smoke observations.

- The first impression on the use of the building is not necessarily defined by code, but rather a common observation such as single-family dwelling, hotel, retail store, and so on.

- After arrival, the first-due decision maker should follow up the initial 180° size-up with an expanded 360° evaluation.

Basic construction considerations (360° size-up)

The fire officer who has invested front-loading time will easily pick up visual clues that will help in reading a building in further detail. Throughout this book, we give you many visual clues that link the clue to a specific fire spread or collapse issue. As you walk around a building, these clues should register and help with tactical understanding. At minimum, the 360° building size-up should help to further define the general construction, era, and interior/exterior features.

Prevailing construction method. In most cases, the initial size-up (while rolling up in the apparatus) also provides some visual clues to help classify the methods and materials used for construction (like a Type III block building with flat roof). The 360° allows you an opportunity to verify that impression. Additionally, the 360° offers a chance to amend the first impression. What appeared to be a Type

III flat roof building from Side A can be amended when you see that Side C is really wood frame—or an alteration of a different construction method. We again mention that the hybrid and alternative methods of construction are an increasing challenge for firefighters. The 360° building size-up helps you spot clues that you are dealing with a hybrid.

Era considerations. We hope the era consideration information in chapter 5 convinced you that we have to tailor our tactics based on the time period that a building was constructed (and/or altered). Building integrity and our safety depend on it! To review, those eras are:

- Pre-WWI historic
- Pre-WWII industrial
- Post-WWII legacy
- Engineered lightweight (1990 and later)

Recall that the building era reflects the methods, materials, codes, and engineering principals that were prevalent at that time. Fire spread and collapse issues differ in each era and thus, our tactical solutions must take those into account.

It is recommended that to assist you in determining the "era" of building construction that you familiarize yourself with the specific features that are prevalent in your jurisdiction. Consider the following examples, which you should be able to expand upon:

- Modern appearing townhouses were constructed during the lightweight engineered era.
- Dwellings with centrally located masonry fireplaces were constructed during the historic and industrial eras.
- Dwellings with attached or adjoining garages with small doorways (originally built for Model A automobiles) were constructed prior to the industrial era.
- Commercial buildings with arched, bridge truss, or sawtooth roofs were constructed prior to the post-WWII legacy era.
- Unreinforced masonry buildings were constructed prior to pre-WWII industrial.
- Buildings with exposed 2 × 4 rafter tails indicate lightweight truss construction from the engineered lightweight era.

Exterior features. Exterior features provide some of the more reliable clues for traps and challenges that firefighters may face for a given building. There are the big-ticket clues like spreaders and anchors, unreinforced masonry construction, parapets, bracing, signage, facades, add-on fire escapes, pilasters, additions, and so on. Exterior features such as facades, false mansards, cantilevers, and cornices are notorious for falling off a building being attacked by fire. Additionally, the exterior features help paint the picture for access and egress options—not just doors and windows, but also for the type of roof and size/amount of utilities. Specific to utilities, the 360° size-up can provide a heads-up on utility challenges or other hazards like cell phone repeaters, solar panels, and the like. Lastly, the exterior features can often help you paint a picture of expected interior features.

Interior features. Many of the exterior features that you find on a 360° size-up help you visualize interior features. Multiple mailboxes and electrical meters can help you recognize that the interior of a building has been subdivided into separate units or occupancy spaces. Additionally, remember that visible construction add-ons to a building can dramatically increase the size of a building as well as alter the original floor plan.

The arrangement of doors and windows can help you visualize the location of bedrooms and living areas in residences, or office space and warehousing areas in commercial structures. Large overhead rolling doors, long windows, and pilasters can suggest that the interior has wide-open spans (increased risk of collapse). Outside utility feeds that seem disproportional to the building can indicate that industrialized equipment will be found inside. As an example, in figure 9–30 the original three-wire service has been replaced by a

The Art of Reading Buildings

four-wire quadruplex service, indicating there are now substantial electrical needs within the interior of this building.

While conducting the 360° building size-up, you will likely make tactical judgments about the hazards that are present and what needs to be accounted for and communicated. Making continual tactical judgments is mostly a good thing, but perhaps a caution is warranted here: Try to keep an open mind and just absorb what the building is telling you during your 360°. The prevailing method, era considerations, exterior features, and interior features are painting the whole picture for you. Once you have completed your 360°, you must act and proceed with a course of action that takes into account what you've discovered. Before we talk about those actions, let's discuss the building status.

Building status (360° size-up)

As applied to size-up, *building status* refers to the likelihood that people are occupying a given structure. By evaluating the status of a building, the first-due officer can begin to formulate a foundation for determining an initial risk analysis for the priority of suppression or search operations. Some fire officers approach each and every building as if it is currently occupied. While noble, it must also be said that many firefighters have been killed in buildings where there were no occupants. In the modern world of managed risk, fire officers should make a judgment regarding the occupancy status of buildings on fire based on visual clues. Generally speaking, a given building's status can be classified in one of four ways: occupied, unoccupied, vacant, or abandoned.

Occupied. Occupied buildings are defined as buildings that are occupied or have a *high probability* of being occupied during an incident. Visually, the most obvious clue that a building is occupied is the presence of people exiting, occupants outside the building, credible assurance from neighbors (be careful with this one), and those waiting at windows to be rescued! When obvious clues are not readily apparent, the fire officer can use clues that suggest the building is lived in or is open for business. Open doors or windows, interior lights illuminated, running equipment or HVAC systems, and functional cars parked in the lot/driveway (fig. 10–3) are some examples of clues that a building is likely occupied.

Fig. 10–3. A parked car next to this building can be a credible clue it is occupied.

Using the time of day to make a judgment regarding the occupied status of a building has been a staple of size-up training. This teaching has some merit for commercial buildings that have posted business hours, but is probably dubious for residential structures. As a rule, all residential buildings that appear lived-in should be considered occupied until reliable information proves otherwise.

When a building has been judged occupied, the fire officer must look at smoke and fire conditions and make a further judgment about survivable spaces. The condition of windows can help in making the survivable space judgment (more on this later in this chapter).

Unoccupied. Unoccupied buildings are buildings that are normally occupied but the occupants are *likely to not be in the structure currently*. Most often, commercial properties that have defined business hours are those that will be unoccupied during off-hours or holidays. To help verify that status, the fire officer should couple the time of day (or holiday) with an empty parking lot and locked doors. As an example, the building in figure 10–4

is a commercial retail building, there is a small closed sign in the front door, and there are no cars parked near the building. The chance of occupants inside this building is remote.

Fig. 10–4. Visual clues indicate this commercial building is likely unoccupied.

Vacant. The primary difference between a vacant building and an abandoned building is vacant buildings are likely to still be in an acceptable condition in terms of structural integrity and marketability. A vacant building is likely to be secured (doors/windows) and may have "For Sale" or "For Lease" signs displayed on the front of the building (fig. 10–5). Other clues that a building might be vacant include the absence of contents and window treatments, removed signage, and an empty parking lot.

Fig. 10–5. Visual clues suggest this building is vacant and secure.

Abandoned. Abandoned buildings are those that have outlived their usefulness, fallen into disrepair, and show signs that the owner has basically given up on the building. Boarded-up windows and doors (fig. 10–6), vegetation overgrowth, vermin infestation, and obvious structural degradation are all visual clues that may lead you to classify a building as abandoned.

Fig. 10–6. Abandoned buildings have typically outlived their usefulness and present numerous challenges for firefighters.

Abandoned buildings present numerous challenges for firefighters. These challenges include:

- Dangerous interior conditions including holes, collapsed structural elements, and exposed utilities
- High potential for criminal-related hazards, including arson, illegal lab production (hazmat waste), and gang activities
- Potential for unlawful occupants such as vagrants and squatters

Some fire departments have developed programs to address abandoned buildings that include marking or placarding them. A common marking system includes posting a large white "X" across a red background to denote buildings that are too dangerous (structurally) for interior firefighting operations. These placards are usually accompanied by a "No Trespassing" warning to others that the building will not be searched by firefighters. A red placard with a single white diagonal slash indicates to firefighters that the structure may be

sound but interior operations should be limited to a quick search for squatters by permission of the incident commander only.

The combination of basic building considerations (method, era, and exterior and interior features) and status (occupied, unoccupied, vacant, or abandoned) will help you form a strong impression of the challenges that the incident will present. The size-up, however, needs to consider two more factors: what is burning and the fireground clock.

What is burning?

Of all of the factors that should be included under the heading of important size-up considerations, the determination of what is burning should be of paramount importance, as every firefighter should be well aware that every second a fire burns it is weakening a structure and enhancing a collapse of the burning structure. Therefore, the first-due decision maker needs to make a judgment as to what is actually burning and whether the structural elements of the building are being attacked by fire or heat.

In previous sections of this book we have discussed the strengths and weaknesses of various types of construction such as metal, masonry, wood, and other types of building materials. Unfortunately, the building industry has not consistently used these materials in conventional configurations and has continued to develop alternative methods and materials (lightweight construction) to construct buildings that have proven to collapse—often catastrophically—in faster time frames than buildings constructed prior to the 1960s. Therefore, a modern size-up should consist of determining "what is burning" from the perspective of conventional or lightweight structural materials as follows:

- Conventional materials generally consist of reinforced masonry, large steel members, wood members of 2 × 6 in. or larger, and so on. Additionally, the connection points consist of substantial materials such as nails, metal plates/bolts, substantial welds, and other similar methodologies. The end result can be slower burn and collapse time frames (and often predictable results) as compared to lightweight construction.

- Lightweight materials and construction techniques include the use of lightweight steel, wood members of 2 × 4 in. or smaller, truss configurations, and connection points of staples, gusset plates, glue, and other similar methods.

The focus of assessing what is burning during a size-up is attempting to define the remaining relative strength of the structural members that are being weakened by fire and determining the amount of time that is left to safely conduct your intended operation. It is important to remember that in some cases it will be necessary to determine the type of construction (that is being weakened by fire) from the interior of the building. That is a primary reason why initial attack teams should also carry a pike pole (or other similar tool) to pull a ceiling when entering the building. Additionally, determining what is burning and the amount/extension of fire from the interior of a building is information that must be relayed to an exterior IC if present.

Supplementary considerations to remember when evaluating conventional vs. lightweight construction include the following baseline time frames:

- If fire is exposing lightweight structural members for more than 5 minutes, rethink interior and/or roof operations.

- If fire is exposing conventional structural members for more than 15–20 minutes, rethink interior and/or roof operations.

Remember, if your initial size-up indicated lightweight construction and it turns out to be conventional construction—that is in your favor. However, if your initial size-up indicated conventional construction and it turns out to be lightweight construction—that is not in your favor. These baseline time frames can help with the last building size-up input: the fireground clock.

Fireground clock

When the term *fireground clock* is mentioned, some of the initial questions that come to mind are: How advanced is the fire? How long did it take someone to become aware of the fire and report it? What was the dispatch time? What was the turn-out and response time? While these questions are certainly applicable to fireground operations, there is another fireground clock that is also applicable to fireground operations. That clock is defined as *once personnel arrive on scene, how long will it take to make a visible impact on the fire?* Notice this definition did not focus on the extinguishment of fire, but the time it will take personnel to perform the following tasks:

- Exit the apparatus, making sure the complete personal protective equipment (PPE) is ready.
- Conduct a 360° or more defined size-up.
- Deploy an evolution (water delivery system).
- Force entry into a building.
- Move the water delivery system to apply an effective fire stream on the fire.

This definition is significantly different from the first definition that focused on the time that elapsed prior to companies arriving on scene. Granted, some of the items on our fireground clock can be accomplished simultaneously (with adequate staffing), but remember that fires are not static but dynamic. That means when personnel arrive on scene, the fire will not wait until they put it out, but will continue to burn and weaken a building until the fire is extinguished! Therefore, until a fire is extinguished, it will continue to weaken a structure.

During suppression operations, there are three factors that are working against fireground personnel:

- Gravity wants to flatten the building.
- The fire is weakening the building as each second ticks.
- Water weighs 8.35 lb per gallon.

So the perception of fireground time once personnel arrive on scene is a combination of the time it will take to apply the first drop of water in addition to the time it will take for extinguishment! Obviously, the time it will take to make a visible impact on a fire will vary from incident to incident, but typically the fireground clock can give a general idea of how much time is available to accomplish an intended operation.

As an example, assume the company you are assigned to is the first company to stop in front of a recently constructed two-story wood frame single-family dwelling that is typical of some newer residential structures in your district. Additionally, your initial size-up indicates fire that is visible from a second floor window as well as heavy smoke and some fire visible from several attic vents (fig. 10–7). Because you are familiar with this type of structure, your initial building construction size-up indicates that the fire in the attic is consuming lightweight roof truss construction and the fire is not static but dynamic (it will keep burning until you put it out). Applying our fireground clock definition, you assume that it will take you approximately 4 minutes to get off your apparatus, get your PPE ready to go, perform a 360° evaluation, force the front door, and accomplish a basic stretch of an initial attack line into the structure.

Fig. 10–7. Upon arrival, it is observed that fire is visible from a second floor window as well as several attic vents and is also consuming lightweight truss roof construction.

At this point, three considerations should command your undivided attention: (1) the 4 minutes to start attack just spotted the fire an additional 4 minutes, (2) the 4 minutes given to this fire also means 4 minutes has elapsed without mitigating the fire (which will continue to burn and weaken the structure), and (3) the national average for the collapse of lightweight wood trusses when they are exposed to fire is around 5 to 7 minutes. Therefore, does this simple scenario give adequate time to enter the structure and extinguish the fire before collapse of the lightweight truss construction? Answer: probably not!

Additionally, remember that if fire is showing upon arrival to a structure fire, consideration should be given to how long the fire was burning before arrival (unknown) and add that to the fireground clock implementation time (known). The point of our example is simple: The combination of building size-up, our read on smoke and fire conditions, and consideration of the fireground clock has created a negative risk scenario for firefighters—collapse is likely. For decision making, we must then weigh the survivable spaces left in the building and opt to support rapid room searches (if appropriate) and then resort to a defensive nature.

Closely related to the fireground clock is the judgment on *rate of change*. The 360° building size-up provides an opportunity to gauge smoke changes. Simply applied, is the smoke getting better or worse, in seconds or in minutes? This judgment about the rate of change, combined with the fireground clock, can help the fire officer understand the time window that is available (or *not* available) for safe interior operations.

In summary, the building size-up factors or inputs are as follows:

1. Upon arrival, make a first impression of the building size and use group for your initial radio report.

2. Conduct a 360° size-up to refine basic building considerations (method, era, exterior/interior features, etc.).

3. Judge the building status (occupied, unoccupied, vacant, or abandoned).

4. Determine what is burning and whether conventional or lightweight structural elements are being attacked.

5. Factor in the fireground clock and rate of change.

Remember, size-up is a *continuous* mental evaluation. It never really ends until you wrap your incident and head back to the firehouse. Your size-up efforts set the stage to make better tactical decisions. In the next section we address the six tactical challenges that exist for buildings.

Quick summary

- The 360° building size-up includes a focused evaluation of building construction considerations, the building status, and the fireground clock.

- Building construction considerations include the prevailing methods used, era considerations, exterior features, and interior features.

- Judging the building status as occupied, unoccupied, vacant, or abandoned can help the decision maker with the risk analysis used to formulate tactical priorities.

- Determining what is burning entails the perspective of evaluating the strengths and weaknesses of conventional vs. lightweight construction and the amount of time left to safely accomplish an intended operation.

- The fireground clock includes the time it takes an arriving crew to prepare for engagement (PPE, size-up, deployment of hose and tools) and the time it takes to make a impact on the fire. The building is undergoing attack by fire as the fireground clock elapses.

THE SIX TACTICAL CHALLENGES FOR BUILDINGS

Firefighters responding to a reported structure fire need to think of the building as the *host*—the building is hosting the incident. Our job is to quickly tell what kind of host that building is going to be, and we need to design and implement tactical solutions to make the event conclude. Rare is the perfect host that provides an open, easy, and gracious event with no agenda. Like people, the building has certain tendencies, secrets, and agendas that present tactical challenges for the firefighters who are trying to bring a very unfortunate event to a safe ending. Some challenges are easy to overcome and some require significant effort. Regardless, most buildings share the classic six tactical challenges. In no particular order, the challenges are fire spread, collapse, forcible entry, ventilation, search, and specific hazards.

Fire spread

By nature, fire wants to spread out and up. The internal geometry of a building doesn't always let the fire do what it wants. In actuality, we know that three factors dictate fire spread within a building:

- The internal geometry of the building (ceiling height, floor plan, compartmentalization)
- The fire load (contents and the building itself) and its arrangement
- The heat (exhaust) and air (intake) flow paths

Building preplanning helps us understand how each of the three factors influence fire behavior. When we respond to fire in a building we've never seen before, we rely on our pre-incident study (like this book) and our ability to read the building and smoke to predict fire spread.

The tactical solution for preventing fire spread is generally tied to the act of putting water on the fire at a rate that exceeds that of energy release. While this is a good starting place, the smart fire officer looks past visible fire and applies the rule of six sides to predict the spread of fire and related priorities. Here is how it works. Assume you have a fire on the second floor of a three-story building as depicted in fig. 10–8:

- The fire on the second floor is comprised of *four* sides (two sides, front, and rear). The primary consideration would be the main body of fire, potential extension to the side, rear, and front walls, search and ventilation as necessary.
- The third floor (top) would be viewed as the *fifth* side. The primary consideration would be extension from the second floor, search and ventilation as necessary.
- The first floor (below) would be viewed as the *sixth* side. Although downward extension could be a possibility, salvage would likely be a higher priority (fire goes up and water goes down).

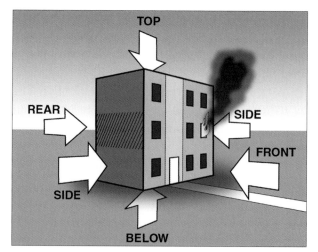

Fig. 10–8. A structure fire has six sides.

Drawing on acquired building construction knowledge to predict where the fire might have gone undetected (or is going) is the hallmark of a good fire spread tactician.

Collapse

Throughout this book, we try to include as many collapse issues as we can squeeze in. Most of these issues we've learned through history and experience. On scene, the first-arriving decision maker must make a judgment on the collapse potential of a building or parts of a building. Typically, fire officers use visual warning signs to make that judgment. These warnings include:

- Sagging floors and roofs
- Bowing or bulging walls
- Cracks
- Settling noises
- Doors out of plumb or jammed
- Water flowing out of the building that doesn't match water going in
- Fire impingement on trusses
- Amount of fire and time it has been burning
- Structural material that is being burned

While this list is important to learn, we feel most are *late* warning signs. There are dozens (if not hundreds) of investigative reports where firefighters have reported that the building "collapsed without warning." A better process to predict collapse is by using a mental one (as opposed to late visual signs). As Francis Brannigan said, "The warning is the brain, in your ability to understand how buildings are built and how they react in a fire."[2] Later in this chapter we offer a few suggestions on how to mentally process collapse judgments, but for now we'll add that judging collapse potential is an algorithm of building construction method, fire and heat exposure, and elapsed time.

Forcible entry

The art of forcible entry has been well documented in many fire service books. The point here is not to teach you how to force any given door or window—hopefully you've learned many techniques. Rather, we include forcible entry as a challenge that the exterior features portion of a 360° size-up will help paint. Using your forcible entry knowledge and size-up info, you can prioritize entry options that help minimize the impacts of the fireground clock. Using your understanding of building construction, you may find that making a door out of a window (or making a door where none exists) may be quicker than trying to tackle a door with serious security hardware. We also must add that any challenge to get *into* a building should also be balanced with a challenge to be able to get out of a building (*escape in a timely manner*) if a firefighter or victim removal situation arises. Therefore, forcible entry should be viewed as a two-part operation that consists of forcible entry and forcible exit! As a thought-provoking side note, we normally require a minimum of two ladders to a roof for personnel operating above ground. Why don't we require a minimum of two ways to exit a building?

Ventilation

The tactical challenge of ventilation is closely related to that of fire spread. Reading the building and reading smoke go hand in hand to help determine what the fire is doing and what it is about to do. As you read both, it is critical to gain an understanding of the exhaust and air flow paths. Inadequate exhaust flow paths sets the stage for explosive smoke ignition and rapid fire spread—especially with today's low-mass, synthetic fire loads. As a general rule, it's best to increase the exhaust flow path size near the fire. The building size-up and your understanding of construction methods (materials, roof types, and window options) will help you prioritize choices to increase the exhaust flow path.

Don't forget to consider the fireground clock and rate of change when making tactical ventilation decisions. For example, you may read the fire and smoke and determine that the best solution to control heat and smoke flow paths for advancing interior crews is that of a large offensive heat hole on the roof near the fire. The fireground clock

suggests that a rooftop heat hole will take 10 minutes to accomplish. The rate of change indicates that the fire and smoke is getting worse each minute (despite the fire attack). Logically, the fireground clock and rate of change has taken away this roof venting option.

Be aware that as this book is being written there are new studies being conducted by the Underwriters Laboratories (UL) and the National Institute of Standards and Technology (NIST) on the effects of horizontal ventilation on fire behavior in legacy and contemporary residential construction, with some surprising results that may modify and/or change some long-held beliefs. Briefly, the studies suggest that initial ventilation efforts are likely to make things worse, and that rapid cooling of turbulent smoke and visible fire conditions are likely to make things better or, at minimum, slow the fire rate of change. These studies are noted at the end of this chapter in the section "Resources for Further Study."

Search

In a simpler time, firefighters *valued* that every building (and every room within) was searched unless it was physically impossible to achieve (fully involved or already collapsed). Without being judgmental, that search value was established based on a history of fighting predictable fires in very resilient buildings with simple floor plans. Smoke was a non-issue for entering firefighters back then—we had our self-contained breathing apparatus (SCBA), and if we could tolerate the heat, it was "safe" to be there (higher smoke ignition temperature). Two things happened since that simpler time: Society changed the fire (high heat release rates and explosive, low-ignition temperature smoke) and society changed the building (engineered lightweight). We still value the need to search what we can, but we must do so with particular attention to survivable spaces (and firefighter safety). The 360° building size-up can help you determine the survivable spaces. Heat-cracked windows that are issuing high-velocity smoke can indicate a room that is not survivable—the human life threshold is 300°F for 1 minute.[3] (Retired Chief John Coleman addresses this issue from a time perspective in his books *Incident Management for the Street-Smart Fire Officer* and *Searching Smarter*. Both are listed in "Resources for Further Study" at the end of this chapter.)

Prioritizing search is an exercise of combining most of our building size-up factors: building use, building size, building status, what is burning, and the fireground clock. Each of these factors has to be weighed with the resources that are immediately available. For example, a large, occupied, multistory, multifamily dwelling fire with rapidly deteriorating fire and smoke conditions should cause you to prioritize search and rescue tactics. You know that it would take several companies 10 to 15 minutes to search the survivable spaces and it will take another two companies 5 to 10 minutes to make an interior knockdown on the fire. If four companies are not immediately available, you may opt to prioritize fire spread tactics (reset and slow the rate of change) to buy time for the search (hit it hard from the exterior).

Specific hazards

Compiling a list of potential building hazards that can affect tactics is probably futile. There are, however, several hazards that seem to always affect tactical choices. They include:

- The use and storage of hazardous materials
- Evidence of hoarding
- Multiple overhead power lines
- Challenging access/egress conditions
- Previous fires and collapses within the building
- Aging and lack of maintenance
- Windowless buildings
- Multiple entry levels or sloping grade configurations

In addition to the specific hazards listed above, we feel there are some significant game-changing hazards worthy of mention. They include the base of operations, green movement, facades, access/egress, and name/utilities.

Base of operations. In chapter 7 we considered the importance of ensuring the stability of your base of operations, which is simply defined as "ensuring what you are standing on (roof or floor) will safely support your weight until you exit the building!" Quite often in the desire to put first water on a fire we tend to assume the grade floor of a building—where we normally conduct initial entry operations—is stable and safe enough to support our weight until our operation is completed. However, with the proliferation of I-joists that are becoming commonly used in floor and roof construction in addition to the increasing use of lightweight steel, the mass of conventional lumber has been replaced with lightweight wood and/or metal construction that can easily catastrophically fail when subjected to the weight of firefighters and their equipment, even when the base of operations looks normal from a visual observation or TIC reading. So, when operating above a fire, either above a cellar/basement, on floors above a grade floor, and/or on a roof, take the necessary time to ensure your base of operations will support your weight.

Green movement. Although the green movement is an emerging program that has and will continue to influence building construction, it presents a new set of size-up hazards that were not an initial size-up consideration a few years ago. Some facets of the green movement cannot be initially seen—such as more efficient insulation and windows, advanced framing techniques, and cool roofs that can collectively contribute to more rapid flashovers. Other hazards that may be readily seen are not as well understood, such as roof mounted solar panels, roof cellular antennas with accompanying electrical feeds, and atriums. These elements create a new set of hazards that not only must be included in an initial size-up but must be emphasized as hazards that every firefighter needs to be familiar with. As explained in chapter 9, an example of a current hazard is roof mounted cellular antennas that are common on many buildings and capable of creating serious injury to firefighters who inadvertently stand in the receiving and transmission paths.

Facades. Of all the items that we recommend you look for in your initial size-up, a facade (some call them a false mansard) is one of the most visible attributes of a building that can be readily identified from the street. Although we discussed facades in chapter 9, the impending hazards that false mansards are capable of creating when exposed to fire bears repeating. Unfortunately, facades are commonplace on some dwellings and numerous commercial buildings. Fortunately, facades are easy to identify, as they are an exterior style of construction that is often quite large and constructed on the front and sometimes the sides of a building. Always consider that facades present at least three primary hazards:

- Many facades are constructed from lightweight construction.

- They are a common attic on the exterior perimeter of a building, are likely common to an adjoining attic, and usually lack fire-stopping and sprinklers.

- They are capable of collapsing over the front access/egress opening (front door) and capable of blocking the movement of firefighters, severing water supply to interior personnel, and burying personnel near the front of a building. As an example, refer to figure 9–24.

If a facade presents any of these hazards, consider using an entrance that is not under it (e.g., a rear door).

Access/egress. In the previous section we briefly looked at some forcible entry considerations. However, let's revisit forcible entry from the perspective of being able to get into and out of a building in an acceptable amount of time. Many firefighter injuries and deaths are caused by not being able to exit a building in a timely manner. Therefore, an initial size-up should include the prospect that personnel who will be committed to the interior of a building are able to satisfactorily

enter and exit the building. This often includes much more than opening/forcing a front door for a rapid deployment of an attack line! Buildings that are prime candidates for this type of analysis and subsequent removal of egress obstructions are vacant buildings that are secured with plywood/OSB type materials, buildings with security bars over the doors and windows as in figure 9–3, and so on. Always ensure that interior personnel are able to exit a building without having to depend on their primary means of access, which is typically the front door!

Name/utilities. In many cases, one of the easiest yet applicable size-up indicators of building contents, layout, and potential hazards is the signage on the front of the building. Residential buildings always have an invisible sign on the exterior that reads "people inside," and, depending on the type of building, can be an important indicator of the expected occupant load. Commercial buildings can be a bit more elusive but will vary from no signage to an exterior name that indicates the type of business and likely interior contents, fire load, expected fire behavior, and potential floor plan. In combination with exterior signage, the number and size of utilities can also indicate potential interior hazards that are noteworthy. Utilities that are more numerous and/or larger than typical residential or standard commercial utilities (as an example, refer to fig. 9–14) than would be expected for a common commercial building indicate interior processes that can be larger and/or more hazardous to interior personnel than processes that are typically encountered. Do not let the excitement of a structure fire and the challenge of deploying the first attack line replace simple size-up indicators that have the potential of clarifying the hazards of the host building you are about to engage.

Now, as a quick review and using the preceding considerations as a resource, let's apply our suggested initial size-up templates and assume you are the first arriving officer who has just stopped in front of the building in figure 10–9 that displays the following:

Fig. 10–9. This building initially appears to be a three-story residential building.

- Three-story residential building. Basement status unknown.
- The building appears to be from the pre-WWII era, as the roof overhang is beginning to collapse, the exterior shingles are likely asbestos, windows are in-line indicating balloon frame construction, and the corbels are likely decorative.
- The window over the porch has been covered over and now supports a heating appliance duct.
- The attic has been converted into multiple living spaces. Notice the dormers on the left and right sides are original while the front dormer is not original.
- The two metal boxes on the right side of the entryway contain 13 mailboxes! This would be noticed upon close inspection.

As this building allows a 360° size-up, let's walk down the right side and see what we did not see from the front by looking at figure 10–10:

- In addition to the front entrance, there is a side entrance to the main building.
- Of particular importance, there are two additions to the rear of the main building—one with a gable roof and another with a flat roof.
- Both of the additions appear to have the same siding as the main building—asbestos shingles.

The Art of Reading Buildings

- Multiple electrical meters are visible on the addition with a gable roof.
- The main building and the two add-ons are likely conventional construction.
- Lastly, this building will present a sizeable search/rescue challenge depending on the type of fire.

Fig. 10–10. From the side and back, the building in figure 10–9 appears quite different from the front size-up.

There you have it—the process of sizing up your next building. You may be thinking that our list is pretty long and quite a challenge to remember, let alone apply at a fire. The key is to create an *unconscious competence* through repeated practice. Formal training helps—and provides the foundation for the competence. The unconscious part comes when you consciously apply the size-up processes during routine or day-to-day building visits that are unencumbered with incident handling. Great company officers and battalion chiefs take advantage of any building visit to linger about the building and exercise their size-up skills. It doesn't matter if the visit is for a meeting, shopping for the crew meal, or running other errands. The learning (and practice) never ends. Your authors are no longer assigned to suppression duties, yet our teaching trips around the world provide great opportunities to discover all manner of buildings and keep our building size-up skills proficient. You can achieve the same.

The building size-up process outlined here is used to develop the rapid street-read guides in the next section of the book. We hope you find them useful as study guides and templates to help you form your own perspective on reading buildings. That is the key—develop *your own* method to get a read on your next building response. To finish this chapter, we want to share some perspectives on how to use your building reading skills to triage and predict collapse of burning buildings.

Quick summary

- Most buildings present six tactical challenges: fire spread, collapse, forcible entry/exit, ventilation, search, and specific hazards.

- Fire spread is influenced by interior geometry, fire loading, and heat/air flow paths.

- Visual collapse warning signs like sagging roofs and floors are considered late warning signs. They are important to know and apply, but a better approach is to use a mental process to predict collapse using building construction and fire degradation knowledge.

- The exterior features that are discovered as part of a 360° size-up will help paint the forcible entry challenge. All forcible entry efforts work against the fireground clock. Using building construction and material knowledge, fire officers may decide that enlarging windows or breaching walls is quicker than forcing a heavily secured door.

- Ventilation challenges are closely related to fire spread issues. The management of heat flow paths is essential to fire control. In most cases, it is best to enlarge the heat flow path near the fire and rapidly apply water to flames and turbulent smoke. Building construction knowledge helps in making the best decisions for increasing heat flow paths. The decisions must take into account rate of change and the fireground clock.

- Search challenges are influenced by building use, size, status, and the fireground clock. As part of the 360° building size-up, a judgment of interior survivable spaces needs to be made. Heat-cracked windows emitting high-velocity smoke is a sign of a nonsurvivable space.

- Depending on resources, firefighters may choose to prioritize fire spread tactics in order to buy time for searches.

- Specific building hazards are myriad. Some significant challenges include hazmat, sloping grades, overhead electrical lines, and hoarding. Game-changing hazards include your base of operations, the green movement, facades, access/egress options, and building name/utilities.

- The building size-up process can become second nature through repeated practice during nonincident building visits. The practice never ends.

PERSPECTIVES ON BUILDING TRIAGE AND PREDICTING COLLAPSE

The first-due decision maker (usually the first-arriving company officer) uses his or her reading building skills for rapid recognition of potential traps and to understand the tactical challenges that a building may present. Once initial operations commence, that first-due company officer will have to focus on the safe accomplishment of tasks. The company officer must maintain situational awareness, but more often than not, the completion of the task requires the crew to get up close and personal with the smoke, fire, and interior of the building. The up-close completion of tactical assignments limits the ability to see the big building picture. Someone needs to assume the role of reading the whole building after the first-due has gone to work. In most organizations that responsibility is automatically assumed by the incident commander (IC). The strategic and stationary nature of the IC responsibilities can limit the ability to continually read the building. For this reason, we support the designation of an incident safety officer (ISO) or IC assistant who has the flexibility to roam the incident scene and continually read the building (as well as the smoke and other hazards).

Those assigned the role of continually reading a building while interior operations are underway also has a responsibility to triage the building and predict collapse. The word *triage* means to sort or prioritize. **Building triage** involves the process of evaluating current and changing conditions and making judgments about the risks and integrity of the various portions of a building. Taken further, the building triage sets the stage for predicting partial, localized, and general collapses.

Building triage and collapse projections are mental processes that use the classic model of identify/analyze/decide. Let's explore triage and collapse predictions a bit more.

Triaging building fires

First-due fire officers are challenged by making "go, no-go" decisions using initial and 360° size-up processes. At best, the size-up investment will result in sound risk management decisions and appropriate tactical accomplishments that lead to a safe and predictable outcome. The reality, however, is that the initial and 360° size-ups are accomplished in a short time frame and rely on surface (exterior) clues to paint an interior picture. All the information to make the best decisions isn't always apparent. First-due decision makers have and will continue to engage in interior operations without the benefit of collecting all the necessary information to ensure our safety (it is a risky business and we willingly accept those risks). This reality has led suppression commanders to develop various checklists and guides to help triage (sort and prioritize) building information and conditions. Using a process of building triage can help

The Art of Reading Buildings

commanders predict hostile events (like rapid fire spread and collapse) and make risk-based changes to the incident action plan that could include rapid withdrawal of interior committed crews.

One example of building triaging is the multifaceted building fire triage and risk management program developed by Los Angeles Fire Department Battalion Chief Joe Castro. (See the feature titled "Joe's Building Fire Triage Program.") Chief Castro has assembled a series of prompts that helps the IC sort through fireground indicators that can be used to consider two primary assessments:

- When to consider not entering a building
- When to consider getting out of a building

The prompts are divided into structural triage, smoke and fire conditions, and insightful tangibles. As you review the feature, note that the prompts include important reminders as well as unanswered information. (We add some clarity to Chief Castro's list in parentheses following the item.)

Using Chief Castro's fire triage program, let's apply his *structural triage* prompts to the building in figure 10–11.

Assume you have just stopped in front of this moderate size, single-story commercial building and it is 2 a.m. Your initial size-up (initially limited to Sides A and B) indicates the building is older as it is in an older, commercial/industrial part of town, the walls (likely of URM) are covered with stucco, the roof appears to be an older gable roof (likely an older truss gable roof with multiple layers of roofing materials) that is flanked by an older flat roof (that although is likely conventional construction, is also covered with multiple layers of roofing materials). The building is obviously vacant (there is a small "Available" sign on the front door and the previous name of the building on Side B has been

JOE'S BUILDING FIRE TRIAGE PROGRAM

STRUCTURAL TRIAGE

- Limited ways in and out
- Can't tell what the building is being used for
- Can't tell where the fire is
- Has the potential to having been burning undetected
- Can't determine the potential floor plan—no prefire intelligence
- Buildings whose construction features frequently result in unexpected fire behavior or rapid collapse

SMOKE CONDITIONS

- Read smoke conditions
- Zero visibility—increases interior risk
- Evaluation of the rate of change
- Best to view all sides of building
- Compare smoke volume to size of building
- Velocity/pressure (turbulent smoke is a flashover warning)
- Density/thickness (smoke is fuel; thicker smoke leads to more severe fire)
- Color (black is heat, dirty white is filtered/distal smoke, brown smoke is heated wood)

FIRE CONDITIONS

- Burn time
- High levels of heat
- No ventilation (inadequate heat flow paths)
- Delays in progress
- Lack of progress

OTHER TANGIBLES

- Out of comfort zone due to size/scope/complexity
- Lack of situational awareness

Fig. 10–11. This is the scenario that is initially viewed by responding resources.

painted over), there is one small entrance door in the front that is flanked by a small display window, and it is unknown if the adjoining building (to the right of the main entrance) is a common occupancy. Assuming there is some smoke visible from this building (particularly the turbine ventilator that is visible on the ridge of the gable roof), you have (or lack) the following structural knowledge:

- Initially, there appears to be one (small) way in and out of the building.
- You have no idea of what the building is used for.
- You are unsure of the fire location.
- Due to the vacant/time of day status, the fire has had the potential of burning for an extended time frame.
- There are no windows on Side B.
- You are uncertain of the interior floor plan (basement?) and contents. It is also uncertain if the adjoining building is common to the subject building.
- The older flat and truss gable roofs are capable of collapse if exposed to fire for an extended period of time. (This can also be enhanced by the likely multiple layers of roofing materials.)

From this simple example that has only used Joe's structural triage and not his smoke conditions, fire conditions, and other tangibles, this building has completely failed the six structural triage prompts. It is easy to see that any of the prompts (individually or collectively) can lead to the order to not enter or to withdraw interior crews, or at the minimum, initiate a cautious approach that is based on the potential of rapid change depending on conditions. The work of Chief Castro is but one example of how a building triage system can help incident commanders and safety officers process incident observations and judgments for the sake of firefighter safety.

The last part of this chapter specifically addresses the mental process of predicting collapse.

Predicting collapse

As with most incident dynamics, nothing is absolute about predicting building collapse. There is no perfect formula. Yet the effort to analyze and predict collapse is absolutely essential to firefighter safety. The book *Collapse of Burning Buildings: A Guide to Fireground Safety*, by retired FDNY Deputy Chief Vincent Dunn, is an excellent starting place to learn *how* buildings and parts of buildings

collapse. This is just one part of the front-loading that needs to take place to help predict collapse. There are other front-loading activities that can help you build your ability to predict collapse:

- Developing the ability to read a building by practicing.
- Reviewing investigative reports from past collapse incidents.
- Making postfire site visits of collapsed buildings.
- Performing prefire planning and site visits.

The process of predicting collapse at an actual incident (especially when crews have been interior committed) becomes a cyclic activity that is akin to an algorithm. The algorithm uses four analytical inputs and one important output:

Input 1: Classifying the construction

Input 2: Determining structural involvement

Input 3: Visualizing and tracing loads

Input 4: Evaluating elapsed time

Output: Predicting and communicating collapse potential

The four inputs are analyzed and a judgment needs to be rendered about the collapse potential of building or various parts or area of a building. Let's examine these inputs a bit more and show how they can lead to the output.

Input 1: Classifying the construction. This is where the classification approach of *era/use/type/size* becomes valuable. This combination better links the building to strengths and weaknesses as well as fire spread patterns. By classifying the building, you can imagine how the materials and arrangement of structural elements might be impacted by fire and heat. As an example, there is a significant difference between older timber trusses and modern lightweight wood trusses.

Input 2: Determining structural involvement. Determining whether a fire is a contents or structural fire is imperative. Most fire departments use the term *structure fire* to dispatch crews. There may be some benefit in the conservative approach that all fires are structure fires, but the reality is that many fires we go to are merely contents fires. Usually it is a combination of reading the smoke and the read you have on the building that will help you determine if the event is a contents fire or if the structural elements of the building are being attacked by heated smoke and fire. A quick visual clue that the structure itself is being attacked is the presence of dark gray or black smoke venting under pressure from structural seams, eaves, attic vents, foundation ridge boards, and wallboard seams. Also, unfinished wood that is being rapidly heated emits a brownish smoke as it pyrolizes.

Input 3: Visualizing and tracing loads.

Visualizing and tracing loads is a "mind's eye" exercise—like using X-ray vision to see how loads are being delivered to ground. In many ways, you are constructing the building mentally. This exercise is designed to help find the weak link, which typically precipitates the collapse. Historical weak links include the following:

- Structural connections
- Overloading
- Occupancy changes
- Modern trusses
- Void spaces
- Stairs
- Parapet walls
- Large, open span interior spaces
- Facades

Input 4: Evaluating elapsed time. Some structural elements fail as soon as fire (heat) reaches the material. Other materials absorb incredible heat for a long duration before they become susceptible to collapse. Elapsed time as a factor should be brought into the collapse equation, although the time it takes for gravity to overcome the structure during a fire is not predictable. Remember that elapsed time includes two clocks: the prearrival

clock (unknown) and the fireground clock (should be known). A number of variables determine the amount of time a material can resist gravity and heat before failure. Many factors can accelerate the potential collapse time:

- Low material mass, or high surface-to-mass ratio
- Higher Btu development and heat release rates (fire load)
- Alterations (undesigned loading)
- Age deterioration or the lack of care and maintenance of the structure
- Firefighting impact loads (fire stream force, accumulated water, forcible entry and ventilation efforts, weight of firefighting teams)
- Breakdown or loss of fire-resistive barriers like drywall and steel coatings

Output: Predicting and communicating collapse potential. The preceding inputs can help you visualize a collapse scenario for the building involved in the incident. In many ways, the inputs create a series of hits that can take down the building. One of the hits alone can take the building (or part of the building). Likewise, any one input may be a marginal hit but the synergy of all the marginal hits will lead to collapse. This information should be communicated to command in such a way that includes the establishment of a collapse zone or zones and an adjustment to the incident action plan (IAP). Collapse zones should be considered no-entry zones for firefighters. Most fire service texts suggest that a collapse zone is at least one and a half times the height of the element that is anticipated to fall. While this may be appropriate for wood structures, it may not be adequate for unreinforced masonry construction. There are many documented collapses where a falling unreinforced masonry wall has propelled bricks and mortar three times the height of the wall!

As a final note, buildings undergoing construction, demolition, or extensive remodeling are basically losers during fires. Treat them all as collapse-ready and push for a *defensive-only* strategy. Granted, there are times when you can prove reasonable integrity in these building and allow interior operations, but the default should be defensive until that proof is gathered.

Quick summary

- Predicting collapse is not an absolute science but is critical for firefighter safety.
- Building triage and collapse projections are mental processes that use the classic model of identify/analyze/decide.
- An algorithm of inputs and an output can serve as a mental process to help predict collapse.
- The inputs include building classification (era/use/type/size), structural involvement observations, visualizing loads to help determine the weak link, and elapsed time. The output is communication of collapse potentials.
- Buildings under construction, renovation, or demolition should default to a defensive approach.

Photo exercise 3

Building use:

Building size:

Basic construction type:

Building era:

Exterior features:

Interior features:

Specific hazards:

Likely building status:

Chapter 10 — READING BUILDINGS: HOW TO SIZE UP A BUILDING

Photo exercise 4

Building use:

Building size:

Basic construction type:

Building era:

Exterior features:

Interior features:

Specific hazards:

Likely building status:

Photo exercise 5

Building use:

Building size:

Basic construction type:

Building era:

Exterior features:

Interior features:

Specific hazards:

Likely building status:

RESOURCES FOR FURTHER STUDY

- Coleman, John F. (Skip), *Incident Management for the Street-Smart Fire Officer*, 2nd ed., Tulsa, OK: PennWell, 2008.

- Coleman, John F. (Skip), *Searching Smarter*, Tulsa: OK: PennWell, 2011.

- Dodson, David W., *The Art of Reading Smoke Practice Sessions*, DVD, Tulsa, OK: PennWell, 2009.

- Dodson, David W., *The Art of Reading Smoke*, DVD, Tulsa, OK: PennWell Corporation, Fire Engineering books and videos, 2007.

- Dunn, Vincent, *Collapse of Burning Buildings: A Guide to Firefighter Safety*, 2nd ed., Tulsa, OK: PennWell, 2010.

- IAAI and USFA, *Managing Vacant and Abandoned Properties in Your Community*, International Association of Arson Investigators (IAAI) and the United States Fire Administration (USFA) Abandoned Building Project, 2006.

- Marsar, Stephen, "Survivability Profiling," *Fire Engineering*, July 2010 (available at http://www.fireengineering.com/articles/2010/07/survivability-profiling-how-long-can-victims-survive-in-a-fire.html).

- UL research on ventilation can be found at http://content.learnshare.com/courses/73/306714/player.html.

NOTES

1. Numerous NIOSH Firefighter Fatality Investigation Technical Reports recommend that the first-due officer perform a 360° size-up of the building fire environment *prior* to interior engagement in order to help prevent future firefighter fatalities.

2. Dodson wrote this quote verbatim from a slide presentation that Professor Brannigan displayed during a pre-conference seminar on steel construction that was presented at FDIC in Cincinnati (late 1980s).

3. Various sources cite the human life threshold for heat (death) and they differ slightly. The sources include the *NFPA Fire Protection Handbook*, *Journal of Forensic Medicine*, and several NIST technical reports. Most use 250°F for several minutes or less. We chose 300°F for 1 minute as a conservative, inarguable data point that favors potential victims and is still within the tolerance levels of compliant structural PPE.

SECTION 3
RAPID STREET-READ GUIDES

USING THE RAPID STREET-READ GUIDES

11

52 BUILDINGS

From a firefighter's perspective, we believe that there are 52 types of buildings. While this may sound like a foolish statement, there is a foundation for it. In the United States, there are approximately 135 million buildings if you add up commercial, residential, and government buildings.[1] It is estimated that every year, 1.5 million new buildings are constructed and 250,000 buildings are demolished. Additionally, there are untold numbers of buildings that are not counted for many reasons (outbuildings, do-it-yourself structures, hidden ones, etc.). The 52 buildings cited are representative of the 135 million-plus aggregate. So how did we come up with the number 52?

If you recall back in chapter 4, we argued that the five classic types of buildings that are outlined in fire service curriculums are a bit oversimplified and can actually lead to some traps for firefighters. We then suggested that from a firefighter's perspective, it is best to classify buildings using a model of *era/use/type/size*. That suggestion became the starting place for developing the list of 52 representative buildings. We then considered the following:

- Lessons learned from previous building fires
- Tactical challenges that are shared by various groups of buildings
- Regional influences in the U.S. (East coast, West coast, Mid-America, North, and South)
- Features or hazards that separate a building group from others

The formula may not be perfect, but we used it to help develop the rapid street-read guides that are found in this section. The intent of the guides is to provide an informative template for the prevalent types of buildings that may exist in your community. Further, we wanted to create a resource that could be used for building preplanning, fire prevention inspections, company training sessions, and responses to building fires. All told, we hope the guides meet our original goal of creating a useful desk reference that helps firefighters with rapid recognition and practical relevancy for the buildings in their community.

As you read through the guides, it may help to understand the specific methodology that went into the design. Likewise, we'd like to offer some suggestions on how to maximize their use.

DESIGN FEATURES

As more responding personnel (firefighter, driver, officer, chief officer) can learn how to quickly evaluate and incorporate applicable building considerations into fireground operations (based on safe and appropriate operational effectiveness), structural fireground operations will become more of an art rather than an experiment. Creating a quick, applicable building reference was our main design criteria to help you develop your art. To that end, we followed these guidelines:

- Use a quick overall indexing system based on the first-due arrival size-up approach presented in chapter 10 (single-family dwelling, public assembly, etc.).
- Subdivide the index into common styles or uses (colonial, condo, church, etc.).
- Include a photograph that shows a classic example of the building being outlined (some guides have more than one photo).
- Use one page per building. The front of the page describes the building, the back highlights tactical concerns that are common to the building.
- Keep text and bulleted items short, easy to read, and to the point.

Using the guidelines above created editing challenges that may open the door for questions of clarity and omissions. In recognition of this possibility, consider the following:

1. The buildings represented are a snapshot of the classic or most prevalent of that kind.
2. Terminology can be amazingly diverse across the United States. As we traveled across the country conducting research, we discovered that firefighters have developed all manner of slang, jargon, or terms to describe a given building. We tried to use the most common terms.
3. Owners and occupants of buildings can be very inventive and resourceful with alterations. There are likely to be buildings out there that combine key points from multiple street guides or simply don't fit into any of them.
4. The guides are developed to be the "starting line" for understanding a given building. By design, they are a *guideline* (flexible in application) and should not be used as a replacement for directive procedures or policies.
5. We know that we will receive reader inputs and corrections—and we encourage it. Actually, we challenge you to develop your own guides for specific buildings in your district—and share them with us!

USING THE RAPID STREET-READ GUIDES

The idea of rapid street-read guides isn't a new one; creative fire instructors have developed variations of the theme to help with in-house training activities for decades. Likewise, some fire departments have developed specific prefire building surveys for those in their jurisdiction—ranging from very simplistic to incredibly complex. Our original vision was to include the street guides as a generic training tool to help firefighters expand their building knowledge past the five types and bridge the gap between book knowledge and fireground application. We shared a couple of the early versions of the guides to firefighters around the country, and their input was invaluable. We then presented a short class on rapid-reads at FDIC in Indianapolis. The resulting feedback from that class blew our minds—and showed how imaginative and training-hungry firefighters can be. Listed below are some ways you can use the rapid street-read guides.

- As a template for collecting building information for the development of prefire plans
- As a conversation starter for your actual on-site building familiarization activities
- Spontaneous "in-service" company training on those bad weather days*
- To help add a field component to basic fire academy building construction classes or community college fire science curriculums
- As a promotional assessment preparation tool
- For a community risk analysis project (the need for sprinklers, fire stations, staffing, etc.)
- A field guide for improving rapid street reads as you drive through your district
- As a reference tool for the preparation of research projects such as those assigned at the National Fire Academy Executive Fire Officer development courses

(*Pick a guide and match it to the crew's knowledge of buildings in their area and the issues that could arise.)

More than anything, we encourage you to get out of the fire station and get into the buildings that could possibly host your next fire. The information you glean from these visits are certain to help you improve your situational awareness, crew safety, and rapid decision making.

As a final note, we are open to your input on making the guides more usable. We fully expect to revise, add, subtract, and expand the guides in future editions and for other media avenues.

Following is a master index of all the street guides. You will see a *use* category (single-family dwelling, public assembly, etc.) followed by a numbered list of differing examples that fall in that category. Each rapid street-read guide is tabbed with a category acronym and number that coincides with the index.

NOTE

1. The figure was derived from data from the U.S. Department of Commerce, Energy Information Administration, and the U.S. Department of Housing and Urban Development. Figure represents total buildings through 2011.

RAPID STREET-READ GUIDE INDEX

Single-Family Dwelling (SFD)

SFD 1: Colonial and Georgian 295
SFD 2: Victorian/Queen Anne and Cape Cod 297
SFD 3: Craftsman and American Four Square 299
SFD 4: Prairie Style 301
SFD 5: Split Level 303
SFD 6: Modern Lightweight 305
SFD 7: McMansion 307
SFD 8: Manufactured (Mobile) Home 309

Multifamily Dwelling (MFD)

MFD 9: Brownstone 311
MFD 10: Tenement 313
MFD 11: Row Frame 315
MFD 12: Railroad Flat 317
MFD 13: Center Hallway Structure 319
MFD 14: Garden Apartment 321
MFD 15: Project Housing—Low Density 323
MFD 16: Project Housing—High Density 325
MFD 17: Legacy Townhome/Condo/Apartment 327
MFD 18: Lightweight Townhome/Condo/Apartment 329

Main Street Commercial (COM)

COM 19: Pre-WWI Ordinary 331
COM 20: Pre-WWII Ordinary (Taxpayer) 333
COM 21: Industrial/Legacy Strip-Style 335
COM 22: Modern Strip-Style 337
COM 23: Fast Food 339
COM 24: Mega-Box 341
COM 25: Big-Box Store 343

Manufacturing/Warehouse (MANF)

    MANF 26: Block/Masonry 345

    MANF 27: Steel 347

    MANF 28: Concrete Tilt-Up 349

    MANF 29: Wood 351

    MANF 30: Converted Mill 353

    MANF 31: Public Storage—Single Story 355

    MANF 32: Public Storage—Multistory 357

Office Building/Hotel (OFF)

    OFF 33: Pre-WWII—Low Rise 359

    OFF 34: Post-WWII—Low Rise 361

    OFF 35: 21st Century 363

    OFF 36: High Rise—1st Generation 365

    OFF 37: High Rise—2nd Generation 367

    OFF 38: High Rise—3rd Generation 369

Institutional Building (INST)

    INST 39: School 371

    INST 40: Hospital 373

    INST 41: Detention (Jail) Facility 375

    INST 42: Attended Care Facility 377

Public Assembly (PUB)

    PUB 43: Restaurant 379

    PUB 44: Stadium/Arena 381

    PUB 45: Auditorium/Theatre 383

    PUB 46: Meeting Hall 385

    PUB 47: Church 387

Miscellaneous Building/Structure (MISC)

    MISC 48: Pole Barn 389

    MISC 49: Kit Building 391

    MISC 50: Silo 393

    MISC 51: Historical Building—Dwelling 395

    MISC 52: Historical Building—Commercial 397

RAPID STREET-READ GUIDE

SFD 1

BUILDING GROUP: Single-Family Dwelling—Colonial and Georgian

BASIC CONSTRUCTION METHOD

- Conventional wood frame and balloon frame
- Some masonry walls

ERA CONSIDERATIONS

- Historical-era homes are built with full-dimensional, rough-sawn lumber.
- Larger historical-era models have heavy timbers for floor beams and internal columns.
- Those built from 1860–1930 are likely to be balloon frame.
- Older examples have stone or some masonry foundations.
- Rehabilitated examples and those with additions may have some lightweight components.

EXTERIOR FEATURES

- Simple gabled and hipped wood roofs are used; some have slate roofs.
- Freestanding brick fireplaces typically anchor Side B and Side D.
- Windows are typically double-hung, narrow, and aligned.
- Crawl spaces and cellars are expected, basements are rare. Cellar access is usually on the exterior of Side C.

- Pillars are used for ornate entryways that may also support extensive upper balconies.
- Some have combustible wood roofs.

INTERIOR FEATURES

- Living spaces typically front the floor plan with the kitchen in the rear. Rooms are boxed-shaped and symmetrical.
- A central stairway near the front door accesses upper floors. Bedrooms are served by a central hallway.
- Narrow hallways and interior doors are common. Interior doors may have transoms above.
- Attic spaces typically have a natural light window. The space may have been converted into a living or sleeping space. Access to the attic is usually through a pull-down stair in the central hallway or one of the bedrooms.

TACTICAL CONCERNS

Fire Spread: Lath and plaster interior walls help absorb heat and contain room fires. Small rooms help with compartmentalization. Fires that originate and/or spread to combustible void spaces create extinguishment challenges. Most voids are combustible and interconnected. Stripping baseboards and placing inspection holes in suspected stud runs can help detect and extinguish fires in voids. Hollow channels on either side of double-hung windows can hide extension of fire. Ornate eave finishes present overhaul difficulties. Fuel-fired, hot-air furnaces that use large flow-through floor registers to distribute heat also help distribute fire and smoke. Check behind any trim molding exposed to fire. Masonry construction will limit the spread of fire. However, renovations can create voids that enhance the spread of fire.

Collapse: The high-mass, full-dimension lumber nature of the construction can resist collapse in a predictable way. Sagging is a common and usable warning. Most collapse issues deal with aging (deterioration and lack of maintenance) or with alterations and additions to the original structure. Modernization efforts can also increase the collapse threat (removing interior walls to create open spaces). Balconies and exterior ornate features can be expected to fall from the building early. Fireplaces are likely unreinforced masonry construction. Heavy roof dead loads can be expected.

Ventilation: Interconnected combustible voids almost mandate the use of vertical ventilation tactics. Thick hardwoods used for roof decking can dull chain saws quickly. A rotary saw with a combination blade is a viable option. Most rooms have an exterior window, which can aid in horizontal ventilation. PPV should be used with caution in balloon frame buildings.

Forcible Entry: Forcible entry can normally be accomplished by conventional methods. Expect heavy, solid doors and strong locks. Narrow windows can make access/emergency egress difficult. Clearing the sash and frame is essential.

Search: Original floor plans include many small rooms that are symmetrical and easy to search. Main floor rooms flow room to room, whereas a central hallway serves bedrooms.

SPECIFIC HAZARDS

- Some utilities may not be able to be secured from the exterior. Inspect for multiple electrical feeds.
- Georgian-style, mansion sized, Civil War- and WWI-era examples may have secret rooms, tunnels, and passageways that lead to outbuildings.
- Expect heavy roof loads on post-and-beam and balloon-frame buildings
- Knob and tube wiring is an ungrounded system.
- Watch for renovations and combustible wood roofs.

RAPID STREET-READ GUIDE

SFD 2

BUILDING GROUP: Single-Family Dwelling—Victorian/Queen Anne and Cape Cod

BASIC CONSTRUCTION METHOD
- Conventional wood frame

ERA CONSIDERATIONS
- Historical-era homes are built with full-dimensional, rough-sawn lumber.
- Those built from 1860–1930 are likely to be balloon frame or post and beam.
- Older examples have stone or masonry foundations.
- Rehabilitated examples and those with additions may have some lightweight elements.

EXTERIOR FEATURES
- Simple gabled roofs are the norm. Some Queen Anne and Victorian examples have steep pitches and turrets. Slate roofs may be encountered as well as combustible wood roofs.
- Fireplaces are central, freestanding unreinforced brick.
- Wood shake, shingle, and shiplap siding is popular.

- Double-hung windows are typically narrow and aligned.
- Cellars are more prevalent than true basements. Most have an exterior cellar access that may have been covered if an addition was made to the back of the house.
- Expect upper balconies and widow's watch type platforms.
- Covered porches are common on Queen Anne dwellings.

INTERIOR FEATURES
- Living spaces typically front the floor plan with the kitchen in the rear. Rooms are mostly small and symmetrical.
- A central stairway near the front door accesses upper floors.
- Narrow hallways and interior doors are common. Interior doors may have transoms above.
- Larger examples may have hidden passageways and rear stairwells.

TACTICAL CONCERNS

Fire Spread: Lathe and plaster interior walls help absorb heat and contain room fires. Small rooms help with compartmentalization. Fires that originate and/or spread to combustible void spaces (such as balloon frame) create extinguishment challenges. Most voids are combustible and interconnected. Stripping double-hung window moldings and baseboards and placing inspection holes in suspected stud runs can help in detecting and extinguish fires in voids. Ornate cornices, scallops, and eave finishes present overhaul difficulties. Fuel-fired, hot-air furnaces that use large flow-through floor registers to distribute heat also help distribute fire and smoke. Remember that cellar/basement fires will readily spread upward, and renovations can add voids that conceal and spread fire.

Collapse: The high-mass, full-dimension lumber nature of the construction can resist collapse in a predictable way. Sagging is a common and usable warning. Most collapse issues deal with aging (deterioration and lack of maintenance) or with alterations and additions to the original structure. Modernization efforts can also increase the collapse threat (removing interior walls to create open spaces). Balconies and exterior ornate features can be expected to fall from the building early. Anticipate heavy roof loads.

Ventilation: Interconnected combustible voids almost mandate the use of vertical ventilation tactics. Queen Anne examples with steep, multiple rooflines will require multiple roof vent openings. Positive pressure ventilation (PPV) fans can create undesired flow paths in combustible voids. Basement and cellar fires can be especially difficult to ventilate.

Forcible Entry: Forcible entry can normally be accomplished by conventional methods. Narrow windows can make access/emergency egress difficult. Clearing the sash and frame is essential. Expect to encounter heavy and/or solid doors and strong locks.

Search: Original floor plans include many small rooms that are symmetrical and easy to search. Main floor rooms flow room to room, whereas bedrooms are served by a central hallway.

SPECIFIC HAZARDS

- Some utilities may not be able to be secured from the exterior. Inspect for multiple electrical feeds.
- Attic fires should be considered a basement/cellar fire first—rule it out.
- Civil War- and WWI-era examples may have secret rooms, tunnels, and passageways that lead to other buildings.
- Anticipate balloon frame construction and knob and tube wiring.
- Asbestos siding may be encountered.
- Queen Anne SFDs are popular choices for conversion to bed and breakfasts or MFDs.
- Attics may be converted to occupant spaces, creating knee walls.
- Be wary of vinyl siding and combustible wood roofs.
- Pulling ceilings below converted attics can be very difficult.

RAPID STREET-READ GUIDE — SFD 3

BUILDING GROUP: Single-Family Dwelling—Craftsman and American Four Square

BASIC CONSTRUCTION METHOD

- Legacy wood frame is most prevalent (Craftsman).
- Load-bearing brick exterior with wood frame interior (Four Square)

ERA CONSIDERATIONS

- Historical-era homes are built with full-dimensional, rough-sawn lumber.
- The 1860–1920 Craftsman can be balloon frame (look for narrow and aligned windows).
- Post-WWII are mostly legacy platform.
- Older examples have stone or masonry foundations.

EXTERIOR FEATURES

- Simple gabled wood roofs with dormer windows are common for Craftsman.
- Simple hip wood roofs with dormer windows are common for Four Square.
- Expect open front porches with large supported overhangs.
- A crawl space or small furnace cellar is more common than a basement. Those with a cellar or basement have an exterior access that may

have been covered with an addition to the back of the house. Larger versions will have an interior stair located in the rear of the house. Basement and cellar windows are very small.

- Attached garages are very rare and, where present, are probably not original to the house.
- Shiplap or masonry siding (can include asphalt siding) is used.
- Mid-floor windows indicate the location of an interior stairway.

INTERIOR FEATURES

- Most have a centered front door that divides a living room and dining room. Rooms are mostly small and symmetrical.
- The front door location (center or to one side) is typically aligned to the upper floor access stairway.
- There are no real hallways. First-floor rooms flow room to room and the bedroom doors surround the stairwell.
- The pictured four-square has an unusual cellar access door and stairway window on Side B, indicating an interior remodel.
- Fireplaces are typically centrally mounted or found on opposing sides of the structure.

TACTICAL CONCERNS

Fire Spread: Lathe and plaster interior walls help absorb heat and contain room fires. Small rooms help with compartmentalization. Balloon frame examples have interconnected combustible voids that create extinguishment challenges. Stripping double-hung window and baseboard moldings and placing inspection holes in suspected stud runs can help in detecting and extinguish fires in voids. Fuel-fired, hot-air furnaces that use large flow-through floor registers to distribute heat also help distribute fire and smoke. As a rule, non-balloon-frame examples contain fires very well. Smoke explosion or backdraft potentials exist behind knee walls on either side of dormers. Remember that basement/cellar fires can rapidly spread vertically.

Collapse: The high-mass, true-lumber nature of the construction can resist collapse in a predictable way. Sagging is a common and usable warning. Most collapse issues deal with aging (deterioration and lack of maintenance) or with alterations and additions to the original structure. Modernization efforts can also increase the collapse threat (removing interior walls to create open spaces). The second floor beams of brick (Four Square examples) sit on a ribbon ledge that can be brick or wood. Beware of heavy roof dead loads.

Ventilation: Rooftop heat holes are the preferred way to ventilate fires that have extended to the second floor or into the attic. Holes may have to be placed on either side of dormers to vent knee wall spaces. PPV can be effective for first- and second-floor fires. Basement and cellar fires are especially difficult to ventilate.

Forcible Entry: Forcible entry can normally be accomplished by conventional forcible entry methods. Narrow windows can make access/emergency egress difficult. Clearing the sash and frame is essential. Expect heavy and/or solid doors and locks.

Search: Original floor plans include many small rooms that are symmetrical and easy to search. Main floor rooms flow room to room, whereas a central stair serves bedrooms.

SPECIFIC HAZARDS

- Some utilities may not be able to be secured from the exterior. Inspect for multiple electrical feeds.
- An oil-burning hot-air furnace in the cellar can cause significant smoke spread into the home. These furnaces often have large heat registers in the floors.
- Oil- and gas-fired boilers with radiators in each room are very popular. The steam pipes were often wrapped in a sleeve that contains asbestos (industrial and legacy eras).
- Anticipate balloon frame construction and knob and tube wiring.
- Asphalt and asbestos siding may be present.
- Knee walls in converted attic spaces can be encountered and need to be checked.
- Vinyl siding can be used to replace older exterior wood siding.

RAPID STREET-READ GUIDE SFD 4

BUILDING GROUP: Single-Family Dwelling—Prairie Style

BASIC CONSTRUCTION METHOD
- Legacy platform wood frame

ERA CONSIDERATIONS
- Those from 1950–1970 have 2 × 6 in. exterior wall studs and cut lumber floor beams. Rafters are tied to a ridge beam.
- Those from 1970–1990 may have a nailed-lumber truss roof, although the floors can remain solid cut lumber.

EXTERIOR FEATURES
- Simple gabled wood roofs with low pitches are used for prairie style.
- Simple gable or hip wood roofs are used for split levels and cookie-cutter ranches.
- Large windows, some with creative window shapes are featured (Frank Lloyd Wright style).
- Bay windows are common.
- Expansive eaves and overhangs are used to form porches and patios as part of the roofline.
- Garages are usually attached with interior access.
- Masonry or shiplap siding is used.
- Common roof coverings are composition, rock, and tile.
- Masonry veneers of brick or stone are often used.

INTERIOR FEATURES
- Expect predictable floor plans that can be read easily from the exterior.
- The "great room" concept is introduced in these homes.
- Ranch (single-story) examples with basements have an internal stairway that is located central or near the front entry.
- Stacked (two- to three-story) examples with a basement have an internal stairway that is accessed by a door on the back side (opposite) the second-story access stairs.
- Vaulted ceilings may have exposed, true lumber beams with no attic.
- Interior moldings are less dramatic and smaller than in older SFDs.
- Some lathe and plaster was used in the interior.

TACTICAL CONCERNS

Fire Spread: Drywall is the primary fire spread barrier room to room. Modern HVAC return air and distribution systems are rarely fire-stopped (no dampers). Attached garages are rarely finished with drywall, although the separation wall to the living space must be drywalled (two layers of ⅝ in.). Multistory (stacked) designs are candidates for floor-to-floor fire spread when the fire originates on lower levels (central open stairwell). Hollow-core interior doors were common in the 1970s and can defeat compartmentalization.

Collapse: Legacy platform wood frame with true lumber is collapse resistive in most growth stage fires. Sagging is a common and usable warning and is usually the result of wood consumption during the fire. Multistory examples with bedrooms above the attached garage are more prone to rapid collapse when the garage is involved.

Ventilation: Rooftop heat holes are the preferred way to ventilate single-story fires or fires that have extended to the top floor or into the attic. Horizontal ventilation is often effective because of the large and numerous window options. PPV can also be effectively deployed.

Forcible Entry: Forcible entry can normally be accomplished by conventional methods. Original doors, windows, and locks can be easily removed or defeated. Occupant-installed security devices may slow forcible entry, although breaching walls is usually fast and easy.

Search: Ranch and traditional stacked floor examples are of a simple arrangement and easy to visualize. The chimney nature of the split-level central stair may necessitate the use of ladders to access and search upper level bedrooms.

SPECIFIC HAZARDS

- Most utilities can be secured from the exterior.
- Side C walkout basements and garden levels can be expected on slope-grade examples.
- Grade floor joists may be unprotected and exposed.
- Tongue-and-groove roof sheathing can be found where vaulted ceilings exist.
- Some shiplap siding has been replaced by vinyl siding.
- Original kitchen cabinetry that has been replaced with modern modular units can create hidden voids that extend to the second floor or attic knee walls.
- Lightweight construction can be found, particularly in floor joists and for roof rafters.

RAPID STREET-READ GUIDE

SFD 5

BUILDING GROUP: Single-Family Dwelling—Split Level

BASIC CONSTRUCTION METHOD

- Legacy platform wood frame
- Staggered floor levels may have some balloon-like framing where levels meet.

ERA CONSIDERATIONS

- Those built from 1950–1970 have 2 × 6 in. exterior wall studs and cut lumber floor beams; rafters are tied to a ridge beam.
- Those built from 1970–1990 typically have a nailed-lumber truss roof. However, some lightweight construction may be present.
- Those built from 1990 to present are of predominately lightweight construction.

EXTERIOR FEATURES

- Simple gable or hip wood roofs are used for split levels.
- Large windows and creative window shapes are found (Frank Lloyd Wright style).
- Bay windows are used.
- Garages are usually attached and have interior access.
- Basements are common.
- Double-hung and/or sliding windows are common.
- Masonry veneers of stone or brick can be used.

INTERIOR FEATURES

- Predictable floor plans are read easily from the exterior.
- The "great room" concept was introduced in these homes.
- Tri-levels have a centrally located side-by-side stairwell that serves the upper and lower levels. A basement door and stair can be accessed on the back side of the lower stairwell.
- Bi-levels typically have an entry door opening to a stairway that splits the upper and lower level.
- Vaulted ceilings in legacy examples have exposed, true lumber beams and no attic, whereas lightweight models will have a scissor-like truss and void space attic above the vault.
- Interior drywall is common.

SFD 5

TACTICAL CONCERNS

Fire Spread: Drywall is the primary fire spread barrier from room to room. Modern HVAC return air and distribution systems are rarely fire-stopped (no dampers). Attached garages are rarely finished (with drywall), although the separation wall to the living space must be drywalled (with two layers of ⅝ in.). Split-level designs are notorious for floor-to-floor fire spread when the fire originates on lower levels (via a central open stairwell). Hollow-core interior doors were common in the 1970s and can defeat compartmentalization.

Collapse: Legacy platform wood frame with true lumber is collapse resistive in most growth-stage fires. Sagging is a common and usable warning and is usually the result of wood consumption during the fire. Split-level examples have a structural weak link: the central open stairwell. While most interior and exterior walls are fire-stopped, the studs that form and support the stair walls are rarely fire-stopped. Lightweight construction will rapidly fail when exposed to heat and/or fire. Second-floor occupant areas above car garages are prone to rapid collapse if the garage is involved in fire.

Ventilation: Rooftop heat holes are the preferred way to ventilate fires that have extended to the top floor or into the attic. Horizontal ventilation can be effective because of the large and numerous window options. PPV can be effective for all levels.

Forcible Entry: Forcible entry can normally be accomplished by conventional methods. Original doors, windows, and locks are easily removed or defeated. Occupant-installed security devices may slow forcible entry, although wall breaching is usually fast and easy. Interior hollow-core doors can be easily forced.

Search: Ranch and traditional stacked floor examples are of a simple arrangement and are easy to visualize. The chimney nature of the split-level central stair may necessitate the use of ladders to access and search upper level bedrooms. Due to typical residential floor plans, search patterns are typically simplified.

SPECIFIC HAZARDS:

- Most utilities can be secured from the exterior.
- Side C walkout basements and garden levels can be expected on slope-grade examples.
- Split level examples may present communication confusion when labeling the first floor, second floor, and other levels.
- Side-to-side split levels are visible the front while front-to-back splits are not.
- Anticipate the presence of lightweight construction.
- Watch for zero-clearance fireplaces.

RAPID STREET-READ GUIDE SFD 6

BUILDING GROUP: Single-Family Dwelling—Modern Lightweight

BASIC CONSTRUCTION METHOD

- Lightweight platform wood frame with wood roof and floors
- Basements may have steel cylinder post columns and steel or laminated veneer lumber (LVL) girders to support the main floor.
- Extensive use of lightweight wood trusses, engineered wooden I-beams, and oriented strand board (OSB)

ERA CONSIDERATIONS

- Those built from 1980–2000 are likely built with classic wood frame techniques (cut lumber). Floors and roof use trusses and/or engineered wooden I-beams, exterior walls are framed with 2 × 6 in. studs or twin/tripled 2 × 4s at corners, and interior walls are 2 × 4s.
- Those built from 2000–present may have advanced framing methods (reduced lumber use) and incorporate the use of more engineered lumber such as laminated veneer lumber (LVL), parallel strand lumber (PSL), oriented strand board (OSB), and finger-jointed lumber (FJL).

EXTERIOR FEATURES

- Larger windows, most of e-glass and with double and triple panes, are common.
- Steep gable and hipped roofs are most common. OSB is most prevalent for roof sheathing and composite shingles are common, although light masonry and baked tile covering can also be found. Expect vented attic spaces with well-slotted eaves and numerous rooftop vents or a ridgeline vent; unvented (sealed) attics are rare.
- Exterior wall sheathing is of OSB with a wallboard, stucco, brick or stone veneer, or vinyl finish.
- Basement windows are large enough for occupant escape (deep well).
- Expect to find vertical and horizontal sliding windows.

INTERIOR FEATURES

- Great room concepts are the rule. Expect large open spaces for kitchen, dining, family, and living areas.
- Bathroom areas are likely to align with rooftop sewer vent pipes.
- Most water and waste piping is synthetic rather than copper or iron pipe. Natural gas piping can also be plastic.
- Vaulted and coffered ceilings are common.
- Adhesives are commonly used for structural binding applications.

SFD 6

TACTICAL CONCERNS

Fire Spread: Drywall is the primary fire spread barrier room to room. Modern HVAC return air and distribution systems are rarely fire-stopped (no dampers). Attached garages are rarely finished (drywall), although the separation wall to the living space must be drywalled (two layers of ⅝ in.). Open stairwells and great rooms allow explosive smoke to accumulate in upper areas. Those same features will contribute to extreme fire spread when the outside wind is over 15 mph. The use of adhesives can cause contents to rapidly burn and spread fire. When present, combustible wood roofs can contribute to spreading exposure hazards, and zero-clearance fireplaces can easily spread fire vertically to common areas.

Collapse: Exposed steel columns and girders in unfinished basements and garages fail quickly. Any fire or hot smoke that has captured structural wood areas is cause for concern; expect rapid collapse. Roofs and floors are "assembly-built," which means they can collapse prior to sagging. Higher heat release rates of building contents can cause drywall to fail quicker. Brown or tan smoke coming from roof or floor seams and vents is a collapse warning sign. Overall, the modern lightweight wood building collapses very quickly. Exposed I-beams (basements) can rapidly collapse.

Ventilation: Higher insulation values in newer homes may mask smoke and heat signs from the exterior. Ventilation-limited fires can be expected on arrival. Opening the front door for access may cause air to be introduced into explosive smoke. These homes may benefit from positive pressure attack (PPA) methods on a single room and contents fire (follow PPA protocol). Rooftop ventilation is a risky option due to lightweight trusses and should only be attempted with a well-trained crew and on an uninvolved portion of a roof. (Some departments prefer personnel to work from roof ladders to distribute weight.) Gable end venting may provide a safer alternative but can be dangerous.

Forcible Entry: Forcible entry can normally be accomplished by conventional methods. The front entry door is usually the most secure with multiple locking devices in a solid wood core. Interior doors are likely hollow core. Larger windows increase entry/egress options as ground floor windows are easily converted into doors with a power saw.

Search: A 360° size-up of window status can help prioritize search areas (survivable spaces). Using a window "vent-enter-isolate-search" method can help speed up searches of survivable spaces. Floor plans can vary from typical residential floor plans.

SPECIFIC HAZARDS

- Most utilities can be secured on the outside of the home.
- Solar electrical panels are live until they are covered with a dark salvage cover.
- Access to Side C can be difficult for homes built on sloping grades.
- Vinyl siding can be encountered.
- Adhesives will off-gas formaldehyde as they burn.
- Double-pane windows can resist heat and fire and enhance flashover conditions.
- The presence of solar electrical panels usually indicates that a battery room will also be present.
- Zero-clearance fireplace flue spaces may not be fire-stopped.
- Combustible wood roofs can add to exposure concerns.

RAPID STREET-READ GUIDE — SFD 7

BUILDING GROUP: Single-Family Dwelling—McMansion

BASIC CONSTRUCTION METHOD

- Conventional or lightweight platform wood frame with wood roof and floors

- Large span spaces with steel girders and steel post columns; some exposed beams may appear to be heavy timber but are likely steel with a wood veneer.

- Insulated concrete forms (ICF) and structural integrated panels (SIPs) may be used for portions of the building or the entire structure (look for signs of very thick walls).

ERA CONSIDERATIONS

- The McMansion is a relatively new trend (1990–present). Most are a morphed version of the modern lightweight wood single-family dwelling (SFD). Floors and roof use trusses and/or engineered wooden I-beams.

- The use of engineered lumber (LVL, PSL, OSB, and FJL) can be extensive.

Exterior features

- Multiple rooflines combine gables, hips, and flat roofs. OSB is most prevalent for wall and roof sheathing. Light masonry and baked tile/slate roof coverings dominate. Vented attic spaces have well-slotted eaves and numerous rooftop vents or a ridgeline vent.

- Basement windows are large enough for occupant escape (deep well).

- Multiple garages are common. They may be colocated or separate. Side C may have its own garage for golf carts or landscaping apparatus.

- Decorative masonry veneers (stone and/or brick) are common.

INTERIOR FEATURES

- Unusual or inventive floor plans can be expected. Great rooms can be cavernous. Bedroom areas are likely to be split, with guest rooms in one area and the master suite in a separate wing, floor, or area. A separate guest suite or nanny apartment may exist within the interior and not be discernible from the exterior.

- Vaulted and coffered ceilings are common.

- Natural gas piping can be plastic.

- Basements can include a lap pool, wine cellar, theatre, or other high-end amenities.

TACTICAL CONCERNS

Fire Spread: Drywall is the primary fire spread barrier room to room. Modern HVAC return air and distribution systems are rarely fire-stopped (no dampers). Open stairwells and great room concepts allow explosive smoke to accumulate in upper areas. Those same features will contribute to extreme fire spread when the outside wind is over 15 mph. The large footprint can mask interior fire conditions when viewed from the exterior. A thermal imaging camera may not reveal hidden fire in floors or walls when tile and decorative veneers are used for floors and walls. Large open areas are conducive to rapid fire spread.

Collapse: The oversized room concept of the McMansion makes them susceptible to early collapse once floor and roof elements are attacked by heat. The use of heavy floor tiles and granite counters/accruements adds dead load. Any fire or hot smoke that has captured structural wood areas is cause for concern; expect rapid collapse. Roofs and floors are often assembly-built, which means they can collapse prior to sagging. Crews assigned to interior ops should routinely test the floors to check for integrity.

Ventilation: Higher insulation values and vast floor plans may mask smoke and heat signs from the exterior. Numerous rooms, vaulted ceilings, stairwells, and unusual floor plans present significant ventilation challenges. Positive pressure attack (PPA) can be questionable for all but the smallest fires due to the large square footage and complex floor plans. Rooftop ventilation is a risky option due to lightweight trusses, multiple rooflines, and masonry/tile roof coverings. A combination of natural horizontal ventilation and multiple positive pressure ventilation (PPV) fans will have to be thoughtfully considered.

Forcible Entry: Forcible entry can normally be accomplished by conventional methods. Interior doors can be solid and/or hollow core. The front entry door is likely to be oversized, ornate, and very stout. Larger windows increase entry/egress options as ground floor windows may be converted into doors with a power saw if masonry siding is not present.

Search: A 360° size-up of window status can help prioritize search areas (survivable spaces). The use of large-area search tactics is warranted (tag lines). There are likely to be interior rooms and areas that are distal to exterior escape options.

SPECIFIC HAZARDS

- Most utilities can be secured on the outside of the home.
- Commercial grade or multiple furnaces and boilers are common.
- In-floor radiant heating systems add dead load and collapse concerns.
- Solar electrical panels usually indicate that a battery room will also be present.
- Access to Side C can be difficult for homes built on sloping grades.
- Insulation and modern energy-efficient windows can enhance interior temperatures and flashover conditions.
- High ceilings and large rooms can hold large volumes of smoke, creating a rapid fire spread potential.
- Zero-clearance fireplaces are common and can be numerous.
- Formidable security measures for doors and windows can be encountered.

RAPID STREET-READ GUIDE — SFD 8

BUILDING GROUP: Single-Family Dwelling—Manufactured (Mobile) Home

BASIC CONSTRUCTION METHOD

- Wood frame of lightweight construction, often of 2 × 2 in. framing.
- Larger examples may have aluminum skeleton with 2 × 2 in. framing for interior walls.
- Wood structure built upon twin steel I-beams that serve as a frame for transport axles

ERA CONSIDERATIONS

- These types of structures date back to the 1940s.
- Newer models are often called modular homes and can include two or more transported sections that are connected together on the home site.

EXTERIOR FEATURES

- The steel frame rails are anchored to a foundation of cinder blocks (CMUs), jacks, or other similar materials that allow the building to be relocated if necessary.
- The foundation is typically above ground and is skirted with panels to allow access to utility feeds.
- Siding is typically lightweight wood siding such as T-11 panels. In some cases, wood shiplap or synthetic masonry can be used. Older models may have aluminum or galvanized steel panels as siding.
- Roof structural members are also of lightweight construction and covered with a waterproof membrane material.
- Windows are typically small.
- Underside storage is common.

INTERIOR FEATURES

- Interior layouts are linear with narrow hallways between rooms.
- Interior privacy doors are quite narrow and hollow core.
- Finish can be drywall (minimal thickness) or panel board. Adhesives are also used for bonding agents for many materials.
- There is typically a main entry door on one side and a small emergency door off a bedroom on the opposite side.

TACTICAL CONCERNS

Fire Spread: Fires in these types of structures will rapidly grow in intensity and quickly spread due to the lightweight construction throughout the structure and the common use of flammable adhesives. Radiant heat is initially significant due to the low ceiling height. Flashover can occur in a few minutes or seconds. Floor burn-through is common. Underside storage is likely flammable and will rapidly extend fire under the structure.

Collapse: Collapse can occur in a short period of time due to the construction and relatively small size (as compared to typical single-family homes) of the structure. Multistory or high ceiling models are rare, and where they exist, present a more rapid collapse potential.

Ventilation: Due to the rapid extension of fire and the size of these structures, roof ventilation is normally not an option. Most self-ventilate as the exterior walls and roof burn through. Natural horizontal ventilation is rarely effective due to window size. The use of positive pressure ventilation (PPV) can be useful if applied in a judicious and timely manner to uninvolved areas.

Forcible Entry: Forcible entry can normally be accomplished by conventional forcible entry methods—often with only hand tools due the lightweight nature of the materials that form the doors, door frames, and other elements such as windows.

Search: Search operations are totally dependent of the amount of involvement. Remember, these fires burn hot and fast, so uninvolved areas must be able to allow sufficient time for search operations.

SPECIFIC HAZARDS

- Extensive use of flammable adhesives is expected.
- Smoke from these fires can contain a noteworthy amount of formaldehyde that is used as an adhesive in the wood paneling and for other similar uses.
- Mobile homes in parks are normally in close proximity to other mobile homes, so the potential of rapid exposure problems should be considered (depending on the amount of fire).
- Utility feeds are usually post mounted (detached from the home) with underground conduit that feeds the underside of the home.
- Underside storage is common and normally flammable.

RAPID STREET-READ GUIDE

BUILDING GROUP: Multifamily Dwelling—Brownstone

BASIC CONSTRUCTION METHOD

- Conventional wood and noncombustible construction, conventional flat wood roofs
- Commonly have basements/cellars

ERA CONSIDERATIONS

- They were erected during the late 1800s as private dwellings.
- Brownstones are indigenous to the East coast (New York, New Jersey, etc.).
- Adjoining buildings resulted in rows that can be a block long.
- If different builders were involved, noteworthy differences can be visible in each building.
- Remodels are common and can employ light-weight materials.

EXTERIOR FEATURES

- Brownstones are three to five stories in height with a cellar underneath a basement, and are about 20 ft to 25 ft wide and 60 ft deep.
- The basement is the first floor and the entrance (doorway) to the basement is under the stoop (stairs).

- A partially sunken patio in the front provides access to the doorway under the stoop.
- Exterior walls are typically conventional brick construction.
- Parapet walls above a roof and between occupancies should be visible.
- Division walls were constructed between each occupancy but are likely to have been breached.
- Expect a lack of proper fire-stopping in common attics/cocklofts.
- Wood or metal cornices are frequently found, and some can be elaborate.
- Roof skylights are common over an interior stairwell.
- Windows are older double-hung with associated hazards.

INTERIOR FEATURES

- Balloon frame construction and knob and tube wiring was commonly used.
- The basement contains a kitchen, dining room, and an interior stairway to the second floor and cellar.
- The second floor is accessed by the front stoop (stairs) and contains a front and rear parlor, bedroom, and primary entrance to the building.
- The third and fourth floors contain two bedrooms each.
- Originally, a hot air furnace was located in the cellar with ducting to other portions of the building.
- Dumbwaiters that served several floors were often used. Sprinklers are not common.

MFD 9

TACTICAL CONCERNS

Fire Spread: The interior is comprised of conventional, flammable materials. Fires in the cellar and basement can easily spread vertically through pipe chases, stairways, dumbwaiter shafts, and other openings. Fire can also spread horizontally to adjoining buildings via substandard construction in the division walls, common cocklofts, exterior cornices, and wooden gutters. Vertical extension can also occur in open stairs at the rear of a building and the main staircase. The main stairway is often paneled with combustible wood products.

Collapse: The noncombustible masonry exterior is resistant to fire; however the combustible wood interior is not. Although it will take a sizeable fire for an entire building to collapse, the interior—with its many vertical passageways—is capable of collapsing when attacked by a fast-moving fire. If the interior portions begin to collapse, the brick exterior is also at risk of collapsing. Do not forget the potential of floors over a developed fire (particularly a cellar or basement fire) collapsing, as well as cornices collapsing downward from the front of a building.

Ventilation: If the interior stairway is contaminated, smoke can be cleared by the removal of the roof skylight over the stairway. Fire on lower floors and/or the cellar and basement will quickly rise toward the top portion of a building. It is a top priority to quickly vent the top portion by vertical roof operations (opening roof, skylights, etc.) and/or horizontal operations using appropriate windows. Strip/trench ventilation can be used to limit horizontal extension in attics/cocklofts. Fires in cellars will be difficult to ventilate as there are limited means for ventilation. PPV can be beneficial with proper coordination.

Forcible Entry: On the exterior, barred windows at grade level (front and rear) can be expected, with some bars set into the masonry construction. There is normally a metal grate over the cellar vent, an iron gate on the door into the stoop that accesses the basement, and substantial door/locking devices on the main entrance doorway. In the interior, pocket sliding doors and doors within rooms are not very substantial. However, if a building has been converted into a multiple residential type occupancy, then doors and subsequent locking devices can be very substantial. In most cases, forcible entry operations can be accomplished by conventional methods.

Search: Search operations are governed by the potential occupant load. If a single family is the sole occupant, search operations are simplified as opposed to a converted multiple dwelling type occupancy with multiple occupants/families. Four areas deserve special consideration. One, small bedrooms (front and rear) with a single entrance from a hallway (referred to as a "deadman's" room). Two, rear extensions to a building that may be several stories in height. Three, buildings with a fully dormered attic apartment. Four, the cellar under the basement. Remember that some rooms will be small to moderate in size, and vertical stair shafts can be narrow. Search operations above a fire will be challenging.

SPECIFIC HAZARDS

- Many brownstones have been converted into multiple residential dwellings. In the process, sprinklers may have been installed in lieu of a secondary means of escape.
- Some conversions have resulted in illegal multiple dwellings without fire escapes or sprinklers.
- Common attics/cocklofts should be expected. Look for attic vents and their condition.
- Division walls are often breached and can allow openings for horizontal extension of fire.
- Look for the presence of a fully dormered attic apartment and/or an added rear extension.
- Adjoining buildings can present serious exposure considerations and apparatus placement problems to the rear.
- Expect significant roof dead loads.
- Anticipate the presence of inverted roofs.

RAPID STREET-READ GUIDE

BUILDING GROUP: Multifamily Dwelling—Tenement

BASIC CONSTRUCTION METHOD

- Brick masonry exterior or wood framing with wood exterior and conventional wood roofs
- Commonly have a basement

ERA CONSIDERATIONS

- Due to noteworthy construction differences, tenements can be defined as old law tenements—those built prior to 1901, or new law (or retrofit) tenements—those built after 1901.
- Tenement type buildings are not indigenous to New York City, but the old/new law regulations are.

EXTERIOR FEATURES

Old law summary

- Old law tenements are three to seven stories high, 20 ft to 25 ft wide, and 50 ft to 85 ft deep.
- Roof styles are normally a conventional flat roof or inverted/raised sloping roof over a common attic.
- Fire escapes on both the front and back denote four apartments per floor.

New law (retrofit) summary

- New law tenements are typically six to seven stories high, 35 ft to 50 ft wide, and 85 ft deep. Front cornices (often quite large) are common.
- Buildings with fire escapes on both the front and back have four apartments per floor.

INTERIOR FEATURES

Old law summary

- Cellars have a stairway to the grade floor and floor joists for the grade floor *are exposed* to the cellar.
- There are two to four apartments to each floor. There may be two railroad flats per floor.
- Stairways and stairs *can use combustible materials*.
- There are two means of egress from each apartment (interior stairway and fire escape).
- They may have dumbwaiter chutes that travel from the cellar through the roof.

New law (retrofit) summary

- The cellar ceiling is of fireproof construction with *no* stairway to the floor above.
- The cellar entrance is by way of exterior stairs and there are five to six apartments per floor.
- The interior stairs (grade floor to top floor/roof) are fireproof and enclosed in partitions of fireproof construction.
- A second means of egress to occupants is either by another stairway or an exterior fire escape.
- Newer new law buildings can have elevators that run from the cellar to the top floor, with a roof bulkhead. After 1929, these buildings were called apartment houses.

TACTICAL CONCERNS

Fire Spread: The open stairway from the cellar to the grade floor in old law buildings can allow the vertical spread of fire up the stair shaft to the top of the building. Also expect fire spread due to combustible materials in the vertical stair shafts, dumbwaiter shafts, pipe chases, substandard construction, elevator shafts if present, common attics, stacked closets, structural steel I-beams that provide a substantial void within the web of the beam, exterior cornices, voids that extend vertically through all floors, and *throats* that are between various types of large multiple buildings (e.g., H, E, O, U, double H) and duplex apartments.

Collapse: The numerous voids and construction configurations (common attics, exposed floor joists to the cellar, etc.) will expedite the rate of failure. Additionally, steel I-beams that are exposed to heat/fire can warp and collapse, often allowing large portions of a building supported by the I-beams to also collapse.

Ventilation: Stair shafts(s) can be vertically ventilated by opening the bulkhead door on the roof, and contaminated hallways and rooms can be horizontally ventilated by natural and/or PPV operations. Roof ventilation can be accomplished by a suitable variety of opening bulkhead doors, bulkheads over dumbwaiter shafts, skylights that will help to ventilate the top floor hallway (common to the skylight), and opening the roof. Roof ventilation teams should be alert to the presence of an inverted/inverse roof that can feel spongy without being exposed to fire and multiple layers of roofing materials. Ventilation of a contaminated cellar can be challenging and should be accomplished through openings to the exterior of the building, not the stairway into the grade floor of a building.

Forcible Entry: Substantial wood doors and/or metal clad doors in metal frames in addition to multiple locking devices are common. Some windows are barred by protective gates, particularly those windows that are common to fire escapes. Although conventional forcible entry methods can work, expect delayed times and/or the possibility that extreme security measures may require the consideration of other means of entrance such as entering from a fire escape, walls between apartments, or some other means. The exterior entrance to the cellar in new law buildings will likely be of substantial construction with formidable security considerations. These forcible entry concerns are particularly true in East coast buildings

Search: The sizeable number of people in these structures can easily require an intensive and coordinated commitment of resources for search operations. However, the size of each unit/occupancy is small to moderate and the floor plans are typically command and residential in nature. Access is limited to each unit from a central hallway. Coordinated search operations may require that the numerous units to be searched be visibly identified as (1) being searched, (2) needs to be searched, and (3) search completed in order to avoid repetition.

SPECIFIC HAZARDS

- Differentiate between old law and new law buildings.
- The status of the vertical stair shaft(s) and hallway(s) is a key initial consideration.
- Determine how the floors and apartments are designated, as 3E could be on the third or fifth floor.
- Determine the type of stairways that are present (e.g., transverse, wing, and isolated).
- Anticipate renovations, settling of a building, and poor workmanship.
- Common attics/cocklofts should be expected. Look for attic vents and their condition.
- Exterior cornices can present a formidable collapse hazard when exposed to fire.
- Stairs are often steel risers with marble/slate treads and can collapse with little warning when heated or suddenly cooled by water.
- Buildings that are not on level ground will have differences in height between the front and back.
- Look for the presence or absence of sprinklers and elevators.
- Rear access is typically poor at best and may only be accomplished through the building.
- Always consider the possibility of a wind-driven fire on the windward side.

RAPID STREET-READ GUIDE

BUILDING GROUP: Multifamily Dwelling—Row Frame

BASIC CONSTRUCTION METHOD

- Conventional wood framing unless remodeled with lightweight materials
- Conventional wood roofs, either with a pitch or flat
- Perimeter foundations and basements

MFD 11

ERA CONSIDERATIONS

- Most were constructed between the 1800s and early 1900s.
- They were built in rows that can consist of numerous buildings. Significant differences can be visible in each building.
- Expect conventional wood and some masonry construction. Roofs are of conventional construction.
- Remodels are common and can employ lightweight materials.

EXTERIOR FEATURES

- Recognized by individual attached residential occupancies in rows that may be over 20 buildings in length.
- Height can vary from two to five stories and are often 20 ft to 30 ft wide and 40 ft to 60 ft deep.
- Exterior can be comprised of conventional brick masonry, brick veneer, wood siding, and some vinyl.
- Roof styles are normally a sloped flat, gable, or hip roof. Most roofs are over a common attic/cockloft.
- Division walls were constructed between each occupancy but are likely to have been breached.
- Expect a lack of proper fire-stopping in the attics/cocklofts.
- Common cornices (between occupancies) are frequent.
- Depending on a specific area, retail stores may be found on the first floor.
- Division walls above a roof may be visible on some buildings.
- Some buildings use dormers for the top floor.

INTERIOR FEATURES

- Basements/cellars are common and may run under multiple buildings with no separation.
- Balloon frame or braced frame construction was used along with knob and tube wiring (unless upgraded).
- Common cocklofts/attics potentially extend over buildings in a row.
- Expect to find remodeled interiors and resultant voids.
- Older buildings can be expected to use lath and plaster, whereas remodels use drywall.
- Dumbwaiter and light shafts are common. Sprinklers are likely not present.
- Scuttle on roof allows access into the top floor.
- Newer versions commonly use lightweight construction.

TACTICAL CONCERNS

Fire Spread: These types of buildings can be considered large rectangular boxes of dried lumber that are capable of producing copious amounts of fire, in all directions, and in a relatively short amount of time. Fire can easily spread vertically in pipe chases, light/air shafts, auto exposure via rear and front windows, balloon frame construction, from basements/cellars upward, and in voids created by modern remodels. Horizontal extension is a primary concern in common attic/cocklofts. Once a well-developed fire creates copious amounts of smoke in a common attic/cockloft, it may be difficult to find the involved building. Fire can also easily spread horizontally in the exterior common cornices, between walls separating occupancies, in common basements/cellars under adjoining occupancies, and from adjoining roofs. Lightweight floor remodels will allow rapid extension of fire between the flooring, and ducting from the original heating plant to registers throughout a building can also easily extend fire.

Collapse: Due to the age and type of construction of these buildings, the potential of collapse should be a measurable concern. Heavy fire in a common attic/cockloft will rapidly weaken ceiling and roof structural members. Brick veneer on the exterior of a building is subject to collapse when the wood framing behind the masonry is exposed to fire. Rear walls have been known to collapse in one section, and floors over a well-developed fire (e.g., basement/cellar) can be a prime candidate for collapse. Wood/metal truss construction in the floors and roof structural members from remodels is a prime candidate for collapse when exposed to fire.

Ventilation: Horizontal and/or PPV is normally effective and easily used for cross ventilation of contaminated occupancies. Vertical roof ventilation can be effectively used for top floor occupancies in multi-story buildings (remember that slate type roofs will require additional time), and opening penthouses and/or bulkheads can effectively ventilate stair shafts. Skylights and light wells can also be used for vertical ventilation.

Forcible Entry: Forcible entry can generally be accomplished by conventional methods as entry is simplified by typical doors and windows that are found in the average residential type of occupancy (although older buildings can have substantial doors). Although conventional forcible entry methods can work well, expect multiple locking devices (some with dead bolts) in entry doors as the close proximity of people in these occupancies often result in increased security concerns. This is particularly true in East coast buildings.

Search: The potential number of people in these structures can easily require a larger resource commitment for search operations. The size of each unit/occupancy is dependent on the size of the building; however, the floor plans are typically residential in nature. Access to grade floors is obtained by front/rear doors. Access to basements/cellars and floors above grade level may be limited due to narrow stairways. Search operations can be minimized during the day depending on observed factors but are generally maximized at night.

SPECIFIC HAZARDS

- Some row frame buildings can be very large, over 100 years old, made entirely of wood, and contain numerous occupants.
- Large buildings can present a significant search challenge.
- Always try to determine the presence of a remodel and the resultant use of lightweight construction.
- Common attics/cocklofts are a standard. Look for attic vents and their condition.
- Although division walls are often visible, they are often breached.
- Exterior brick veneers are decorative only and are subject to collapse if fire exposes the wood framing. Buildings in close proximity can present considerable extension problems.
- Vinyl siding (from retrofit modifications) will readily burn and also emit toxic smoke.
- Look for the presence or absence of sprinklers. Some newer units do have sprinklers, depending on applicable codes that are area specific.
- Anticipate inverted roofs on older buildings.

RAPID STREET-READ GUIDE

BUILDING GROUP: Multifamily Dwelling—Railroad Flat

MFD 12

BASIC CONSTRUCTION METHOD

- Brick masonry exterior, wood framing with wood exterior, conventional roof of wood
- Perimeter foundations and basements

ERA CONSIDERATIONS

- Railroad flats made their initial appearance in New York City around the middle of the 19th century, and are most common in New York, San Francisco, and their neighboring areas.
- Railroad flats are also common in brownstone type buildings.

EXTERIOR FEATURES

- Buildings are normally narrow and can vary from single-family residential type to those that are five to six stories in height.
- The exterior finishes can be comprised of conventional masonry, brick veneer, or wood siding. Remodels can use other materials, such as vinyl siding.
- Roof styles are normally a sloped gable or hip roof over a common attic, and flat roofs (normally over a common attic/cockloft) are also common.
- The placement of the common hallway in single-family type residential buildings may be on the side of the building and visible.
- In larger buildings, the short hallway and associated stairwell per floor is not visible from the exterior.
- They often back up to a similar building on an opposing street.
- Some single-family dwellings can contain railroad flat floor plans.

INTERIOR FEATURES

- Railroad flats/apartments are configured with a series of rooms connecting to each other in a line (linked together like railroad cars) that are served by a common hallway.
- A short center hallway that serves four flats per floor is accessed by a vertical stairway. Some older buildings may have trash chutes and other similar vertical voids.
- Normally, the bedroom areas are toward the front with the kitchen toward the back.
- There are two entrances to each flat.
- Kitchens and bathrooms are stacked on top of each other in multistory buildings.
- These buildings are prime candidates for remodeled interiors and resultant voids.
- Individual HVAC units typically serve each occupancy.

TACTICAL CONCERNS

Fire Spread: Fires in basements can spread vertically very rapidly. This is enhanced by openings/voids such as the vertical stairway and pipe chases for stacked kitchens and bathrooms that allow for the spread of fire/heat/smoke upward into upper flats and the cockloft/attic and roof. If fire and/or smoke is able to extend into the vertical stairway, it will rapidly spread upward and contaminate each landing above the point of origin. Fire on the top floor can spread into the attic/cockloft above, which can then rapidly spread horizontally due to the presence of a common attic/cockloft. If division walls are present, they must be verified for security. Cornices can horizontally spread fire. A one-room fire will quickly contaminate all rooms in a flat.

Collapse: It will take a sizeable fire for an entire building to collapse, particularly in masonry buildings. However, remember that unreinforced masonry construction is not as resistant to collapse as conventional brick construction and remodels will likely use lightweight materials that readily burn and fail at a faster rate than conventional materials. Whether a building is single story or multistory, common attics are usual and are subject to collapse when exposed to fire (a contributing factor is numerous layers of roofing materials that have built up over the years). Cornices are a willing candidate for collapse when weakened by fire.

Ventilation: Horizontal and/or PPV can be effective and easily used for cross ventilation of contaminated occupancies. Particular attention must be directed toward the hallways, as keeping these areas clear is essential. It is also important to quickly vertically ventilate a contaminated stair shaft in multistory buildings. Vertical ventilation for contaminated flats can be effectively used for one-story structures and for top floor occupancies in multistory buildings.

Forcible Entry: Entry is simplified by the typical doors and windows that are found in these occupancies. Although conventional forcible entry methods can work well, anticipate advanced and/or multiple locking devices (some with dead bolts) in entry doors as the close proximity of people in these occupancies often result in increased security concerns. However, once inside a flat, sliding/pocket type doors are common and are not difficult to force.

Search: As the individual rooms are somewhat small, the floor plans should be known in advance. The rooms in each flat are in line, so they can be rapidly searched in an orderly manner. In multistory buildings, there are typically four flats per staircase landing. It is normally best to enter a flat through the door nearest the staircase landing, as it should enter the kitchen area and not be blocked as is common with the other door that enters the bedroom and/or living room areas. Search operations can be conducted in linear fashion from the kitchen to the bedroom/living room area.

SPECIFIC HAZARDS

- Remember that each flat has two entry doors. In multistory buildings, the entry door by the staircase enters the kitchen and is normally not blocked. The other entry door enters a bedroom/living area and is often blocked.
- The stacked bathrooms and kitchens in multistory buildings provide vertical passageways that often travel into the cockloft/attic and above the roof.
- Common attics/cocklofts are standard. Look for attic vents and their condition.
- If the vertical stairway becomes contaminated with fire/smoke, it will quickly travel upward and contaminate floors above the point of origin.
- Rear access to multistory buildings is typically poor at best.
- Always try to determine the presence of lightweight renovations.
- If division walls are visible, they are often breached.
- Cornices can be a collapse hazard to fireground personnel.
- Determine the presence or absence of sprinklers.

RAPID STREET-READ GUIDE

BUILDING GROUP: Multifamily Dwelling—Center Hallway Structure

MFD 13

BASIC CONSTRUCTION METHOD

- Configured with residential or commercial units that are accessed by interior center hallway(s)
- Wood/metal frame, masonry of concrete block or brick construction, brick veneers, and poured-in-place concrete with either wood or metal roofs
- Concrete slab, perimeter foundation, and some basements

ERA CONSIDERATIONS

- Pre-1960 buildings are primarily conventional construction comprising the walls, floors, and roofs.
- Post-1960 buildings have embraced light-weight construction due to cost effectiveness.

EXTERIOR FEATURES

- These buildings are easily identified by the presence of center hallways with windows at each end.
- Floor plans can be either residential or commercial.

- Buildings are configured from one story to multiple stories in height, including some high-rise buildings.
- As these structures have been built for many years and encompass a wide range of building types, the exterior finish can be a variety of materials such as stucco, wood/vinyl siding, concrete, concrete block, brick veneer, and brick masonry.
- Depending on the type of structure, roof construction will encompass all types and configurations.

INTERIOR FEATURES

- Interior units will vary between residential and commercial type occupancies and will also vary in size and complexity.
- Interior residential or commercial units are accessed by an interior center hallway.
- Older buildings that have been remodeled will be a mixture of old and new construction that is often accompanied by voids that were not an original part of the building.
- Common attics will be present in many buildings and, depending on the type of building, can be extensive.
- Lath and plaster is used in older buildings and drywall and/or dropped ceilings is used in newer buildings.
- Expect to find truss floor construction in newer buildings.
- Depending on the age and configuration of a building, sprinklers and/or standpipes may be present.
- HVAC units vary between a central system and individual units for each occupancy within a building.

TACTICAL CONCERNS

Fire Spread: A primary concern for the spread of fire is the center hallway configuration. If fire is able to extend into a hallway, it can quickly become a horizontal avenue for the spread of fire and can also hamper occupant egress and firefighter access. This is why the status (clear or contaminated) is a primary consideration. Fire on the grade floor of single-story buildings can easily spread into an attic (which can be a common attic). Fire on the top floor can easily spread into the attic above and is only slowed by lathe & plaster or drywall type materials. If fire spreads into a common attic it will allow rapid horizontal spread of fire to both ends of the building. Remember that multistory buildings with lightweight construction often have lightweight construction between the floors that will allow rapid extension of fire between the flooring. Also, if division walls are present, they may be breached and must be verified for security.

Collapse: Generally, it will take a sizeable fire in conventional construction for an entire building to collapse. Concrete is the most fire resistive, conventional masonry is somewhat less resistive, and wood/metal construction is the most prone to collapse when exposed to fire. Wood/metal truss construction in the floors and roof structural members is a prime candidate for collapse when exposed to fire. Some of these buildings commonly used a brick veneer on the exterior that is subject to collapse when the wood framing behind the masonry is exposed to fire.

Ventilation: Horizontal and/or PPV can be effective and easily used for cross ventilation of contaminated occupancies by using blowers to clear hallways and then individual units/occupancies. PPV can also be used to pressurize vertical stair shafts and horizontal hallways. Vertical ventilation can be effectively used for one-story structures and for top floor occupancies only. Opening penthouse doors when present can also clear common vertical stair shafts

Forcible Entry: Depending on the type of structure, forcible entry can generally be accomplished by conventional methods as entry is simplified by typical doors and windows. Remember that commercial type structures often employ more intensive security measures than residential type structures. Expect multiple locking devices with dead bolts in entry doors.

Search: A common configuration with these structures is individual units or occupancies that are accessed by a center hallway. Therefore, if only individual units or occupancies are contaminated and need to be searched, the number of units/occupancies dictates the type and difficulty of search operations. Additionally, if the hallways are contaminated, this will add to the difficulty of search operations. Remember that some type of order to search operations when multiple units/occupancies are involved will include visually designating rooms that have not been searched, are being searched, and need to be searched to help avoid duplication.

SPECIFIC HAZARDS

- The status of the hallway(s) is a major indicator of the type and extent of fire. If the hallways are clear, it is likely a room and contents fire with possible vertical/horizontal extension. If the hallways are contaminated, the fire has spread into the hallway that suddenly becomes a horizontal passageway for the extension of fire and the potential entrapment of occupants on the contaminated floor(s).

- Fires that have captured the central hall on upper floors will likely require significant resources to ladder windows for search and rescue.

- Look for the presence of lightweight construction.

- Anticipate the presence of a common attic.

- Although division walls are often present, they are frequently breached and can allow openings for horizontal extension of fire.

- Exterior brick veneers may be decorative only.

- Sprinklers and/or standpipes may be present.

- Vinyl siding will readily burn, emit a toxic smoke, and can create significant exposure problems.

- Consider the presence of zero-clearance fireplaces and elevators.

RAPID STREET-READ GUIDE

BUILDING GROUP: Multifamily Dwelling—Garden Apartment

BASIC CONSTRUCTION METHOD

- Residential or commercial units that open to the exterior of a building; *no interior hallways for unit access*
- Wood frame is most predominant, with some masonry of concrete block or brick construction
- Concrete slab or perimeter foundations

MFD 14

ERA CONSIDERATIONS

- Pre-1960 buildings are primarily conventional construction comprising the walls, floors, and roofs. Some use brick veneer, and older buildings use conventional brick construction.
- Post-1960 buildings have embraced lightweight construction. Brick veneers are still common.

EXTERIOR FEATURES

- Garden apartments are easily identified by units that open to the exterior of a building and a lack of center hallways.
- The front of each unit is normally configured with a door and one or more windows.
- Buildings are normally configured from one to three stories in height. One to two stories is the most prevalent.
- The exterior finish will vary (stucco, wood or vinyl siding, concrete, concrete block, etc.).
- Roofs are typically sloped gable or hip but can be flat. Older roofs are constructed from conventional lumber and likely have numerous layers of roofing materials (composition) or other materials such as tile, corrugated metal, etc.
- Stairways and landings are easily recognized as they serve the upper or lower units.
- Division walls projecting above a roof will be visible on some buildings.

INTERIOR FEATURES

- Residential or commercial units face the exterior of a building and are accessed by landings and stairways.
- Individual units are normally simplistic and consist of smaller residential type occupancies (e.g., motels, apartments, etc.) or office type units.
- Lath and plaster is used in older buildings and drywall is used in newer buildings.
- Common attics can be expected.
- Sprinklers and/or smoke detectors may or may not be present.
- Individual HVAC systems are normally used for each unit.
- Expect to find a single wall between older occupancies and offset studs between occupancies in newer buildings.
- Expect to find lightweight truss roof and floor construction in newer buildings.

TACTICAL CONCERNS

Fire Spread: Fire on the grade floor of single-story buildings can easily spread into an attic, which is likely a common attic. Fire on the grade floor of multistory buildings can be hampered by the floor construction above the fire. Fire on the top floor can easily spread into the attic above. If fire spreads into the attic, a common attic will allow rapid horizontal spread of fire to both ends of the building. Additionally, if division walls are present, they may be compromised or breached and must be verified for security. Remember that multistory buildings with lightweight construction often have lightweight construction between the floors that will allow rapid extension of fire between the flooring.

Collapse: It will take a sizeable fire for the entire building to collapse. Concrete is the most fire resistive, conventional masonry is somewhat less resistive, and wood/metal construction is the most prone to collapse when exposed to fire. Wood/metal truss construction in the floors and roof structural members is a prime candidate for collapse when exposed to fire. Remember that some of these buildings use a brick veneer on the exterior that is subject to collapse when the wood framing behind the masonry is exposed to fire.

Ventilation: These buildings have multiple avenues that can be used for ventilation operations. Horizontal and/or PPV can be effectively used for cross ventilation of contaminated occupancies. The fact there are no enclosed stairwells and hallways greatly simplifies horizontal ventilation considerations. Vertical ventilation can be effectively used for one-story structures and for top floor occupancies only.

Forcible Entry: Grade floor units are easily accessible, and exterior stairs to landings simplify access to upper floors. Typically there are no rear doors, so access to the interior is simplified by either a window next to the entry door and/or a door (wood) that is similar to those found on single-family dwelling structures. Multiple locking devices and dead bolts are common. Conventional forcible entry methods can work well.

Search: Search operations are normally simplified by the small to moderate size of each unit. The floor plan is dependent on the type of occupancy (residential or commercial). Easy access is limited to the front of each unit; however, some units feature large windows at the rear and front. Search operations are minimized at night in commercial occupancies and maximized at night for residential occupancies.

SPECIFIC HAZARDS

- Lightweight construction is common.
- Common attics spread fire horizontally.
- Division walls are often breached and can allow openings for horizontal extension of fire.
- Offset studs and walls can be found in newer construction. This will allow horizontal and vertical fire spread between the walls that separate individual occupancies.
- Exterior brick veneers are for decoration only.
- Asphalt siding (also referred to as gasoline siding) will readily burn.
- Vinyl siding will readily burn and also emit a toxic smoke.
- A lack of sprinklers is common.
- Check for basements. Access to basement units may be not be visible from the front of the building. Check the sides and rear.
- Elevators are not typically present.

RAPID STREET-READ GUIDE

BUILDING GROUP: Multifamily Dwelling—Low Density

MFD 15

BASIC CONSTRUCTION METHOD

- Poured-in-place concrete, masonry of concrete block or brick, wood/metal structural members with wood/metal framing
- Lightweight materials of wood, metal, and/or lightweight concrete
- Concrete slab, perimeter foundation, and basements

ERA CONSIDERATIONS

- These occupancies are recognized by containing multiple dwelling units in one or more buildings (that are grouped together) in low-cost areas. The defining feature is a more relaxed or widely spaced type of housing as opposed to high-density low-cost housing, which places as many occupancies as possible into the least amount of space.
- Older buildings typically use conventional construction and materials such as concrete, masonry, wood, and metal for the walls, floors, and roofs. Newer buildings use an increasing amount of lightweight methods and materials.

EXTERIOR FEATURES

- Exterior finishes can be comprised of conventional masonry, brick veneer, wood/vinyl siding, and some poured-in-place concrete. Lightweight synthetic panels are often used on newer low-cost construction.
- Roof styles are normally a sloped gable or hip roof over a common attic. Flat roofs (often over a common attic) are also common.
- Center hallways are the typical configuration for multistory buildings, but some garden apartment configurations can be found in smaller multistory buildings.
- Multistory buildings may have exterior patios or balconies for some or all units.

INTERIOR FEATURES

- Normally consists of individual residential units that are accessed by exterior doorways or a center hallway.
- Older buildings can be expected to use lath and plaster, whereas newer buildings use drywall.
- Expect to find some older buildings with remodeled interiors and resultant voids.
- Some offset studs/walls between adjoining occupancies are found in newer buildings, and single walls in older buildings.
- Depending on the building, individual HVAC units serve each occupancy or a central HVAC unit is used.
- Sprinklers may be present, depending on the age of a building.
- Older buildings may have lift or elevator devices that have been added to the building to meet ADA or occupant needs. Newer buildings are likely to have more modern elevators.
- Some older buildings may have trash chutes and other similar vertical voids.

TACTICAL CONCERNS

Fire Spread: Fire on the grade floor of single-story buildings can easily spread into the attic, which is likely a common attic. If a fire extends past a room of origin in a single-story garden apartment type building, the fire will primarily extend to the exterior of the building. Fire on the grade floor of multistory buildings can be hampered by the floor construction above the fire (stronger and thicker than lath/plaster and/or drywall). However, if fire extends into a center hallway, the hallway will become a horizontal channel within the building for the extension of heat, smoke, and fire. Fire on the top floor can easily spread into the attic above. If fire spreads into the attic, a common attic will allow rapid horizontal spread of fire to both ends of the building. Also, offset studs between adjoining walls will willingly spread fire horizontally and vertically.

Collapse: Newer buildings constructed from lightweight materials can readily burn and will fail at a faster rate than conventional buildings. Wood/metal truss construction in the floors and roof structural members is a prime candidate for collapse when exposed to fire. Remember that some of these buildings use a vinyl siding that can enhance exposure problems and emit a toxic smoke. Brick veneer on the exterior of a building is subject to collapse when the wood framing behind the masonry is exposed to fire.

Ventilation: Horizontal and/or PPV can be effective and easily used for cross ventilation of contaminated occupancies. Ventilation operations will be more extensive for contaminated hallways/stair shafts, but can be ventilated with PPV. Hallways/stair shafts are not a concern with garden apartment type occupancies. Vertical ventilation can be effectively used for one-story structures and for top floor occupancies only in multistory buildings.

Forcible Entry: Forcible entry can generally be accomplished by conventional methods, as entry is simplified by typical doors and windows that are found in the average residential type of occupancy. Although conventional forcible entry methods can work well, expect multiple locking devices (some with dead bolts) in strong entry doors as the close proximity of people in these occupancies often result in increased security concerns. This is particularly true in East coast buildings as opposed to West coast buildings.

Search: The increased number of people in these structures (as opposed to typical single-family dwellings) can easily require a larger resource commitment for search operations. However, the size of each unit/occupancy is small to moderate and the floor plans are typically residential in nature. Easy access is limited to the front of each unit above grade floors and the front/rear of many single-story structures. Search operations can be minimized during the day depending on observed factors but are generally maximized at night.

SPECIFIC HAZARDS

- Always try to determine the presence of lightweight construction.
- These types of buildings are typically found in low-income areas and can present additional safety concerns toward fireground personnel.
- Offset studs and walls can be found in newer construction.
- Common attics are standard. Look for attic vents and their condition.
- Although division walls are often present, they are frequently breached.
- Exterior brick veneers are for decoration only and are subject to collapse if fire exposes the wood framing.
- Asphalt siding (also referred to as gasoline siding) will readily burn.
- Vinyl siding will readily burn and also emit a toxic smoke.
- Look for the presence or absence of sprinklers. Newer units may have sprinklers, depending on applicable codes for that area.
- Multiple buildings in close proximity and/or remote to roadways can present apparatus placement problems.

RAPID STREET-READ GUIDE

BUILDING GROUP: Multifamily Dwelling—Project Housing—High Density

MFD 16

BASIC CONSTRUCTION METHOD

- Wood or metal structural members, some poured-in-place concrete, masonry of concrete block or brick
- Lightweight materials of wood, metal, and/or lightweight concrete
- Concrete slab, perimeter foundation, and basements

ERA CONSIDERATIONS

- These types of buildings share a common characteristic of providing affordable housing (also known as socialized/public housing) in single or multiple buildings that place a large number of people in close proximity. These buildings have been built for many years and in many different configurations. Some buildings that are often associated within this category are large multistory buildings that can be of older construction and newer lightweight multistory buildings that have been specifically built for this use.

EXTERIOR FEATURES

- Multiple dwelling units in one or more buildings (that are grouped together) in high-density areas.
- Exterior finishes can be comprised of conventional masonry, brick veneer, some wood/vinyl siding, and some poured-in-place concrete. Lightweight synthetic panels are often used on newer buildings.
- Roof styles are normally flat, sloped gable, or hip roof, all of which can be over a common attic.
- Center hallways are the typical configuration for multistory buildings.
- Older buildings typically have exterior fire escapes and standpipe connections.
- Division walls above a roof may be visible on some buildings.

INTERIOR FEATURES

- Individual residential units are often accessed by doorways from a center hallway configuration.
- Older buildings can be expected to use lathe and plaster, whereas newer buildings use drywall.
- Some offset studs/walls between adjoining occupancies are found in newer buildings, and single walls in older buildings.
- Expect to find some older buildings with remodeled interiors and resultant voids.
- Elevators are normally present, and their design will depend on the age of the building.
- Standpipe connections of various types are located in hallway landings. Sprinklers may be present.
- Fire doors are used in some buildings but are often blocked open.
- Basements are common. Some are divided into lockable storage areas using fencing.
- Depending on the building, individual HVAC units serve each occupancy or a central HVAC unit is used.
- Some older buildings may have trash chutes and other similar vertical voids.

TACTICAL CONCERNS

Fire Spread: If fire is able to extend into a center hallway, it will become a horizontal channel within the building for the extension of heat, smoke, and fire. Fire in hallways on lower floors of multistory buildings can extend vertically if fire doors are not present or are blocked open. Fire on the top floor can easily spread into the attic above. If fire spreads into the attic, a common attic will allow rapid horizontal spread of fire to both ends of the building. Offset studs between adjoining walls will willingly spread fire horizontally and vertically. Remodels will likely have voids that can easily spread fire. Storage area basement fires can rapidly extend upward.

Collapse: It will take a sizeable fire for an entire building to collapse, but newer buildings constructed from lightweight materials can readily burn and will fail at a faster rate than conventional materials. Concrete is the most fire resistive, conventional masonry is somewhat less resistive, and wood/metal construction is the most prone to collapse when exposed to fire. Wood/metal truss construction in the floors and roof structural members is a prime candidate for collapse when exposed to fire. Brick veneer on the exterior of a building is subject to collapse when the wood framing behind the masonry is exposed to fire, and fire in a common attic will quickly weaken roof structural members.

Ventilation: Horizontal and/or PPV can be effective and easily used for cross ventilation of contaminated occupancies (using entrance doors and windows). Ventilation operations will be more extensive for contaminated hallways/stair shafts but can be ventilated with natural ventilation and/or PPV. Remember that it will be necessary to open penthouse/bulkhead doors to vertically ventilate a stair shaft. Vertical ventilation can be effectively used for top floor occupancies in multistory buildings.

Forcible Entry: Entry is simplified by typical doors and windows that are found in the average residential type of occupancy. Although conventional forcible entry methods can work well, expect multiple locking devices (some with dead bolts) in entry doors as the close proximity of numerous people in these types of occupancies often result in increased security concerns. This is particularly true in East coast buildings as opposed to West coast buildings. Main entrance doors can be formidable.

Search: The increased number of people in these structures (as opposed to typical multifamily dwellings) and in close proximity to each other can easily require a larger resource commitment for search operations. The size of each unit/occupancy, however, is small to moderate and the floor plans are typically residential in nature. Easy access is limited to the front of each unit on the above-grade floors and the front/rear of occupancies on the grade floors. Search operations in these structures can be a considerable challenge depending on the size and extension of fire.

SPECIFIC HAZARDS

- Due to the number and close proximity of people, expect challenging rescue considerations.
- These types of buildings are typically found in low-income areas and can present additional safety concerns toward fireground personnel.
- Always try to determine the presence of lightweight construction.
- Offset studs and walls can be found in newer construction.
- Common attics are standard.
- Although division walls may be present, they are often breached.
- Hallway fire doors may be blocked open.
- Exterior brick veneers are for decoration only and are subject to collapse if fire exposes the wood framing.
- Look for the presence of sprinklers and standpipe connections.
- Multiple buildings in close proximity and/or remote from roadways can present apparatus placement problems.
- Expect large-scale electric utility rooms and altered utility distribution equipment in large buildings.

RAPID STREET-READ GUIDE

BUILDING GROUP: Multifamily Dwelling—Legacy Townhome/Condo/Apartment

MFD 17

BASIC CONSTRUCTION METHOD

- Mostly brick masonry and conventional wood frame, some concrete block and poured-in-place concrete
- Conventional wood floor and some concrete over conventional wood flooring
- Typically conventional wood roof construction and some conventional metal roof construction
- Concrete slab, perimeter foundation, and basements

ERA CONSIDERATIONS

- The defining difference between these structures built prior to 1960 and the same type of structures constructed after the 1960s is the use of lightweight construction in newer buildings and the predominance of conventional construction in the older buildings. Many older commercial buildings have been converted to residential apartment and condominium type structures, resulting in a mixture of conventional and lightweight construction.

EXTERIOR FEATURES

- Range from simple two- to four-story structures with several units to numerous adjoining units that comprise one common structure (e.g., row buildings).
- The exterior is normally of brick masonry construction or wood siding on all-wood structures.
- Cornices, corbels, and other decorative type exterior features are often used.
- Roof is usually of conventional or lightweight construction with multiple layers of roofing materials.
- Some of these buildings incorporate adjoining occupancies that share a common attic.
- Unreinforced masonry construction can be identified by its visible characteristics.
- Basements are common depending on the area of the country.

INTERIOR FEATURES

- The majority of these buildings use the floor plan of an interior hallway configuration, where each unit is accessed from a center hallway.
- Multiple units normally share a common attic.
- Depending on the age, lath and plaster is the standard interior finish for older buildings and drywall for newer buildings and/or renovations.
- Older elevators are common in larger buildings and smoke detectors/sprinklers may not be present.
- Standpipes may be present depending on the size and type of building.
- There is generally no HVAC system.

TACTICAL CONCERNS

Fire Spread: Although these buildings are rather simplistic from the perspective of dwelling type configurations and the construction is generally older and of conventional materials, fire can rapidly spread in vertical stair shafts, pipe chases, and particularly within common attics. Many of these structures used full-dimension wood and conventional masonry construction, which can resist fire in a superior manner compared to lightweight construction. However, structures that use wood exterior siding can rapidly burn, particularly when covered by asphalt siding. Lastly, a fire in a basement can rapidly travel vertically within unprotected channels. Exposed floor joists over an unfinished basement increases the collapse hazard.

Collapse: Conventional construction (walls, floors, and roofs) is not prone to early collapse when exposed to fire. Conventional wood and masonry construction is also not prone to early collapse. However, many of these buildings use brick veneer on a wood frame that can easily collapse if the wood frame is weakened by fire. Remember that if a fire is on the top floor, it can easily extend into an attic and cause an early collapse of a wood or metal roof, particularly when the roof has numerous layers of roofing materials that impose a significant dead load.

Ventilation: Ventilation operations are similar to single-family residential homes in that either horizontal and/or vertical ventilation can be beneficial, depending on the location of fire. Horizontal ventilation is enhanced by windows in each room and can often be assisted by PPV. If present, hallways normally terminate into vertical stair shafts that may exit a roof through a penthouse. Although vertical ventilation can be useful for fires on the top floors of these buildings, expect heavy construction and numerous layers of roofing materials that will require appropriate resources and time for completion.

Forcible Entry: Conventional forcible entry methods are normally sufficient as doors and windows are similar to common residential doors and windows. Most older windows are plate (annealed) glass, although newer-construction and remodels likely use double-pane/energy-efficient type windows. It is not uncommon for entry doors to be solid wood and secured by multiple locking devices, which can hamper forcible entry operations. Some of the larger complexes can have on-duty security personnel who should be available to assist with entry/exit considerations.

Search: Search is not much different from conducting a search in a residential occupancy as the floor plans are often similar. In many cases, the size of each occupancy and individual rooms are small to moderate, which can be a benefit to search operations. Larger buildings can present a significant occupant consideration as well as occupants above a fire in multistory buildings. Remember that some buildings will have numerous adjoining occupancies that may also need to be searched depending on exposure considerations.

SPECIFIC HAZARDS

- Common attics are normally found over adjoining occupancies. Read the attic vents.
- Asphalt siding (also known as gasoline siding) can be expected to rapidly burn and fail when exposed to fire.
- Older exterior cornices (if present) may be flammable and will not support the weight of an aerial device or personnel.
- Expect many older buildings to have been renovated with lightweight construction and voids that were not present in the original building.
- Fires on the top floor can present a hazard by extending into a common attic.
- Always check for a basement.
- Well-developed fires can cause collapse of exterior brick veneers.
- The older buildings normally lack sprinklers.
- Older elevators may be present in some large buildings.

RAPID STREET-READ GUIDE

BUILDING GROUP: Multifamily Dwelling—Lightweight Townhome/Condo/Apartment

BASIC CONSTRUCTION METHOD

- Light wood or metal framing, and truss floor and roof construction
- Some brick veneer, wood, vinyl, and stucco exteriors
- Concrete slab or perimeter foundation and some walkout basements

MFD 18

ERA CONSIDERATIONS

- These types of buildings have universally embraced the use of lightweight (engineered) construction in almost every aspect.

EXTERIOR FEATURES

- Whether listed as apartments, condominiums, or townhomes, the construction, floor plans, and resultant hazards are similar.
- Sizes vary from simple two- to four-story structures that can incorporate several units (modular) to numerous units that comprise one common structure.

- The exterior is normally finished with stucco, T-11 type siding, vinyl siding, or brick veneer.
- Roof construction consists of some type of lightweight wood or metal construction. Fake dormers, decorative gable vents (nonfunctional), and foam cornices are often part of the roof construction.
- The presence of stairways to individual units can indicate the number of units within a structure.
- A complex of buildings can be difficult to size-up and reach with resources from on-scene apparatus.
- Underground parking is often used in larger buildings.

INTERIOR FEATURES

- Garden apartment style floor plans do not have a center hallway configuration. However, some large buildings have center hallways that serve the units within a building.
- The interior normally consists of interconnected rooms that are similar to single-family dwellings.
- Drywall is the standard interior finish.
- Zero-clearance fireplaces are common.
- Adjoining units often use offset studs, which can be considered a modern form of balloon frame construction.
- Multiple units can share a common attic.
- Mechanical and utility-areas (laundry, storage) are typically located in lower levels of the complex.
- Modern elevators and enclosed interior vertical stair shafts are common in larger buildings. Smoke detectors are common in both older and newer buildings.
- Sprinklers are normally present in newer buildings and/or municipalities with strict codes.

TACTICAL CONCERNS

Fire Spread: Although these buildings are rather simplistic from the perspective of dwelling type configurations and small units to moderate sizes, fire will rapidly spread in lightweight construction, particularly within common attics, open truss type floors, and vertically in zero-clearance fireplaces. Fire can also easily spread in voids within soffits, dropped ceilings, and within center hallways (if this floor plan is utilized). A fire in an automobile in an underground parking area can create severe exposure problems. Vinyl siding can create noteworthy exposure problems to the area of the fire building and other buildings in close proximity that also have vinyl siding.

Collapse: Lightweight construction (walls, floors, and roofs) is prone to early collapse when exposed to fire, particularly when glue has been used as a bonding agent for connection points in trusses. Remember that if a fire is on a top floor, it can easily extend into an attic and cause an early collapse of a truss roof. Brick veneer on a wood frame can also easily collapse if the wood is weakened by fire.

Ventilation: Ventilation operations are similar to single-family residential homes as either horizontal and/or vertical ventilation can be beneficial, depending on the location of fire. Horizontal ventilation is enhanced by windows in each room and can often be assisted by PPV. Although vertical ventilation can be useful for fires on the top floors of these buildings, it is not normally beneficial for floors below the top floor. Roof ventilation operations on lightweight roofs is not recommended in an area where trusses are exposed to fire.

Forcible Entry: Conventional forcible entry methods are normally sufficient as doors and windows are not commercial grade. Most doors are lightweight and windows are likely plate (annealed) glass, although newer buildings likely use double-pane windows. Some of the larger complexes can have on-duty security personnel who should be available to assist with entry/exit considerations.

Search: Search in these buildings is not much different from conducting a search in a residential occupancy, as the floor plans are often similar. In many cases, the size of each occupancy and individual rooms are small to moderate, which can be a benefit to search operations. Remember that it is not uncommon for occupancies above ground to have only one means of exit, so ground ladders may need to be implemented. Also, remember that exposed lightweight trusses offer minimal time prior to collapse.

SPECIFIC HAZARDS

- Lightweight construction rapidly burns and fails when exposed to fire!
- Common attics are widespread. Read the attic vents.
- Exterior cornices constructed of rigid foam (if present) are flammable and will not support the weight of an aerial device or personnel.
- Look for the presence of offset studs in adjoining occupancies.
- Fire can easily spread vertically in zero-clearance fireplaces.
- Fires that originate on balconies or the top floor can rapidly spread into attics spaces and lead to early collapse—even when the occupant spaces have fire sprinklers.
- Well-developed fires can cause collapse of exterior brick veneers.
- Anticipate below ground escape windows when walkout basements are present.
- Consider the use of ground ladders for above ground occupancies with a single means of entry/exit.
- Vinyl siding is flammable and emits a very toxic smoke.

RAPID STREET-READ GUIDE

BUILDING GROUP: Main Street Commercial—Pre-WWI Ordinary

BASIC CONSTRUCTION METHOD

- Ordinary load-bearing brick or stacked stone walls
- Timber floor beams and wood roof
- Perimeter stacked stone foundations and rock pads for central columns (where present)

COM 19

ERA CONSIDERATIONS

- Some examples may still have water-soluble sand and lime mortar.
- Most have heavy timber floor beams. Some have central heavy timber columns.
- Fire-cut floor beams emerged in this era.

EXTERIOR FEATURES

- First floor businesses often have dwelling spaces on upper floors.
- Front wall is curtain-style, all other exterior walls are load bearing.
- Narrow, arched windows have keystones or solid stone lintels.
- Bridge truss and authentic mansard roofs look similar from the street level. Decorative cornices/parapets are common for flat roof examples.
- Cast iron columns and lintels are used for glass storefronts (pictured).
- Access to under-building crawl spaces and utility tunnels are typically made from exterior cellar-like doors.
- Presence of spreaders, anchors, and/or rafter tie plates are common and indicate that efforts have been made to prolong structural stability or give earthquake protection.

INTERIOR FEATURES

- Open stairwells are used for upper-floor access.
- Narrow center hallways for upper floors and simple floor plans are common.
- Expect open first floor with tin ceiling coverings.
- Interior doors are likely to have transoms.
- Cellars and/or basements are not uncommon.

TACTICAL CONCERNS

Fire Spread: Expect rapid fire spread through the open area first floor. Open stairwells and center hall construction will cause rapid fire spread through common areas on upper floors. Fire can communicate through multiple and connected combustible voids. Individual room fires on upper floors are typically contained, but transom openings above hallway entry doors will accelerate fire spread where present. Fires in mansard roofs are difficult to extinguish due to multiple voids. While simple, the upper floor layout will present difficulties for hoselines due to distances, corners, and multiple small spaces.

Collapse: The pre-WWI ordinary commercial building is high mass and can typically withstand the effects of fires contained in small rooms. However, some of the most spectacular collapses have occurred in pre-WWI ordinary construction buildings. Fire-cut (self-releasing) floors can cause a general collapse, and sagging floors are a typical warning. The use of cast iron for columns and lintels can cause the curtain wall supported above the cast iron to collapse with little visual warning. Alterations to load-bearing walls and floors create significant weaknesses that are accelerated by heat and fire. Decorative cornices and parapets fail easily with the slightest roof sag. Aging and deterioration of brick mortar creates higher collapse risk. The presence of anchors and rafter tie plates indicate that the building has been strengthened to resist collapse (look for uniform spacing of anchors on the exterior wall). The presence of tie-rod ends and spreaders on the exterior wall indicates a rapid collapse potential; when an interior fire heats the steel tie-rod, it will elongate and allow floor beams to pull away from the wall.

Ventilation: Roof ventilation should be a high priority if the fire has captured combustible voids (likely). Stairwell penthouses may assist ventilation of open, center hall configured upper floors. The size and interior geometry can present challenges for the use of PPV fans.

Forcible Entry: Forcible entry can normally be accomplished by conventional methods as the doors and frames are typically solid wood. Security hardware on storefront windows and entry doors can require power saws or hydraulic spreaders.

Search: First floor open spaces may require large area search techniques. Stock, furnishings, kitchens, and arrangement variables can also be challenging on the first floor. Open stairwells and center hallways are also smoke/heat/fire pathways that can hamper search and rescue. Window rescues are typical for upper floor fires. Upper floor plans are typically simple and repeated.

SPECIFIC HAZARDS

- Multiple electrical feeds are usually found on Side C.
- Steam heat, water, and sewer systems are usually under the building and are accessed by an exterior cellar door.
- Utility tunnels and underground passageways can cause fire spread to adjacent exposures.
- Expect numerous poke-through holes created by years of utility and infrastructure upgrades.
- Anticipate multiple layers of roofing materials.
- Sprinklers may not be present.
- Cellars/basements can be encountered, and they present a formidable challenge due to storage, vertical fire spread paths, and limited access and ventilation options.
- Utilities are old (with associated hazards) unless they have been upgraded.

RAPID STREET-READ GUIDE

BUILDING GROUP: Main Street Commercial—Pre-WWII Ordinary (Taxpayer)

BASIC CONSTRUCTION METHOD

- Mostly unreinforced and conventional brick masonry, full-size wood, and some rough-cut lumber
- Perimeter foundations or concrete slab
- Basements are common and can traverse multiple stores
- Common cocklofts/attics are typical.

COM 20

ERA CONSIDERATIONS

- Primarily constructed from the late 1800s to the early 1900s, the first floor is typically the commercial occupancy and the second and/or third floors are the residential occupancies. Basements are also incorporated below grade level and often contain a noteworthy fire load. Although these buildings are old and constructed from conventional materials, they are prime candidates for remodels that can change the original floor plans and construction materials (notice the blocked second-floor windows in the bottom photo).

EXTERIOR FEATURES

- The sizes of these occupancies most often range from simple two to three stories. All can incorporate a basement, which adds an additional story to the building.
- Some buildings are a block long; multiple adjoining taxpayer type buildings can appear to be a block long.
- Cornices and corbels are often used.
- Roofs are typically flat or sloped and use conventional wood construction that is covered with multiple layers of roofing materials.
- Unreinforced masonry construction and roof rafters/floor joists will often be strengthened by rafter tie plates that are visible from the exterior.
- Access to the residential occupancies is normally through a door/interior stairway on the street level and/or a door/exterior stairway on one side of the building.

INTERIOR FEATURES

- The first floor is commercial, while the floors above are residential or office space.
- Basements can contain noteworthy amounts of storage.
- Lath and plaster wall and ceiling coverings are standard. Some tin ceilings can be found in first floor occupancies. Sprinklers are not typical, and, where present, are likely to be a retrofit.
- Interior hallways and stairways can be narrow.
- Large display windows of plate (annealed) glass on the grade floor are common.
- Common cocklofts/attics can extend the full length of these buildings.

TACTICAL CONCERNS

Fire Spread: Although these buildings are rather simplistic from the perspective of a single-story occupancy that is often over a basement, a fire in a basement can rapidly spread vertically to the floor above. Similarly, a fire in the first floor occupancy can spread to the overhead common attic/cockloft and rapidly spread horizontally. Because of adjoining occupancy configurations, there can be an exposure problem with occupancies on either side of the fire occupancy.

Collapse: It will take a sizeable fire and time for collapse to occur in most of these buildings as they are of substantial construction. However, unreinforced masonry construction is prone to collapse when exposed to prolonged fire due to its inherent weaknesses. Another potential area of collapse is the front portion of these buildings. A wood or cast iron beam is often used as the supporting member (over the display windows) for the parapet wall above, both of which can fail and allow the entire front portion of a building to collapse outward into the street. Remember, if fire extends into a common attic/cockloft, collapse of the roof depends on the size of structural members that are normally conventional construction, but the roof will have a sizeable dead load from numerous layers of roofing materials.

Ventilation: Fires in a basement are slow to be ventilated due to few openings; often there is only one opening—the stairway—for basement access/egress. In many cases, PPV can be used to enhance natural ventilation if there are sufficient openings. Horizontal and/or PPV can be used on the grade floor by cross ventilating contaminated areas from front to back (or the reverse). Roof ventilation that consists of either offensive operations or defensive operations for an extending attic/cockloft fire can be useful. However, cutting a roof will be slowed by conventional construction and multiple layers of roofing materials.

Forcible Entry: Forcible entry can generally be accomplished by conventional methods. The display windows are usually plate (annealed) glass and can be easily broken, but will break in large shards that can fall outward. Expect heavy security on the front of some of these buildings (such as scissor gates, overhead doors, multiple locking devices, etc.). The rear door is also of heavy construction and may be secured with an interior security bar. (Look for a double pair of carriage bolt heads on the door.)

Search: Search operations in a basement with fire will be difficult and dangerous at best. Search operations on the grade floor commercial(s) will be simplified by the floor plan (often rectangular with few partition walls), but can be degraded by a noteworthy fire load. Access/egress to the front and rear of grade floor occupancies can be helpful to search operations.

SPECIFIC HAZARDS

- When exposed to fire, unreinforced masonry construction can collapse outward at least twice its height.
- Masonry parapets are also prone to collapse outward when exposed to fire.
- Exterior cornices are likely flammable and will not support the weight of an aerial device or personnel.
- The name displayed on the exterior of the business is an indicator of the interior contents.
- Fire can easily spread in a common attic/cockloft.
- These occupancies are prime candidates for remodels.
- Consider forcing entry to occupancies on either side of the original fire occupancy if extension into a common attic is suspected.
- If a large advertising sign is mounted on the roof, determine whether the roof has been strengthened to adequately support the sign (see photo).
- Expect multiple/altered utility shutoffs on the rear of the building. Apparatus access may be restricted in alleyways.
- Anticipate the presence of inverted roofs.

RAPID STREET-READ GUIDE

BUILDING GROUP: Main Street Commercial—Modern Strip-Style Store

BASIC CONSTRUCTION METHOD

- Load-bearing CMU exterior walls (Sides B, C, and D) with lightweight steel, wood, or composite trusses
- Side A steel and/or wood columns and girders to support trusses
- Foundation usually slab on grade unless a storage basement is included
- Tenant spaces separated by light steel or wood framed stud partition walls

COM 22

ERA CONSIDERATIONS

- Those built from the late 1970s to 1990s are CMU exterior walls with parallel chord steel bar truss roofs.
- The 21st century examples are engineered lightweight. They may still have CMU walls but utilize a lightweight trussed flat roof (wood chords with a stamped metal web is most common).

EXTERIOR FEATURES

- Side A is usually glass with customer entry doors for each tenant. Side A will likely include a cantilevered canopy or column-supported overhead facade (pictured).
- Rooftop HVAC and exhaust vents can be hidden by decorative skirts, false mansards, or parapets.
- Expect to find no windows on Sides B, C, and D. Steel access (delivery) doors are located on Side C, and are rarely used for customer access.

INTERIOR FEATURES

- Individual retail spaces or tenants (units) are divided by lightweight partition walls.
- The rear of each unit is also partitioned to create stock and office spaces for staff.
- The products/services that individual tenants offer will dictate interior arrangement and potential fire load. Attention to the individual signs on the facade or entry door can help with size-up for interior features (physician's office, hair salon, coffee shop, auto parts retail, etc.).
- Small restaurant tenant spaces will also include a partitioned kitchen/prep area. This introduces fixed cooking equipment (including a grease hood), commercial-sized coolers, and stationary food prep counters.
- Common attics over most or all occupancies are normal.

TACTICAL CONCERNS

Fire Spread: Strip stores are subdivided into tenant units by simple partition walls that should be considered temporary (they can be torn down and moved to reallocate space for larger or smaller tenants). These walls rarely extend to the roof, which creates a shared void space above a suspended ceiling. Fire can easily spread from one unit to another. It is paramount to check for fire extension on adjoining units (remember to check the adjoining unit that stands to receive the most loss first). Fires that originate in or behind occupancy signs can easily capture the entire facade or canopy. Not all codes require the facade to be separate from the roof structure; be sure to check for fire above the dropped ceiling of interior spaces. Fire intensity and smoke production are obviously influenced by the fire load that each unit holds. (Read the tenant signs!) Additionally, remember that division walls do not typically extend into exterior facades, which means they are not fire-stopped!

Collapse: The primary collapse threat is that of the roof and Side A facade or canopy. Most examples have a significant dead load consisting of multiple HVAC units, cooking hoods, vents, and communications equipment. Newer buildings with lightweight trusses are especially dangerous. Firefighters can expect a rapid collapse of the roof when fire and heat enter a suspended ceiling space of newer buildings. Often, the collapse is limited to the areas above and immediately adjoining the fire unit. The threat of a domino style collapse past the involved unit is rare but can be thwarted by cooling the trusses in adjoining, uninvolved unit spaces.

Ventilation: The modern strip-style store is easily ventilated using common tactical practices—with a few caveats. In most cases, it is dangerous to perform rooftop ventilation directly over the fire due to the lightweight nature of the trusses. A proven approach to ventilation typically includes the creation of a primary exhaust flow path for the involved unit (consider outside wind direction and take out the front window if appropriate). This is followed by adding defensive ventilation for uninvolved, adjoining units (PPV, defensive roof strips, etc.).

Forcible Entry: Forcible entry can normally be accomplished by conventional methods for the main customer entrance doors. Most are glass with an aluminum frame. Rear service/supply doors are typically all steel and are secured with commercial-grade security hardware and/or interior bars that require powered equipment for forced entry. Breaching CMU walls can be accomplished, although it is often time consuming and discouraged.

Search: The need to prioritize search is rare as customers/staff are likely to self evacuate during fires. The only exception could be staff members who become trapped in the rear office or storage area. Rear access doors are typically fortified for security.

SPECIFIC HAZARDS

- Gas and electrical utilities can be secured on the building exterior, although they may require the forcing of locks.
- Dumpster fires that are next to the building can extend into the roof structure if an eave or other opening exists.
- The rear of the building may include auxiliary equipment or waste/recycle collection areas.
- Common attics allow horizontal fire extension.
- Division walls may partition an attic but not exterior facades.
- Signage can help indicate interior contents, arrangement, and fire load potential.
- Facades over entry/exit areas can be very dangerous.
- Consider not using ground ladders on facades.

RAPID STREET-READ GUIDE

BUILDING GROUP: Main Street Commercial—Fast Food

BASIC CONSTRUCTION METHOD

- Load-bearing CMUs with wood or steel roofs or lightweight steel or wood framing with trusses
- Usually slab on grade unless a storage basement is included

ERA CONSIDERATIONS

- Legacy examples are likely to be CMU walls with wood beam roofs.
- Engineered lightweight may still have CMU walls but utilize a lightweight truss roof of metal or wood.

EXTERIOR FEATURES

- Large windows are used in the customer seating area; few to no windows are included in the cooking/storage areas (perhaps with a drive-up window).
- Rooftop HVAC and exhaust vents are hidden by decorative skirts, false mansards, or facades.
- Rear service areas are often fenced and contain exterior refrigeration equipment, dumpsters, and/or waste cooking oil vats.
- Drive-thru traffic arrangements vary but most allow 360° apparatus access.
- Exterior facades typically have a rear-mounted ladder for roof access.

INTERIOR FEATURES

- Interiors are usually divided into three areas: order counter and customer seating area, prep/cooking area, and restroom areas.
- A small office at the rear of the building is common. Likewise, a stairway to a basement storage area can be found in the rear of the building.
- Older examples have fixed cooking equipment (including a grease hood), commercial-sized or walk-in coolers, and stationary food prep counters.
- Newer examples have open cooking areas with drop-down utilities to allow for modular (and moveable) cooking equipment and appliances.
- Suspended ceilings are common and create a void above the ceiling.

COM 23

TACTICAL CONCERNS

Fire Spread: Fires that originate in cooking equipment are usually contained by an appliance sprinkler/suppression system. Older examples may have antiquated suppression systems that are defeated by vertical spread of fire in ducting that terminates on the exterior (roof or rear exhaust vents). Fire can also start in electrical and gas appliances and can spread in a similar fashion. Fires in the dining area are not common but can easily spread due to the openness of the room and lack of partition walls. Soffits, facades, and other decorative modifications must be checked for hidden extension of fire in concealed voids. Remember that fire can also easily spread in voids above suspended ceilings.

Collapse: The primary collapse threat is that of the roof. Most examples have a significant dead load consisting of multiple HVAC units, cooking hoods, and exhaust vents. Newer buildings with lightweight trusses are especially dangerous. Partition walls between the seating and cooking areas are rarely load-bearing. Firefighters can expect a rapid and general collapse of the roof when fire and heat enter the suspended ceiling space of newer buildings. Fire spread into decorative facades can cause localized collapse.

Ventilation: Roof ventilation should only be considered for legacy era types. PPV fans will likely be needed for horizontal ventilation of the cooking area and restroom areas. The seating area is much easier to vent.

Forcible Entry: Forcible entry can normally be accomplished by conventional methods for the main customer entrance doors. Most are glass with an aluminum frame that are typically secured by Adams Rite cylindrical locks. Rear service/supply doors are typically steel and are secured with commercial-grade security hardware and/or interior bars.

Search: The need to prioritize search is rare as customers/staff are likely to self evacuate during fires. The only exception could be staff members who become trapped in the rear office or storage area.

SPECIFIC HAZARDS

- Gas and electrical utilities can be secured on the building exterior, although they may require the forcing of locks.
- Dumpster fires that are next to the building can extend into the facade. Many fire codes require that dumpsters and waste cooking oil holding systems be located away from the building.
- Large indoor or outdoor children's play areas can create a considerable fire load as well as a resulting smoke that is extraordinarily toxic/flammable due to the use of synthetic materials.
- Sprinklers may or may not be present, depending on the age of the building.
- Rapid collapse of the roof and facade cannot be overemphasized in lightweight constructed fast-food buildings.
- Roofs with facades can typically be accessed by a rear-mounted ladder.

RAPID STREET-READ GUIDE

BUILDING GROUP: Main Street Commercial—Mega-Box

COM 24

BASIC CONSTRUCTION METHOD

- Mostly fire resistive (Type I) that combine reinforced concrete and coated steel
- Some load-bearing CMU walls
- Roofs of lightweight steel parallel bar trusses with numerous skylights, domes, and finishes
- Slab-on-grade foundation and footers on peripheral stores
- Central core with a reinforced concrete basement foundation for delivery tunnels and utility conveyance

ERA CONSIDERATIONS

- Legacy examples are likely to be reinforced concrete. Additional wings built onto the original mall may be constructed using CMU walls where appropriate occupancy separation exists.
- Newer malls use mostly coated steel to meet fire resistivity requirements.

EXTERIOR FEATURES

- Big-box anchor stores surround a central core (mall) for smaller retail shops.
- Anchor stores have exterior customer entry doors, whereas smaller stores may only be accessed from the central mall area. Some larger restaurants may also have exterior entries.
- A central receiving/loading dock is shared by smaller tenants. Each anchor store may have an independent receiving dock.
- Expect a flat roof for anchor stores. The central mall area may have multiple light wells, skylights, or other features as part of a flat roof.
- Most HVAC systems are on the roof, although a large chiller unit may be found on site.

INTERIOR FEATURES

- The central mall is usually designed with a grand avenue with spur hallways. Some spurs open to the parking lot and some are dead ends that include restrooms, staff offices, and emergency-only exits.
- Older malls incorporate a central facility utility plant for boilers, water, and electrical needs (usually in a basement). Central receiving and stock distribution is colocated near the utility plant.
- Newer malls are likely to have multiple utility service and stock receiving areas.
- For malls with multiple floors, expect to find escalators, open stairways, and elevators for customers and freight.
- Food courts are common.

TACTICAL CONCERNS

Fire Spread: Mega-malls rely on an automatic fire sprinkler system to suppress fires. The system is designed to have multiple zones and includes multiple risers. Compartmentalization of fires to a given area is accomplished using a combination of fire walls, fuse-link doors, and ventilation system dampers as well as the fire sprinkler system. Fire spread issues exist with the utility conveyance system, waste/recycle collection areas, and suspended ceiling spaces. If fire extends to a metal bar joist roof, it can travel between the metal decking and roof composition.

Collapse: Fire resistive buildings also have a certain resistance to collapse. Collapse threats do exist for suspended ceilings in individual retail stores and for the exposed atrium trusses in the central mall. In a fire, bar-joist roofs are candidates for collapse.

Ventilation: The combination of individual stores, dead end corridors, cavernous atriums, and long distances to exterior doors make ventilation a challenge. Smoke removal from smaller fires can often be accomplished with the help of the building engineer using the facility's ventilation system. Newer malls typically have a fire command center that allows fire officers to control various building systems. Tactical ventilation for uncontrolled fires is difficult and usually requires the removal of skylight or atrium glass and PPV fan relays.

Forcible Entry: Forcible entry can normally be accomplished by conventional methods for the main customer entrance doors. Most are glass with an aluminum frame with Adams Rite circular locks. Emergency exit doors should include panic hardware that is easy to force. Stock receiving areas with overhead rolling doors require power saws and well-trained firefighters to defeat. Retail stores that are accessed by the central mall may not have actual doors, but rather roll-down security gates. Most are key-activated, with an electric motor that raises and lowers the gate. These can be defeated but generally require power tools and special training. Security personnel, mall staff, and building engineers may have more rapid solutions for gaining entry into given spaces that are well secured.

Search: The need to prioritize search is rare as customers/staff are likely to self evacuate during fires. Multiple entry/exit doors on each side of the building should assist in occupant egress routes. Most malls have 24/7 security services whose employees may help responders target areas that need to be searched. The use of large area search techniques may be required for larger stores where visibility has been hampered.

SPECIFIC HAZARDS

- Gas and electrical utilities can be secured on the building exterior, although they may require the forcing of locks. Multiple gas and electrical feeds are common, extensive, and large. Expect high voltage and high pressure service feeds.

- Congested parking lots and evacuating occupants can hinder arriving apparatus. Mega-box facilities that include multiplex movie theatres will exacerbate this challenge. Fire departments are encouraged to preplan apparatus access areas and practice arrival options during times of crowd congestion prior to actual incidents.

- Legacy-era malls are likely to have asbestos in insulation, pipe casings, and as part of the spray coating of steel structural elements. Asbestos can be released when firefighters tear into these components for overhaul.

- Anticipate the extension of fire between the metal roof decking and roof composition if fire exposes a metal bar joist roof. Likewise, always check for fire extension above suspended ceilings.

- External storage areas are prime locations for fire.

- Interior displays can pose a specific combustible hazard (i.e., automobiles, etc.)

- Food courts can present restaurant-type hazards.

RAPID STREET-READ GUIDE

BUILDING GROUP: Main Street Commercial—Big-Box

BASIC CONSTRUCTION METHOD

- Load-bearing CMU exterior walls with interior steel columns to support lightweight steel parallel bar trusses
- Tilt-up concrete walls with interior steel columns to support lightweight steel parallel bar trusses
- Slab-on-grade foundations and footers

ERA CONSIDERATIONS

- The big-box store is mostly a modern engineered lightweight building.
- A few legacy examples exist that are CMU with solid steel girders supporting steel bridge trusses.

COM 25

EXTERIOR FEATURES

- Expect a large, expansive footprint with 360° apparatus access.
- Exterior walls can range from 20 ft to 40 ft high, and decorative facades are common.
- Windows are on the storefronts only.
- Customer access doors are usually glass and aluminum. Emergency evacuation doors surround the building. The rear or side of the building will have several shipping/receiving overhead roll-up doors.
- Most HVAC systems are on the roof, although a large chiller unit may be found on Side C.

INTERIOR FEATURES

- Two general interior arrangements can be found: (1) A wide-open floor plan with high-rack displays and storage as part of the customer area, or (2) a large showroom style floor plan with a separate high-rack storage area that is partitioned from customers.
- Both variants have a separate receiving area for incoming goods.
- Some examples include a mezzanine-like office area built as a "building within a building" for staff offices and a staff break area. These should be of noncombustible construction, although wood framing can be found.
- Most have exposed roof trusses. Flat, metal bar joist roofs are the most common. HVAC ducting, lighting, and signage are typically hung from the trusses.
- A fast food, bank, or other convenience service tenant may have a sublet space inside that is located against an exterior wall.

TACTICAL CONCERNS

Fire Spread: The big-box store relies on an automatic fire sprinkler system to suppress fires. The system is designed to have multiple zones and typically includes an overhead (truss) protection grid as well as a high-rack storage grid. Fires that surpass the capabilities of the sprinkler system are rare but are cause for great concern. High fire loads, deep-seated storage rack challenges, and long hose lay lengths typically challenge attack crews. Risk managers consider the big-box store an *engineered loss*, meaning that a fire or other event that cannot be controlled by automatic devices will result in a total loss of the contents and building. If fire extends to a metal bar joist roof, it can travel between the metal decking and roof composition.

Collapse: The primary collapse threat is that of the high-rack storage units, either by fire or water saturation (sprinkler activation). High-rack storage units can be considered vertical trusses with cantilevers—they are very strong but have design limits that can be exceeded by water saturation or the application of heat (800°F). Large-scale fires that attack the roof truss system should be considered an instant collapse threat. Failure of roof trusses can cause wall stresses. Walls that are no longer supporting a roof should be considered unstable. Shoring of walls (using rakers) is mandatory prior to overhaul and fire cause investigations.

Ventilation: The cavernous nature of big-box interiors renders most fire department ventilation techniques inadequate. For those occasions where an interior operation needs ventilation, some consideration can be given to removing rooftop light diffusion panels (where they exist) and setting up PPV fan relays. Although few exist, the use of an apparatus-mounted, ultra high-volume fan may prove effective.

Forcible Entry: Forcible entry can normally be accomplished by conventional methods for the main customer entrance doors. Most are glass with an aluminum frame with Adams Rite circular locks. Emergency exit doors should include panic hardware that is easy to force. Stock-receiving areas with overhead rolling doors require power saws and well-trained firefighters to defeat.

Search: The need to prioritize search is rare as customers/staff are likely to self evacuate during fires. Multiple emergency exit doors on each side of the building should assist in occupant egress routes, although a panic-stricken stampede for the storefront egress can produce a mass-casualty event. After-hours stockers are common for busier stores, as some companies schedule stock deliveries at night. In most cases, these staffers will be waiting for the fire department's arrival and can help account for all workers.

SPECIFIC HAZARDS

- Gas and electrical utilities can be secured on the building exterior, although they may require the forcing of locks. Multiple gas and electrical feeds are common, extensive, and large. Expect high-voltage and high-pressure service feeds..
- Congested parking lots and evacuating occupants can render the storefront inaccessible to responding apparatus. Fire departments are encouraged to preplan apparatus placement options for various types of incidents in big-box buildings.
- Anticipate fire extending between the metal roof decking and roof composition if fire exposes a metal bar joist roof.
- Expect to encounter large utility feeds.
- Food courts can present restaurant-type hazards.

RAPID STREET-READ GUIDE

BUILDING GROUP: Manufacturing/Warehouse—Block/Masonry

BASIC CONSTRUCTION METHOD

- Walls of brick or concrete block
- Roofs of metal or wood, often of light-weight materials
- Concrete slab foundations

ERA CONSIDERATIONS

- Pre-1960 buildings commonly used either brick or concrete block with conventional wood roofs (bowstring roofs were common) or flat metal roofs of open web bar joists. Some unreinforced masonry buildings (pre-1933) can be found with either conventional wood roofs or flat metal roofs. Some wood panelized roofs can be encountered in the West.
- Post-1960 buildings used conventional reinforced brick construction and concrete block with lightweight roofs of open web metal bar joist construction. Some wood panelized roofs can be encountered. Arched roofs are normally not found after 1960.

EXTERIOR FEATURES

- Depending on the size and age of a building, many types of roofs can be used, including flat, sawtooth, gable, and arched (bowstring or rigid arch truss). Metal deck roofs are commonly found on the East coast and wood roofs are common on the West coast. Bowstring roofs are very common in many areas of the country.
- It is not uncommon for additional buildings to be attached to the original building for additional space without having to modify the original building. The attached buildings are often a lesser grade of construction.
- Depending on the occupancy, large display windows may be present.

INTERIOR FEATURES

- Large open areas are common with these types of buildings as the openness favors storage and manufacturing processes.
- A lack of partition walls is common.
- Small office areas are common and are used for records, reports, management of the business, and other purposes. In many cases, the office area is a building-within-a-building and maybe of dubious construction.
- Sprinklers may or may not be present.
- Mezzanines may be present.

TACTICAL CONCERNS

Fire Spread: The spread of fire can be rapid and intense due to open floor plans, the size of some buildings, and primarily the type of contents. In most cases, not only will the contents readily burn, but so will wood structural components of a building. Metal structural components will soften and then fail between 800°F and 1,000°F. The rapid spread of fire within flammable contents can be enhanced by the type of storage that is in close proximity to other flammable contents (racked or tiered storage, etc.). The presence of operative sprinklers can retard the spread of fire. Suspect a lack of sprinklers on older buildings unless they have been retrofitted.

Collapse: Collapse is totally dependent on the type of construction, size of the building, type of contents, and presence or lack of sprinklers. It will take a sizeable fire and time to affect the stability of concrete block walls; however, if a roof does collapse, the concrete block walls will be freestanding and weaker than their original configuration. Conventional brick walls are more prone to collapse than concrete block walls, and unreinforced brick walls can readily collapse if a roof collapses. Buildings with unreinforced masonry construction should be considered extremely dangerous when exposed to fire. Collapse of any roof depends on the size and type of roof. The large wood timber roofs will resist fire for longer periods of time than metal roofs or wood roofs of smaller conventional and/or lightweight lumber.

Ventilation: These buildings can normally support horizontal, vertical, and/or PPV as is most appropriate for the type/size of building and extent of fire encountered. Horizontal ventilation will often be limited to minimal doors such as a single standard size door on one end of a building and multiple overhead loading doors on other ends of a building (depending on size). Vertical ventilation can be a viable operation, particularly in wood roofs, but depending on the size of building may be a resource- and time-intensive operation to have a meaningful effect. If a particular building is equipped with glass and/or plastic skylights, these can be quickly removed for ventilation operations. As a building approaches the large warehouse size, PPV will require fans capable of large CFM capability.

Forcible Entry: Forcible entry is usually limited to either a standard size interior door into an office area or overhead loading doors. Some overhead doors will have a standard size door inserted as part of the larger unit. Some windows may be present if the building houses processes that require openable windows for ventilation, but windows are usually not in abundant supply. Challenging and time-intensive entry operations should be anticipated, as security is a major consideration in these types of commercial buildings.

Search: Unless there are known trapped occupants, search should not be a primary concern as the occupants are normally ambulatory. Obviously, a rapid escalation of fire can dramatically change this consideration; trapped occupants often seek refuge in the office area (building within a building). Additionally, if a search is necessary, the inherent risk should be evaluated and personnel must be proficient in large area searches as they can be difficult and time consuming.

SPECIFIC HAZARDS

- Always look for clues that give an indication of the contents and hazards, such as signage on the exterior of the building, for any particular storage outside a building, and the size of the utilities.

- As entry and exit openings can be in short supply, and the size of these buildings can be formidable, consider making as many large entry and exit openings as feasible.

- Mezzanines should be anticipated.

- Unreinforced masonry construction should be considered very dangerous (due to collapse) when exposed to fire conditions.

- Buildings that are not equipped with operative sprinklers pose an additional risk.

- The larger the building, the more difficult suppression operations will likely become. This applies to forcible entry, attack, and ventilation operations.

- Searches in large buildings can be challenging unless personnel are proficient in large area searches.

- Collapse hazards are presented by burned rack/tiered storage (particularly during overhaul operations).

RAPID STREET-READ GUIDE

BUILDING GROUP: Manufacturing/Warehouse—Steel

BASIC CONSTRUCTION METHOD

- Structural members of metal beams that form a skeleton for wall and roof coverings (often "kit" buildings such as Butler or Murphy red-beam)
- Wall sidings typically of corrugated steel, aluminum, or fiberglass panels
- Roofs of metal decking over metal structural members (some use wood sheathing)
- Concrete slab foundations

ERA CONSIDERATIONS

- Pre-1960 buildings use thicker and stronger structural materials.
- Post-1960 buildings use lighter and smaller structural materials, and some fibrous materials for roof decking.

EXTERIOR FEATURES

- Exterior walls can appear as corrugated metal panels, but can be corrugated fiberglass. Various types of wood such as plywood can also be used as a covering.
- Roofs are typically of a gable or flat configuration.
- These buildings can vary widely from a medium size to large buildings that are hundreds of feet in length and width.
- In some cases, masonry veneers are used for aesthetic purposes.

INTERIOR FEATURES

- Interior configurations can fall into three categories: (1) open floor plans used for storage, (2) floor plans used for manufacturing processes, and (3) a combination of manufacturing and storage.
- Typically a small to medium size area is reserved for office/administration space.
- The buildings are often equipped with sprinklers.
- The interior contents can vary widely. This also holds true for the flammability of the contents.
- Hazardous materials can be encountered in some buildings.
- Some buildings will have interior insulation panels for the walls and/or roof.

MANF 27

TACTICAL CONCERNS

Fire Spread: The extension of fire is dependent on the size of a building, contents, and the presence of functional partition walls and sprinklers. Partition walls will tend to compartmentalize areas within these buildings and reduce interior open areas. Fire spread is also dependent on the type of manufacturing processes (if present) and the type of storage, such as piled, stacked and/or rack-tiered materials. A lack of sprinklers can contribute to an enhanced fire spread.

Collapse: Depending on the amount of fire, steel, aluminum, and fiberglass wall panels can readily fail, particularly fiberglass panels. Steel structural members will weaken and become a candidate for collapse when heated to 800°F to 1,000°F. Corrugated roof panels are also prone to rapid collapse when exposed to sufficient heat and/or fire.

Ventilation: Horizontal ventilation is usually limited to doors and windows, which can be minimal in these buildings. PPV can be beneficial but is dependent on the size of a building and the ability to use strategically placed openings. Roof ventilation is enhanced by removing plastic/fiberglass skylights (if present) and/or opening the roof sheathing, which can be wood, metal, or fibrous type materials. Roof ventilation can be time intensive depending on the size of the building. Remember that activated sprinklers near roof ventilation openings can negate vertical ventilation. Roof ventilation personnel must know what type of roof they are on.

Forcible Entry: Forcible entry is normally limited to either the standard interior door(s) to the office areas (small opening) or overhead loading doors (much larger). Security is a concern in these buildings that is reflected in the substantial doors and windows. Proper training techniques for overhead doors is essential for efficient operations.

Search: Search operations can be challenging due to the amount of floor space and the need for personnel who are proficient in large area searches. Remember that occupants are normally ambulatory and are not usually lost in commercial building fires.

SPECIFIC HAZARDS

- The signage on the exterior of a building can be a good indicator of the interior contents.
- The sizes of utilities (electrical, gas, and water) are also a good indicator of noteworthy interior processes.
- Exterior hazmat placarding and exterior storage of palletized goods may offer additional clues to interior processes.
- Mezzanines can be found in these buildings and are normally located above the office door and not the overhead loading door areas.
- Roof ventilation on large buildings will require resources, time, and large and/or many openings to create effective vertical ventilation. Corrugated metal roofs are not recommended to support roof ventilation personnel.
- Security measures can contribute to lengthy forcible entry operations.
- Activated sprinklers can minimize or negate interior contaminates from vertically exhausting through roof ventilation openings. In this case, the sprinklers must be temporarily shut off to allow vertical ventilation to be effective.
- Be wary of burned racked/tiered storage and building access/egress under mezzanines.

RAPID STREET-READ GUIDE

BUILDING GROUP: Manufacturing/Warehouse—Concrete Tilt-up

BASIC CONSTRUCTION METHOD

- Walls of tilt-up concrete panels, either poured on site or prefabricated
- Roofs of metal or wood
- Concrete slab foundations

ERA CONSIDERATIONS

- Pre-1960 buildings commonly use concrete tilt-up panels with conventional wood roofs (bowstring roofs are common) or flat metal roofs of open web bar joist. Wood panelized roofs are very common on the West coast. The concrete panels of the buildings are often more substantial in their connection to the slab foundation than their newer counterparts. As a point of interest, if a concrete tilt-up building has a conventional wood roof such as a gable or arched, it likely was built prior to 1960.

- Post-1960 buildings use concrete tilt-up panels with lightweight metal and wood roofs (panelized). Conventional arched roofs are normally not found after 1960.

EXTERIOR FEATURES

- Depending on the size and age of a building, many types of roofs can be used, including flat, gable, and arched (bowstring or tied truss). Metal deck roofs can commonly be found on the East coast and wood roofs are common on the West coast. Bowstring and rigid arch roofs on pre-1960 buildings are very common in some areas of the country. Flat roofs are the most common.

- In some cases, additional buildings are attached to the original building for additional space without having to modify the original building. The attached buildings are normally not concrete panels and are often a lesser grade of construction.

INTERIOR FEATURES

- Large open areas are common with these types of buildings as the openness favors storage and manufacturing processes.
- A lack of partition walls is common.
- Small office areas are common and are used for records, reports, management of the business, and other uses.
- Mezzanines are often found above the office area, which is accessed through a standard size door and not above the overhead loading door areas.
- Sprinklers are often present in these buildings.
- Depending on the occupancy, manufacturing processes and rack/tiered storage are common.

MANF 28

TACTICAL CONCERNS

Fire Spread: The spread of fire can be rapid and intense due to open floor plans, the size of the building, and the type of contents. In most cases, not only will the contents readily burn, but so will the wood structural components used for the roof. Roof metal structural components will soften and then fail between 800°F and 1,000°F. The rapid spread of fire within flammable contents can be enhanced by the type of storage that is in close proximity to other flammable contents (racked-tiered storage, etc.). The presence of operative sprinklers can retard the spread of fire. Suspect a lack of sprinklers on older buildings unless they have been retrofitted.

Collapse: Collapse is totally dependent on the type of roof construction, size of building, type of contents, and presence or lack of sprinklers. It will take a sizeable fire and time for the concrete tilt-up panels to be in danger of collapse. However, if a roof does collapse, the concrete panels will be freestanding and weaker than their original configuration and can collapse outward. Collapse of any roof depends on the size and type of roof. The large wood timber roofs will resist fire for longer periods of time than metal roofs or wood roofs of smaller conventional and/or lightweight lumber. Mezzanines that are subjected to water from sprinklers are prone to sudden collapse.

Ventilation: These buildings can normally support horizontal, vertical, and/or PPV as ventilation options depending on the building and extent of fire encountered. Horizontal ventilation will often be limited to minimal doors such as a single standard size door on one end of a building and multiple overhead loading doors on other ends (depending on size) of a building. Vertical ventilation can be a viable operation, particularly in wood roofs, but depending on the size of building, may be a resource- and time-intensive operation. If a particular building is equipped with glass and/or plastic skylights, they can be quickly removed for ventilation operations. PPV is totally dependent on the size of a building, and as a building approaches the large warehouse size, PPV will require fans capable of large CFM capability.

Forcible Entry: Forcible entry is usually limited to either the standard size door into an office area or overhead loading doors. Some windows may be present, such as in buildings with specific interior processes that require openable windows for ventilation, but windows are usually not in abundant supply. Challenging and time-intensive entry operations should be anticipated as security is a major consideration in these types of commercial buildings. Personnel should be proficient in forcible entry techniques for overhead doors.

Search: Unless there are known trapped occupants, search should not be a primary concern as the occupants are normally ambulatory. Obviously, rapid fire escalation can dramatically change this consideration. Additionally, if a search is necessary, the inherent risk should be evaluated and personnel must be proficient in large area searches as they can be difficult and time consuming.

SPECIFIC HAZARDS

- Always look for signage on the exterior of a building and for any particular storage outside a building, as these can be indicators of the interior contents and processes.
- The size of the utilities can be an excellent indicator of any unique hazards within a building.
- As entry and exit openings can be in short supply, and the size of these buildings can be formidable, consider making as many large entry and exit openings as feasible.
- Mezzanines should be anticipated and are subject to collapse when sprinklers have been activated.
- Buildings that are not equipped with operative sprinklers pose an additional risk.
- The larger the building, the more difficult suppression operations will likely become. This applies to forcible entry, attack, and ventilation operations.
- Search in large buildings can be challenging unless personnel are proficient in large area searches.
- Be wary of rack/tiered storage collapse issues if exposed to fire.
- Consider the dangers of access/egress under mezzanines.

RAPID STREET-READ GUIDE

BUILDING GROUP: Manufacturing/Warehouse—Wood

BASIC CONSTRUCTION METHOD

- Walls of wood frame with wood and/or corrugated metal siding
- Substantial wood structural members, and some large timber wood trusses for the roof
- Concrete slab foundations

ERA CONSIDERATIONS

- Pre-1960 buildings used a heavy grade of wood for structural members. Large timber wood trusses were common.
- Post-1960 manufacturing/warehouse buildings made from wood are rare due to cost constraints and/or code requirements. Where found, smaller dimensional lumber and trusses for structural members will be common.

EXTERIOR FEATURES

- Depending on the size and age of a building, many types of roofs can be used, including flat, monitor, gable, and arched (bowstring or tied truss). Metal deck roofs can also be found but wood roofs seem to be more common. Bowstring and rigid-arch roofs are very common in many areas of the country.
- It is not uncommon for additional buildings to be attached to the original building for additional space. The attached buildings are often a lesser grade of construction (pictured above).
- Even though some of these buildings can be quite large, their walls are often covered by corrugated metal siding and/or plywood/OSB type materials.

INTERIOR FEATURES

- Large open areas are common as the openness favors storage and manufacturing processes.
- A lack of partition walls is common.
- Small office areas are common and are used for records, reports, management of the business, and other purposes.
- Sprinklers may or may not be present.
- Expect storage and manufacturing processes.

MANF 29

TACTICAL CONCERNS

Fire Spread: The spread of fire can be rapid and intense due to open floor plans, the size of the building, and the type of contents. The rapid spread of fire within flammable contents can be enhanced by the type of storage that is in close proximity to other flammable contents (racked-tiered storage, etc.). When exterior siding of plywood type materials has been used, it provides little resistance to fire. Wood roofs will readily burn, but the older wood timber roofs will last longer than their newer lightweight counterparts. The presence of operative sprinklers can retard the spread of fire. Suspect a lack of sprinklers on older buildings unless they have been retrofitted.

Collapse: Collapse is totally dependent on the type of construction, size of building, type of contents, and presence or lack of sprinklers. It will take a sizeable fire and time for collapse to occur in the older wooden walls. The large wood timber roofs will resist fire for longer periods of time than metal roofs or wood roofs of smaller conventional and/or lightweight lumber. Smaller buildings with more conventional roofs will fail depending on the size and duration of fire in a particular building. Watch for overloaded mezzanines.

Ventilation: Horizontal ventilation will often be limited to minimal doors such as a single standard size door on one side of a building and multiple overhead loading doors on other ends (depending on size) of a building. Vertical ventilation can be a viable operation, particularly in wood roofs, but may be a resource- and time-intensive operation to have a meaningful effect. If a particular building is equipped with glass and/or plastic skylights, these can be quickly removed for ventilation operations. Roof ventilation operations on metal deck roofs is not normally recommended. PPV is totally dependent on the size of a building, and as a building approaches a large warehouse size, PPV will require fans capable of large CFM capability. Monitor roofs with openable windows/vents can provide vertical ventilation.

Forcible Entry: Forcible entry is usually limited to either the standard size door into an office area or overhead loading doors. Some windows may be found in buildings that house specific interior processes that require openable windows for ventilation, but windows are generally not in abundant supply. Challenging and time-intensive entry operations should be anticipated as security is a major consideration in these types of commercial buildings. Personnel should develop forcible entry techniques for overhead doors.

Search: Unless there are known trapped occupants, search should not be a primary concern as the occupants are normally ambulatory. Additionally, if a search is necessary, the inherent risk should be evaluated and personnel must be proficient in large area searches as they can be difficult and time consuming.

SPECIFIC HAZARDS

- The size of the utilities can be an excellent indicator of any unique hazards within a building.
- Always look for signage on the exterior of the building and for any particular storage outside a building as these can be indicative of the interior contents and processes.
- As entry and exit openings can be in short supply, and the size of these buildings can be formidable, consider making as many large entry and exit openings as feasible.
- Buildings that are not equipped with operative sprinklers pose an additional risk.
- The larger the building, the more difficult suppression operations will likely become. This applies to forcible entry, attack, and ventilation operations.
- Search in large buildings can be challenging unless personnel are proficient in large area searches.
- Overhead doors require specific forcible entry techniques.

RAPID STREET-READ GUIDE

BUILDING GROUP: Manufacturing/Warehouse—Converted Mill

BASIC CONSTRUCTION METHOD

- Heavy timber and timber floor beams and columns
- Exterior walls of load-bearing brick
- Flat, conventional wood roof
- Perimeter concrete foundation with internal concrete footer pads for central columns
- Rarely have basements

ERA CONSIDERATIONS

- Converted historic-era examples may have a steel post and beam substructure backing the brick exterior walls.
- Converted industrial-era examples are more likely to have steel reinforcement only for selected heavy timber components.

EXTERIOR FEATURES

- In rehabbed brick buildings, energy-efficient windows and modern material doors are common.
- Some buildings have first floor businesses, with dwelling spaces on upper floors.
- Modern construction additions are used for new stairs and/or elevators (pictured above).
- The presence of bolt ends and anchor plates on the exterior wall indicates a steel-reinforced substructure.
- Decorative cornice/parapets are common.

INTERIOR FEATURES

- Vast open spaces may have been partitioned into individual apartments or condos accessed by a center hall. Top floor units may be two stories with lofts and mezzanine arrangements.
- Modern stairs and elevators may be added.
- First floor businesses are accessed by a central mall.
- The buildings may have been retrofitted with sprinklers and modern HVAC systems.
- Central hallways are common.

MANF 30

TACTICAL CONCERNS

Fire Spread: The large, strong, and open spaces of heavy timber construction are ideal for conversion to a multi-tenant apartment or condominium building. To meet code, an automatic fire sprinkler system and standpipe system must be added. The greatest fire spread threat comes during the conversion or renovation period; expect rapid fire spread throughout and tremendous heat release rates. Because of the expansive space, master streams are unable to penetrate the interior. This can lead to a fire that could burn for hours, resulting in collapse. Once finished, the converted mill is highly compartmentalized and protected with sprinklers. Fire spread is usually influenced by drywall and other fire-stop methods added during the conversion (e.g., fire doors, dampers, etc.). High-rise fire attack protocol may be warranted. Fire spread can also be enhanced by vertical stair shafts, elevator shafts, and common hallways.

Collapse: Mill buildings are quite stout and collapse resistive for all but the fully involved fire. Likewise, the greatest collapse threat comes when the mill building is undergoing conversion, as exposed, aged, and temporarily braced structural elements are prone to rapid failure. Once converted, the mill building is actually more collapse resistant as steel reinforcement has been added to load-bearing walls and internal beams and columns. Add to this the compartmentalization and automatic sprinkler system. Many converted mills have a completely new roof of lightweight steel or composite (wood/steel) trusses and lightweight sheathing and membrane. Like all lightweight truss systems, a collapse threat exists. Fires on the top floor can collapse the trusses or may weaken cornice connections, leading to collapse.

Ventilation: The use of high-rise apartment ventilation tactics is the ventilation option of choice. PPV fans used to pressurize stairwells can help. Exhaust flow paths can be established by removing downwind windows. Top floor fires may benefit from defensive holes. (They should not be located above the fire as lightweight trusses are likely.) Where present, stair and elevator penthouses may assist in ventilation efforts by pressurizing these shafts.

Forcible Entry: Forcible entry can normally be accomplished by conventional methods for individual living units. The multitude of interior doors and locking devices are likely of modern materials. Security hardware on ground floor businesses may require power saws or hydraulic spreaders.

Search: First floor open spaces may require large area search techniques. Stock, furnishings, and other arrangement variables can also be challenging on the first floor. Upper floor plans are often simple and repeated unit to unit. Smoke conditions in exit corridors may cause occupants to await rescue near exterior window openings.

SPECIFIC HAZARDS

- Extensive remodeling has most likely consolidated utility shutoffs to a central location within the building. Expect modern, commercial-style equipment.
- Backup generators and uninterruptable power supply battery banks may be present.
- Occupant amenities such as swimming pools may be found.
- Original building features such as smokestacks, water tanks, and boilers may have been left in place to add character or serve a purpose.
- Windows may have been replaced with tempered glass or energy-efficient windows.
- Using outside windows to count floors for orientation may not be accurate due to loft-style interior units. Fire departments are encouraged to preplan converted mill buildings and develop floor plans for incident referencing.
- Elevators and HVAC systems will tend to be older.

RAPID STREET-READ GUIDE

BUILDING GROUP: Manufacturing/Warehouse—Public Storage—Single Story

BASIC CONSTRUCTION METHOD

- There are three different types:
  - Concrete block with conventional or lightweight construction
  - Walls of wood frame with conventional and/or lightweight construction and plywood and/or metal siding
  - Lightweight metal frame construction with plywood materials and/or corrugated metal siding
- Concrete slab foundations

ERA CONSIDERATIONS

- Pre-1960 buildings are wood frame construction with wood siding.
- Some use poured-in-place concrete and concrete block with light construction materials.
- Post-1960 buildings are typically concrete block and lightweight construction, metal interior/exterior, and wood exterior and lightweight interior construction.
- Wood/metal frame buildings typically have plywood type materials or metal panels for exterior walls.

EXTERIOR FEATURES

- These buildings typically use minimal construction such as T-11 siding, metal studs, and/or lightweight construction.
- Some buildings were built with concrete block walls, but normally are of lightweight construction.
- Attached units are a single-story row of various lengths with secure doors of various sizes for each unit.
- Locks on each door are the property of each renter.

INTERIOR FEATURES

- Individual compartmentalized rooms of various sizes are used.
- Unknown contents will vary within each unit. Contents can consist of hazardous storage, such as ammunition, flammable items, and hazmat items.
- Partition walls between rented spaces may not go to the roof, leaving an open void that spans multiple individual units.

MANF 31

TACTICAL CONCERNS

Fire Spread: Rooms in common single-story buildings often use partition walls that do not travel from the foundation to the roof. This lack of separation (at the top) will easily allow fire spread between numerous units and can result in a fast spreading fire as well as weakening large sections of a lightweight roof. Partition walls are often made from drywall that will not resist fire for long periods. The type of fire is dependent on contents.

Collapse: Collapse is dependent on the type of construction. Concrete block and conventional construction are not prone to rapid collapse and can offer a measurable amount of fire resistance. However, common single-story row buildings are often of minimal wood and/or metal construction and will be prone to early collapse, particularly the roof.

Ventilation: Ventilation is limited as there is typically only one door per unit. Therefore, ventilation is limited to opening the access door to an involved unit, an operation that can be dangerous. Roof ventilation is not normally recommended due to a lack of substantial roof construction. When utilized, roof ventilation operations should be defensive (flank the burning area) as opposed to ventilating directly over the fire.

Forcible Entry: Forcible entry is simplified by easy access to the door locks that are normally a single padlock. Although the locks are the property of the renter, their attachment to the door is not usually substantial and can be removed with conventional forcible entry methods.

Search: Search is not a primary concern as these buildings/units are principally used for storage.

SPECIFIC HAZARDS

- The primary hazard comes from the variety of unknown contents that can vary widely, including flammable materials, hazardous materials, ammunition, and anything else that renters would want to store in a secure location. Illegal disposal of chemicals and waste has been found in storage units.
- These buildings can rapidly burn due to the common use of lightweight materials.
- Expect a common void between each unit at the top of the partition walls.
- These buildings are not usually sprinklered.
- Be careful of opening a door to an involved unit before adequate extinguishment operations.

MANF 31

RAPID STREET-READ GUIDE

BUILDING GROUP: Manufacturing/Warehouse—Public Storage—Multistory

BASIC CONSTRUCTION METHOD

- Either concrete block or poured-in-place concrete walls with conventional or lightweight interior construction
- Concrete perimeter or slab foundation
- May have basements

ERA CONSIDERATIONS

- Pre-1960 buildings are poured-in-place concrete or concrete block construction and may be converted multistory commercials of concrete and conventional construction.
- Post-1960 buildings are typically concrete block and lightweight construction, metal exterior, and lightweight interior construction.

MANF 32

EXTERIOR FEATURES

- Older multistory buildings are frequently converted commercial buildings and often have few entry doors. However, some buildings have windows on each floor (see picture above).
- Older multistory buildings can be of substantial construction such as concrete or block and may also have a substantial roof.
- Entry doors are likely substantial.
- Locks on each door are the property of each renter.

INTERIOR FEATURES

- Individual compartmentalized rooms are of various sizes.
- Unknown contents will vary within each unit.
- Interior hallways are common—expect some dead-end hallways.
- Contents can consist of hazardous storage, ammunition, and other flammable materials.
- Locks, which are the property of each renter, can vary widely.
- Sprinklers may or may not be present in these buildings.

TACTICAL CONCERNS

Fire Spread: Rooms in converted multistory commercials are normally well compartmented with minimal voids. If the rooms are sealed floor to floor, fire spread is largely dependent on the type of access door to each room. These are usually not fire rated and can vary widely from heavy to minimal construction. If fire is able to extend into a hallway, the hallway will become a horizontal channel for fire and its by-products. A lack of sprinklers will enhance the spread of fire. Light smoke showing can indicate a deep-seated fire that is difficult to locate due to extensive compartmentalization. Expect extremely smoky conditions for most fires.

Collapse: Collapse is dependent on the type of construction. Poured-in-place concrete, block, and conventional construction are not prone to rapid collapse and can offer a measurable amount of fire resistance. The strength of the roof is dependent on the type of construction.

Ventilation: Ventilation is limited as these buildings typically have only one door per unit. Involved rooms within multistory buildings can create heat and smoke problems in the hallways that can be ventilated by horizontal and/or PPV ventilation operations, but will take resources and time. Dead-end hallways will be difficult to ventilate.

Forcible Entry: Forcible entry is simplified by easy access to the locks on the doors, which are normally a single padlock. Although the locks are the property of the individual renter, their attachment to the door is not usually substantial and can be removed with conventional forcible entry methods. Forcible entry to the building can prove challenging, as the exterior doors/locks can be substantial.

Search: Search is not a primary concern as these buildings are principally used for storage. However, multistory buildings with interior hallways that are contaminated may need to be verified for a lack of occupants.

SPECIFIC HAZARDS

- The primary hazard comes from the variety of unknown contents, such as flammable materials, hazardous materials, ammunition, and anything else that renters would want to store in a secure location. Illegal disposal of chemicals and waste has been found in storage units.
- The center hallway design can complicate ventilation and/or search operations.
- A lack of sprinklers can compound an existing problem.
- Smoky conditions, numerous compartments, and dead-end hallways are a dangerous formula that can lead to lost firefighters, air-management issues, and getting trapped.

RAPID STREET-READ GUIDE

BUILDING GROUP: Office Building/Hotel—Pre-WWII—Low Rise

BASIC CONSTRUCTION METHOD

- Either poured-in-place concrete, concrete block, or conventional or unreinforced masonry construction
- Interior structural members of conventional wood
- Concrete slab or perimeter foundations, some include basements

ERA CONSIDERATIONS

- They were built prior to World War II and constructed of heavy grade conventional materials.
- Unreinforced masonry construction was popular until 1935.
- Basements are frequently found.
- External fire escapes and sprinklers may be present.

EXTERIOR FEATURES

- Varying sizes of these occupancies range from simple two-story office buildings to multistory buildings up to about six stories.
- Roof construction normally consists of metal deck built-up construction or heavy timber/conventional wood construction. Expect numerous layers of roofing material on either type of roof.
- Conventional cornices are often used and can be quite large.
- Check the rear and sides of the building for fire escapes.
- There are numerous windows.

INTERIOR FEATURES

- These buildings are prime candidates for renovations.
- Lath and plaster walls and ceilings are common; some have tin ceilings.
- The interior normally consists of offices that are common to a center hallway.
- The interior environment of the building is typically heated by a central system (steam or forced hot air) and A/C is provided by retrofit methods (individual wall/window units or a central system).
- Older elevators and external fire escapes are widespread.
- Basements are common and can be quite large/extensive.

OFF 33

TACTICAL CONCERNS

Fire Spread: Although these buildings use a standard floor plan of offices that are accessed from a center hallway, fire can spread by the following primary avenues: (1) auto exposure, (2) HVAC ducting if present, (3) poke-through construction from renovations, (4) center hallways, and (5) voids from renovations. Remember that fire in the top floor or roof that exposes an exterior building cornice can easily spread in the void behind the cornice and cause it to collapse, as well as allow the fire to spread to the attic area. Fires that originate in a basement can rapidly spread upward through numerous vertical avenues. Retrofit data and electrical cable distribution trays add additional fire load and fire spread hazards.

Collapse: It will take a sizeable fire and time for collapse to occur in most of these buildings as they are primarily constructed of older conventional materials. In most cases, the roofs are of heavy wood construction that does not readily fail when exposed to heat/fire. Remember that unreinforced masonry construction can readily collapse when exposed to fire and/or the walls or structural members of the roof collapse. Watch for weakened cornices that can easily collapse.

Ventilation: Although the windows are plate (annealed) glass, they are numerous and can be easily used for natural and/or PPV horizontal ventilation in contaminated areas. Pressurized air from PPV blowers can be supplied to vertical stair shafts and then to contaminated areas, exhausting the contaminates either horizontally through window openings or up vertical stair shafts to exit a roof. Roof ventilation can be accomplished but should be approached with the perspective that older roofs, although strong, can have numerous layers of roofing materials. Renovations can result in a roof that is constructed from lightweight metal or wood materials. If present, the HVAC system should be turned off as soon as possible and only used if personnel are totally familiar with its operation and capabilities.

Forcible Entry: Forcible entry can normally be accomplished by conventional methods as the doors/windows are of conventional construction. The glass in the windows is normally plate (annealed) glass unless upgraded to tempered or double-pane glass. Some of these buildings have on-duty security personnel who should be available to assist with entry/exit considerations.

Search: A potential search in this type of building should be preplanned as there will be numerous rooms that will need to be accounted for. In most cases, searches will emanate from a center hallway, but must be done in a systematic manner until all affected rooms are searched. The older conventional construction that is used in these buildings can be a benefit due to the fact that conventional construction is more fire resistive than modern lightweight construction, allowing additional search time. Search operations can be enhanced by appropriate information from on-duty security personnel, if present. Occupants above the fire floor may try to escape to the roof—check the top floor of all stairwells.

SPECIFIC HAZARDS

- Unreinforced masonry construction can easily collapse outward in fire conditions.
- Renovations can result in hidden voids and poke-through openings between floors.
- Sprinklers may not be present and elevators are likely old and substandard.
- Older cornices will not support the weight of an aerial device and/or personnel.
- There is a sizeable void behind exterior cornices that may be common to the attic/roof.
- Older cornices can readily fail when exposed to fire and/or heavy streams.
- Consider the safety of using older elevators and external fire escapes.

RAPID STREET-READ GUIDE

BUILDING GROUP: Office Building/Hotel—Post-WWII—Low Rise

BASIC CONSTRUCTION METHOD

- Poured-in-place concrete; concrete block; conventional brick; metal frame with exterior of decorative metal, stucco, glass, or brick veneer; and wood frame with brick veneer
- Roofs vary from older conventional roofs to the more modern lightweight materials
- Concrete slab or perimeter foundation; some basements, particularly older buildings

ERA CONSIDERATIONS

- Post-WWII buildings are a challenging mix of old (built after the war), an intermediate combination of old and emerging new technologies (1960–1985), and modern construction renovations.

EXTERIOR FEATURES

- The buildings range from simple two-story office buildings to buildings up to about six stories.
- From the exterior, these buildings can present a vast array of configurations and materials.
- Roof construction normally consists of conventional wood and/or metal deck built-up construction, with numerous layers of roofing materials on older buildings to lightweight metal and wood structural members covered by composition, rock, membrane materials, and so on.
- The use of flammable decorative cornices or conventional framed cornices is dependent on the age of the building.
- Expect a wide range of window styles. Most can be opened but have a small range of motion.
- Curtain exterior wall construction can be found on newer buildings.

INTERIOR FEATURES

- The interior normally consists of office spaces that surround a central hall or elevator corridor. Circuitous office floor plans were starting to be utilized (open floor plans that circle around a central elevator lobby or access area).
- Suspended ceilings that hide voids above the ceilings are common on newer buildings, and lath and plaster ceilings are common on older buildings.
- Older windows will be openable and use plate (annealed) glass, while newer windows will be fixed and of tempered glass or double panes.
- Central HVAC systems were increasingly being utilized (older buildings may still have central heat but retrofit A/C).
- Elevators are common and sprinklers may be present.
- Older buildings are prime candidates for renovations, including retrofit data and electrical cable trays.
- Many office buildings include a paper file storage area(s) that include large, steel, rolling units.
- Hotels typically have a first floor that includes common-use areas (lobby/lounge, restaurant, conference rooms, etc.) and upper floors with a repeated floor plan of individual sleeping rooms.

OFF 34

TACTICAL CONCERNS

Fire Spread: Older buildings use a more conventional, heavy type of construction that can be more resistant to the spread of fire as compared to the use of lightweight materials in newer buildings. Both circuitous and center hallways can spread fire/heat smoke. The older buildings often have vertical voids/shafts/stairways/cornices that support the spread of fire. The newer buildings have curtain wall construction, HVAC systems, voids above the suspended ceilings, and other fire spread hazards.

Collapse: It will take a sizeable fire and time for collapse to occur in the older buildings due to the conventional type of construction. Newer buildings with lightweight metal/wood construction and masonry veneers will take less time for collapse when exposed to fire/heat. If fire is on a top floor, collapse of the roof depends on the size and type of roof. Renovated buildings that use lightweight materials are subject to faster collapse times than conventional materials.

Ventilation: Although the windows are plate (annealed) or tempered glass, they are numerous and can be used for natural and/or PPV horizontal ventilation in contaminated areas. Pressurized air from PPV blowers can be supplied to vertical stair shafts and then to contaminated areas, exhausting contaminates either horizontally through window openings or up vertical stair shafts that exit the roof. Although older roofs will likely be a conventional type of wood or metal construction, expect a heavy layer of roofing materials. Newer roofs are likely constructed from lightweight metal and some wood materials. If present, the HVAC system should be turned off as soon as possible.

Forcible Entry: Forcible entry can normally be accomplished by conventional methods. Doors should be of conventional construction. However, if glass in the windows is tempered, it will require specific techniques to break.

Search: A potential search in this type of building should be planned as there will be numerous rooms that will need to be accounted for. In most cases, searches will emanate from a center hallway or elevator lobby, but must be done in a systematic manner until all affected rooms are searched. This process can be enhanced by appropriate information from on-duty security personnel (if present). Occupants above the fire floor may try to escape to the roof—check the top floor of all stairwells.

SPECIFIC HAZARDS

- These buildings can present a mix of old and new fire protection features, but most have stairwell standpipes.
- Exterior foam cornices are flammable and will not support the weight of an aerial device. Older cornices also are unlikely to support an aerial device and can hide a void that may be common to an attic.
- Suspended ceilings are common in newer buildings and can subject firefighters to entanglement in the supporting wires if the ceiling fails.
- In newer buildings, fire can spread vertically from several avenues. Older buildings also have many vertical extension routes, but the original construction is stronger than modern lightweight construction.
- Older buildings are prime candidates for renovations.
- In case of fire/smoke, HVAC systems (if present) should be shut off as soon as possible.
- It is best to preplan the use of elevators for fire incidents on an individual building basis.
- Circuitous open floor plans are often finished with modular office cubicles that add tremendous fire load and make search patterns and hose movements difficult.
- Underground parking may be present.

RAPID STREET-READ GUIDE

BUILDING GROUP: Office Building/Hotel—21st Century

BASIC CONSTRUCTION METHOD

- Concrete block or metal frame with exterior of glass, stucco, decorative metal, and brick veneer
- Curtain exterior wall construction in many cases
- Concrete slab or perimeter foundation
- Hybrid construction with noncombustible common areas and wood/metal-framed individual rental or room units for some of the newest buildings.

ERA CONSIDERATIONS

- Post-1960 buildings commonly use metal structural members (skeleton) that are then finished with an exterior covering of decorative metal, glass, brick veneer, or stucco. This type of construction commonly utilizes the methodology of curtain construction. The roofs are usually metal deck with built-up or membrane finish.
- Post-2000 versions are constructed with concrete, CMU, or steel common areas (lobby, stairwells, elevator shaft) with wood-framed sleeping rooms or office spaces.

EXTERIOR FEATURES

- These occupancies range from simple two-story office buildings to multistory buildings of about six stories that do not qualify as high-rise (buildings over 75 ft).
- Many of these buildings use curtain construction with various exterior materials.
- Roof construction normally consists of metal deck built-up construction. In some smaller buildings of several stories, lightweight wood trusses can be used.
- Decorative cornices made of rigid foam and fastened to a building by adhesives are common.

INTERIOR FEATURES

- The interior normally consists of offices that are circuitous to a common hall or elevator lobby. Hotels use a center hall configuration for sleeping rooms.
- Suspended ceilings that hide voids above the ceilings are the standard.
- Virtually all windows are tempered glass and not openable (sealed building).
- Modern elevators and enclosed interior vertical stair shafts are common.
- Wet standpipes and sprinklers are normally present.
- Underground parking of various levels is common.
- Hotels may have restaurant facilities and associated hazards.

OFF 35

TACTICAL CONCERNS

Fire Spread: Although these buildings are rather simplistic from the perspective of offices and hallways, they are renowned for allowing fire to spread vertically by five avenues: (1) auto exposure, (2) curtain construction, (3) HVAC ducting, (4) poke-through construction, and (5) transfer of heat through the floors. Fire can also easily spread in voids above the suspended ceilings and within the center hallways. Remember that fire in the top floor or roof can expose and ignite the decorative foam cornices on the exterior. Sprinklers are normally present. Fire spread from room to room in hotels is typically limited due to compartmentalization and fire suppression systems. Fire can vertically spread up grease ducts in kitchens.

Collapse: It will take a sizeable fire and time for collapse to occur in buildings that are of noncombustible construction and compartmentalized. Hybrid examples are more prone to a rapid collapse. In most cases, the roofs are constructed of lightweight metal trusses supporting a metal deck that can readily fail when exposed to high heat and/or fire. Lightweight wood truss roofs can also readily fail if exposed to fire. Additionally, interior firefighters can be entangled in supporting wires if suspended ceilings fail.

Ventilation: These buildings have multiple attributes that can be used for ventilation operations. Although the windows are tempered glass, they are numerous and can be used for natural and/or PPV horizontal ventilation in contaminated areas. Pressurized air from PPV blowers can be supplied to vertical stair shafts and then to contaminated areas, exhausting contaminates either horizontally through window openings or up vertical stair shafts that exit the roof. Roof ventilation can be accomplished but should be cautiously approached as most roofs are constructed from lightweight metal materials. Lastly, HVAC systems should be turned off as soon as possible and should only be used if familiar with their operation.

Forcible Entry: Forcible entry can generally be accomplished by conventional methods as there are multiple conventional doors. The glass in the windows is normally tempered, which requires specific techniques to break. Some of these buildings have on-duty security personnel who should be available to assist with entry/exit considerations.

Search: Although the frequency of fires in these buildings is very low, a potential search should be planned as there will be numerous rooms that may need to be accounted for. Typically, searches will emanate from a center hallway or elevator lobby, but must be done in a systematic manner until all affected rooms are searched. This process can be enhanced by appropriate information from on-duty security personnel (if present).

SPECIFIC HAZARDS

- It is imperative to preplan the construction methods used for modern office/hotel buildings. They can range from Type 1 to hybrid.
- The exterior foam cornices are flammable and will not support the weight of resources.
- Failed suspended ceilings can subject firefighters to entanglement in the supporting wires.
- It is best to preplan the use of elevators for fire incidents on an individual building basis.
- Circuitous open floor plans are often finished with modular office cubicles that add tremendous fire load and make search patterns and hose movements difficult.
- Hotels with indoor swimming pools will have boiler and chemical rooms.
- In case of fire/smoke, HVAC systems should be shut off as soon as possible.
- Fires on the top floor can present a hazard to exterior foam cornices and roof structural members.
- Fires in underground parking areas can create noteworthy problems.
- Kitchen facilities may be present with associated hazards.

RAPID STREET-READ GUIDE

BUILDING GROUP: Office Building/Hotel—High-Rise—1st Generation

BASIC CONSTRUCTION METHOD

- Poured-in-place concrete, brick masonry, or stone construction exterior walls with iron floor beams
- Roofs vary from concrete to heavy wood timber and/or metal construction
- Perimeter foundations and basements

ERA CONSIDERATIONS

- These buildings were constructed from the late 1800s to the early 1900s, so virtually everything about them is strong and will resist fire in a superior manner.

EXTERIOR FEATURES

- Height is limited to about 20 stories. Iron/steel beams and cast iron columns are used.
- Exterior walls are typically comprised of concrete or masonry.
- Roof construction normally consists of concrete or heavy wood/metal construction. Rooftop penthouses for stairways and elevators are common.
- Conventional facades are common and comprised of heavy wood materials and some cast iron.
- Stone/concrete corbels are used for decorative purposes.

INTERIOR FEATURES

- Offices/living units that are common to a center hallway are the norm.
- Unprotected vertical passageways are likely.
- They were not originally equipped with modern elevators, building communication systems, or other amenities that are used today.
- Return-type stairs are most common. The first floor may be two stories in height and include a central grand stairway that is open to a mezzanine level as part of the first floor.
- Dry standpipes are common; the presence of wet standpipes and sprinklers is questionable.
- Some tin ceilings are found, particularly in the lobby areas.
- Horizontal and vertical passageways tend to be narrower than modern passageways.
- Lath and plaster walls and ceilings are used.
- Windows are openable and elevators are likely old or substandard.
- Originally these had no HVAC systems.

OFF 36

TACTICAL CONCERNS

Fire Spread: Although these buildings feature strong construction, fire can readily spread vertically up numerous vertical passageways such as elevator shafts, open stairwells (which can go from the lowest floor to the top floor), light wells, voids created by plumbing and/or electrical channels, and the open hallways on each floor. Remember that an extending fire in the top floor or roof can expose the decorative cornice on the exterior of a building. The presence of renovations will introduce voids that were not in the original construction.

Collapse: It will take a sizeable fire and time for collapse to occur in most of these buildings, which is not a common occurrence as the construction is based on heavy concrete, masonry, and wood/metal materials. This also holds true for the roof. However, when exposed to fire, the wood floors and cast iron structural members can collapse and may be considered the weak point of this construction.

Ventilation: As the windows are plate (annealed) glass and they are often numerous, they can be easily opened or broken and used for natural and/or PPV horizontal ventilation in contaminated areas. Pressurized air from PPV blowers can be supplied to vertical stair shafts and then to contaminated areas (including hallways), exhausting contaminates either horizontally through window openings or up vertical stair shafts that exit the roof. If the open stairwells become contaminated with heat/smoke, they can be ventilated if the stair shaft exits a roof. Roof ventilation can be accomplished but should be cautiously approached as most roofs are constructed from substantial materials with heavy layers of roofing materials.

Forcible Entry: Forcible entry can normally be accomplished by conventional methods, but anticipate the presence of substantial doors and windows. The glass in the windows is normally plate (annealed) glass that is easily opened/broken if necessary. Some of these buildings have on-duty security personnel who should be available to assist with entry/exit considerations.

Search: Although the fire load is low to moderate and lath/plaster can better withstand fire and heat as compared to drywall, a potential search in this type of building should be planned as there will be numerous rooms that will need to be accounted for. In most cases, searches will emanate from a center hallway, but must be done in a systematic manner until all affected rooms are searched. This process can be enhanced by information from on-duty security personnel (if present).

SPECIFIC HAZARDS

- The exterior cornices can be flammable, will not support the weight of an aerial device, and will also conceal voids behind the cornice.
- Corbels are likely unstable due to their age.
- Fire can easily spread vertically in the open stairwells and access shafts.
- Expect a large boiler room for heat and hot water utilities.
- Expect many retrofit systems (HVAC, plumbing, electrical, data cable trays, etc.).
- Horizontal and vertical passageways are narrower than modern passageways.
- Elevators original to the building will not have modern fire service use features.
- Heavy objects such as water tanks may be found on the roof.
- Stairwells will not be pressurized by an HVAC system.
- Falling glass and debris during fires can endanger those leaving and entering the building.
- Standpipes and fire department connections may not be capable of high gpm or flow pressures.
- Restaurant facilities and underground parking may be present.

RAPID STREET-READ GUIDE

BUILDING GROUP: Office Building/Hotel—High-Rise—2nd Generation

BASIC CONSTRUCTION METHOD

- Constructed prior to WWII
- Stone/brick masonry or poured-in-place concrete
- Roofs of concrete, heavy wood timber, or metal deck of heavy construction
- Perimeter foundation or with basement, depending on the area of construction

ERA CONSIDERATIONS

- The primary difference between these buildings and first generation high-rise buildings is protected steel structural members, noncombustible materials, and a lack of open vertical passageways.
- Due to the era of construction, virtually everything about these structures is basically strong.
- Building heights greatly surpassed first generation high-rises.

EXTERIOR FEATURES

- Structural elements are normally comprised of concrete, masonry, protected steel, and other noncombustible type materials.
- Roof construction normally consists of concrete or metal. Spire shaped roofs may include radio/TV relay antennas.
- Stone/concrete corbels are used for decorative purposes.
- Cast iron columns and beams are common for first and second floor exterior decorative purposes (not structural).

INTERIOR FEATURES

- Expect protected vertical shaft enclosures.
- Return type stairs are likely present for older examples. Scissor-type stairs are common for later examples. Later examples have enclosed stairways that can be considered an area of refuge.
- The interior normally consists of offices/living units that are common to a center hallway.
- Lath and plaster walls and ceilings are used instead of drywall coverings.
- Elevators likely have older characteristics although elevator lobbies on each floor are typically larger to accommodate multiple lifts. It is common to have separate elevators that serve only a given range of floors.
- Most lack HVAC systems, so expect unpressurized stairwells.
- Dry standpipes are common; wet standpipes/sprinklers may also be present.

OFF 37

TACTICAL CONCERNS

Fire Spread: Although these buildings feature strong construction and protected vertical shafts, fire can readily spread up open vertical passageways such as elevator shafts, stairwells that are open (which can go from the lowest floor to the top floor), light wells, vertical voids created by plumbing and/or electrical channels, and the open hallways on each floor. Renovations will introduce voids that were not in the original construction.

Collapse: These buildings are generally considered to be resistant to fire and collapse as the construction is based on heavy concrete or masonry materials. Therefore, it will take a sizeable fire and time for collapse to occur in these buildings and is not a common occurrence. This also holds true for the floors and the roof, which are also of substantial construction.

Ventilation: As the windows are plate (annealed) glass and are often numerous, they can be broken and used for natural and/or PPV horizontal ventilation. Pressurized air from PPV blowers can be supplied to vertical stair shafts and then to contaminated areas (including hallways), exhausting contaminates either horizontally through window openings or up vertical stair shafts that exit the roof. Stack effect may limit the use of vertical ventilation if the distance to the roof openings allows for smoke cooling. Open stairwells can be ventilated if the stair shaft exits the roof. If not, they must be ventilated by exhausting contaminants horizontally at the highest level. Roof ventilation can be accomplished but should be cautiously approached as most roofs are constructed from substantial materials with heavy layers of roofing materials. HVAC systems are normally not present.

Forcible Entry: Forcible entry can normally be accomplished by conventional methods, but anticipate the presence of substantial doors and windows as this is older conventional construction that was the norm before the use of modern lightweight materials. Window glass is normally plate (annealed) glass that is easily opened and/or broken if necessary. Some of the larger buildings have on-duty security personnel who should be available to assist with entry/exit considerations.

Search: A potential search in this type of building should be planned as these buildings can be large and comprised of numerous rooms that will need to be accounted for. In most cases, searches will emanate from a center hallway, but must be done in a systematic manner until all affected rooms are searched. This process can be enhanced by appropriate information from on-duty security personnel (if present).

SPECIFIC HAZARDS

- Although the vertical passageways are protected, fire can easily spread vertically in stairwells that are open and access other shafts.
- Corbels are likely unstable due to their age.
- Horizontal and vertical passageways are somewhat narrower than modern passageways.
- The buildings were not originally equipped with modern elevators, building communication systems, smoke detectors, and other modern amenities.
- Circuitous open floor plans are often finished with modular office cubicles that add tremendous fire load and make search patterns and hose movements difficult.
- Falling glass and debris during fires can endanger those leaving and entering the building.
- Wet standpipe systems on buildings greater than 10 floors may have pressure limiting devices that restrict flow.
- Expect multiple boiler rooms for heat and hot water utilities. Industrial-size electrical transformers and distribution equipment can be found near the ground floor or subgrade.
- Restaurants and underground parking may be present.

RAPID STREET-READ GUIDE

BUILDING GROUP: Office Building/Hotel—High-Rise—3rd Generation

BASIC CONSTRUCTION METHOD

- Constructed after WWII
- Primarily steel skeleton frame attached to reinforced concrete cores with exteriors of decorative metal, stucco, glass, or brick veneer
- Roofs are typically modern lightweight metal/concrete
- Perimeter foundation and some underground parking areas that are reinforced concrete

ERA CONSIDERATIONS

- They are curtain construction with exteriors of decorative metal, glass, tile, and concrete slabs.
- The concrete core normally contains the stairwells, elevators, and electrical/plumbing shafts.
- HVAC systems are the standard.
- Building heights can reach over 100 stories.
- Newest examples include a fire command center.

EXTERIOR FEATURES

- Exterior materials are normally curtain panels of glass, metal, tile, or brick veneer.
- Roof construction normally consists of a metal deck with a built-up or membrane finish.
- Green roofs with extensive plantings and cellular equipment can be found on some of these buildings.
- Decorative cornices made of rigid foam that are fastened to a building by adhesives are common.

INTERIOR FEATURES

- The ground level floor is typically greater than one story in height and may include mezzanines.
- Upper floor plans are typically open and include a center core for elevators, restrooms, and stairwells. Open floor plans are finished out by the users in a multitude of ways (partition walls, open office cubicles, conference rooms, etc.).
- Hotel room floors typically have central hallways leading from the elevator lobby.
- Suspended ceilings that hide voids above the ceilings are the standard.
- Virtually all windows are tempered glass and not openable, as the interior environment of the building is maintained by an HVAC system. Expect pressurized stairwells.
- Modern elevators with firefighter controls are the norm.
- Scissor stairs are more common than return stairs.
- These buildings are normally equipped with communication systems, heat and smoke detectors, sprinklers, standpipes at each floor in the stair shafts, and other features.
- The top floor is often two stories in height and likely to include a swimming pool, restaurant, and/or luxury suites.

TACTICAL CONCERNS

Fire Spread: These buildings are renowned for allowing fire to spread vertically by five avenues: (1) auto exposure, (2) curtain construction, (3) HVAC ducting, (4) poke-through construction, and (5) transfer of heat through the floors. Fire can also easily spread in voids above the suspended ceilings, within the center hallways if they are exposed to extending fire, and via access stairways that allow travel between floors within a common occupancy. Fires in open floor plan office cubicle spaces can be quite intense and will require high gpm flows. Extending fire in the top floor or roof can easily ignite exposed decorative foam cornices on the exterior of a building. Always expect a wind-fed fire for upper floors after windows have failed.

Collapse: It will take a sizeable fire and time for collapse to occur in most of these buildings as they are primarily of fire-resistive construction. However, if a fire is on the top floor, collapse of a roof depends on the size and type of roof, particularly if it is supporting a significant dead load (e.g., a green roof, HVAC equipment, etc.). The attacks on the World Trade Center towers on September 11, 2001, remind us that a catastrophic collapse of a modern high-rise is possible.

Ventilation: Although the windows are tempered glass and are numerous, they can be broken (with proper techniques) and used for natural and/or PPV horizontal ventilation. Pressurized air from PPV blowers can be supplied to vertical stair shafts and then to contaminated areas, exhausting contaminates either horizontally through window openings or up vertical stair shafts that exit the roof. Although these roofs are constructed from lightweight metal materials, roof ventilation is very doubtful. High-rise buildings with modern fire command centers may have HVAC control features that can assist with ventilation. Preplanning and training with these features is encouraged. On-site building engineers can be a useful asset when trying to evaluate ventilation options during fire incidents.

Forcible Entry: Forcible entry can normally be accomplished by conventional methods as there are multiple conventional doors that are likely metal frame and glass (usually entry type doors) and interior metal in metal frame doors. However, the glass in the windows is normally tempered glass. Loading dock doors can be forced but will take time and the necessary expertise to accomplish this operation. Some of these buildings have on-duty security personnel who should be available to assist with entry/exit considerations.

Search: A potential search in this type of building should be planned, as each floor can contain a significant amount of floor area and there will be numerous rooms that will need to be accounted for. Remember that many office type high-rise buildings use the workstation or cubicle concept, which results in minimal partitions and large open areas. In most cases, searches will emanate from a center hallway, but must be done in a systematic manner until all affected rooms/areas are searched.

SPECIFIC HAZARDS

- Dropped ceilings are common and can subject firefighters to entanglement in the supporting wires if a ceiling fails.
- On-demand water heating systems can be found in suspended ceiling spaces for each floor or each room in hotels.
- Expect large (moving van size) HVAC equipment on the roof and in mechanical areas on the ground floor or subgrade levels. Some high-rise buildings have an entire floor used for mechanical equipment at some mid-level point between the ground and top floor.
- Industrial-size electrical equipment can be found in numerous locations. Newer high-rise buildings may have a central control room to manage all building utility functions.
- Circuitous open floor plans are often finished with modular office cubicles that add tremendous fire load and make search patterns and hose movements difficult.
- Consider the possibility of falling glass and debris.
- Underground parking may be present.
- Aboveground wind-driven fires are a risk.

RAPID STREET-READ GUIDE

BUILDING GROUP: Institutional Building—School

BASIC CONSTRUCTION METHOD

- Conventional brick, concrete block, and poured-in-place concrete
- Walls of wood frame with wood siding, vinyl siding, stucco, or brick veneer
- Concrete slab or perimeter foundations

ERA CONSIDERATIONS

- Pre-1960 schools use a heavy grade of wood, some poured-in-place concrete, and some concrete block or brick masonry. Roofs are of heavier construction than modern roofs.
- Post-1960 schools use smaller dimensional lumber and trusses for structural members in wood frame buildings. Concrete block is often found, depending on the size of a building. Brick masonry tends to be a veneer on wood framing. Lightweight materials are becoming more popular. Roofs are often of lightweight materials.
- Industrial- and legacy-era schools may still have combustible wall and ceiling finishes.

EXTERIOR FEATURES

- Schools vary from an elementary school of moderate size to a multistory college complex.
- Poured-in-place concrete, concrete block, and brick masonry are popular for larger buildings. Smaller buildings use wood framing with stucco, metal or vinyl siding, and brick veneer.
- Due to cost considerations, some newer schools are using more lightweight materials.
- Different types of roof construction can be used but are often a flat/sloped configuration. Conventional wood roofs and metal deck roofs are common; wood roofs seem to be more common on the West coast and metal roofs are more common on the East coast.

INTERIOR FEATURES

- Floor plan layouts are either garden style or center hallway configurations for classroom areas. A central or anchored commons area may include staff offices, gymnasium, cafeteria, and/or auditorium.
- Larger buildings have multiple center hallways that access numerous rooms.
- Partitioned walls are in abundance.
- Security personnel are on premises at all times for larger complexes such as colleges.
- There are numerous windows and multiple doorways.
- Older/smaller structures may not have fire suppression sprinklers.

INST 39

TACTICAL CONCERNS

Fire Spread: The spread of fire in these buildings is not a common problem as the interior is partitioned with numerous rooms, sprinklers are common (larger buildings), and on-duty staff is present during normal hours. However, if a fire starts and extends into hallways, it will rapidly spread to other areas served by the hallways. Additionally, fire in masonry/concrete buildings does not accelerate as rapidly as fires in combustible buildings (wood frame, etc.). Suspended ceilings are common and will allow fire above the ceiling to rapidly spread.

Collapse: It will take a sizeable fire and time for collapse to occur in most of these buildings. Poured-in-place concrete and concrete block can resist collapse; however, collapse of a roof depends on the size and type of roof. Conventional wood roofs will resist fire for longer periods of time than metal roofs or wood roofs of smaller conventional and/or lightweight lumber. Smaller buildings with more conventional and/or lightweight wood roofs can readily fail depending on size and duration of fire in a particular building.

Ventilation: Although these buildings have numerous doors and windows, ventilation operations will be assisted or hampered by the size and configuration of a building. Individual rooms can be easily ventilated by vertical, horizontal, and/or PPV. However, as the size increases up to and including large complexes, ventilation becomes more intensive. Contaminated center hallways in larger buildings must be ventilated early to allow access and egress to and from the individual rooms. This can be accomplished by a combination of horizontal ventilation and/or PPV. Roof ventilation is dependent on the type of roof.

Forcible Entry: Forcible entry is enhanced by multiple conventional doors and windows in these buildings. Additionally, the on-duty staff should be available to assist with entry/exit considerations during normal school hours. Older buildings will have more substantial doors and locks, whereas newer buildings may not.

Search: The need for search operations is low as the individual rooms are easily entered and vacated by occupants, and schools regularly conduct fire drill evacuations. Although the frequency of fires in these buildings is low, it will take minimal smoke/disruption to create a measurable amount of panic in the students (particularly younger children). Obviously some type of an immediate dilemma can dramatically change this consideration. If rescue operations are needed, they can be challenging for younger children and should be anticipated.

SPECIFIC HAZARDS

- Younger students are the primary hazard as they often need constant supervision and directions for simple tasks. Fortunately, they are ambulatory.
- Schools typically have a high occupancy load during normal school hours. Evacuation of K–8 schools will likely be orderly with teachers accounting for students. High school and college evacuation may have less accountability and may present congestion issues.
- Expect some special events during nighttime hours.
- Schools often have courtyards or other open areas that are not visible from the outside perimeter of the building.
- Aging boilers and electrical distribution systems can be problematic.
- Expect unique challenges in schools with chemistry labs and research facilities, and even the possibility of radioactive materials.
- Accountability and location of occupants is a key consideration.

RAPID STREET-READ GUIDE

BUILDING GROUP: Institutional Building—Hospital

BASIC CONSTRUCTION METHOD

- Poured-in-place concrete or concrete block or brick masonry
- Concrete slab or perimeter foundations
- Limited use of wood for structural members in older hospitals or small clinics

ERA CONSIDERATIONS

- Most industrial- and legacy-era hospitals are well-built using heavier grade materials and engineered for significant dead loads.
- Modern-era hospitals are typically masonry or concrete for load-bearing walls but may include a lightweight truss roof and lightweight trusses with metal pan and concrete flooring.

EXTERIOR FEATURES

- Depending on the size and age of a building, these types of buildings can vary from a single story to large high-rises.
- Hospitals are usually characterized by lots of windows, a large primary entrance area, and a separate drive-up emergency room entrance.
- Most hospitals have an oxygen storage/generation area located on the exterior. Liquefied oxygen tanks and the oxygen generation and distribution system are usually colocated and should be placarded.
- Many hospitals have a large exterior electrical generator and industrial-size electrical transformers.
- The presence of a rooftop wind sock indicates that a helicopter pad (and helicopter) may be on the roof.

INTERIOR FEATURES

- Large open areas are not common with these types of buildings. Instead, there is an abundance of hallways and partitioned rooms.
- Hospitals typically have plumbed medical gases (oxygen, etc.) to many areas within the building.
- Exam rooms may include large, fixed imaging equipment (X-ray, MRI) that present unique hazards. Hallways are typically congested with mobile or staged exam equipment, rolling beds, and wheelchairs.
- Small office areas are common and are used for records, reports, management, and other administrative needs.
- There is always 24/7 security. However, remember that many patients will have little or no ability to self evacuate.
- Interior finishes are typically noncombustible and durable (lots of tile).
- Fire sprinklers are normally present. Interior corridors typically have self-closing doors that are activated by the fire alarm system to help with compartmentalization.

INST 40

TACTICAL CONCERNS

Fire Spread: The spread of fire will be hampered by the numerous partitioned rooms, but the byproducts can quickly spread via the hallways. In most cases, combustible materials are not excessively abundant; however, the rapid spread of fire can be enhanced by the presence of flammable gases. Remember that these buildings often have radioactive materials and specific chemicals that are used in specialized treatments. Hospitals are normally equipped with operative sprinklers that will retard the spread of a fire.

Collapse: Collapse is totally dependent on the type of construction, size of the building, and presence or lack of sprinklers. It will take a sizeable fire and time for collapse to occur in these buildings, and is not a common problem. Collapse of a roof depends on the size and type of roof. Older wood timber roofs will resist fire for longer periods of time than metal roofs or wood roofs of smaller conventional and/or lightweight lumber. Smaller buildings with more conventional roofs can fail depending on the size and duration of fire in a particular building.

Ventilation: These buildings can normally support horizontal, PPV, and, depending on the type of roof construction, vertical ventilation. Horizontal ventilation will be limited to doors and windows, which are normally plentiful. Vertical ventilation can be a viable operation, particularly in wood roofs, but a challenging operation in metal roofs. PPV is totally dependent on the size of the building and the number of doors and windows that can be used. These buildings typically have center hallways that can be used for PPV. It must be remembered that the larger the building, the more resources and time that will be required for any type of successful and timely ventilation operations.

Forcible Entry: Hospitals operating 24/7 with security personnel rarely present a forcible entry issue for firefighters. When needed, forcible entry is usually limited to either the standard size doors into the office areas or overhead loading doors at ground level. Windows are normally abundant. Some areas with specific interior processes will have limited windows and doorways. Likewise, there may be a storage area for controlled substances that is well secured but can be defeated with powered tools.

Search: Most hospital staffs are trained for a systematic and accountable evacuation plan for patients. Some plans include the movement of patients to an interior refuge area. If a search by the fire department is necessary, a method of cataloging the movement/location of patients as well as identifying rooms that have been, have not been, or need to be searched should be established.

SPECIFIC HAZARDS

- The size of the utilities can be an excellent indicator of any unique hazards within a building.
- Expect the presence of medical gases (such as oxygen), radioactive materials, and hazardous chemicals.
- Hospitals typically contain a large number of computers, communication, and electrical-powered medical equipment that may cause interference to fire department portable radios. Some medical imaging equipment may actually damage FD radios and thermal imagers.
- Search in these buildings can be challenging unless personnel are proficient in prioritizing multi-area searches.
- If patients are moved, document names and their new location.

RAPID STREET-READ GUIDE

BUILDING GROUP: Institutional Building—Detention (Jail) Facility

Note: This guide focuses on jails and not prisons.

BASIC CONSTRUCTION METHOD

- Poured-in-place concrete, concrete block, brick masonry, and concrete tilt-up panels
- In a few cases, wood framing with stucco and/or vinyl siding
- Concrete slab or perimeter foundations
- Flat roofs with minimal openings

ERA CONSIDERATIONS

- Industrial-era buildings are likely masonry with sizable wood for floor and roof supports.
- Legacy purpose-built detention facilities are likely poured-in-place concrete.
- Engineered lightweight-era facilities are typically CMU block with metal pan/concrete floors and steel trusses.

EXTERIOR FEATURES

- Most of these buildings feature a stout type of construction to alleviate security concerns. This means minimal doors, windows, and basically solid exterior walls.
- In many cases, these buildings are surrounded by high fencing topped with barbed wire and/or razor wire for additional security.
- Most exterior doorways have a redundant or secondary interior doorway to enhance security.
- Windows for the detention areas of the facility are for light/ventilation only and are typically too small for an adult to pass through.
- An overhead rolling door or multiple garage doors indicate a "sally port," which is a secured garage area for vehicles that transport detainees.
- In some cases, existing buildings in smaller jurisdictions have been converted into a detention type facility. In this case, the construction may be of a lesser grade than the typical detention facility, although security concerns will still be formidable.

INTERIOR FEATURES

- Most detention buildings are divided into three interior spaces: administrative offices, secure detainee processing area, and a secure cell or holding area.
- The holding area in larger facilities will have an open cafeteria and activity area surrounded by partitioned holding cells.
- Holding areas typically have no combustible materials or furnishing.
- Expect highly secure interior doors that are stout and redundant.
- Sprinklers may be present.
- Security is 24/7.

INST 41

TACTICAL CONCERNS

Fire Spread: The spread of fire should be minimal due to a minimal fire load. Although some furnishings are present, they are also minimal due to the type of occupancy. Sprinklers are usually present, but may not be plumbed to individual cells for obvious reasons. Fire spread can also be minimized by constant security in the building.

Collapse: Collapse is totally dependent on the type of construction, size of building, type of contents, and presence or lack of sprinklers. Collapse is not a common concern in these buildings.

Ventilation: Ventilation options in these buildings is typically limited and can present challenges. Vertical ventilation in concrete buildings is normally not a feasible operation unless the roof is constructed from lightweight materials. Horizontal ventilation will often be limited to minimal doors such as standard size doors. In most cases, PPV ventilation fans will be needed, depending on the size of a building and the number of openings to the exterior of a building.

Forcible Entry: Forcible entry will be a significant challenge but will be minimized by 24/7 on-duty personnel inside these facilities. Forcible entry into the building is usually limited to a standard size door into an office/entry area. A few windows may be present in some buildings but windows are not in abundant supply. Challenging and time-intensive entry operations should be anticipated as security is a major consideration in these buildings. If necessary, forcible entry inside the building will be noteworthy without personnel equipped with appropriate keys and security access.

Search: Unless there are known trapped occupant/detainees, search should not be a primary concern as inmates will likely have been moved to a safe place of refuge and are normally ambulatory under supervised conditions. Obviously a rapid change in conditions can dramatically change this consideration. Additionally, if a search is necessary, the inherent risk should be evaluated.

SPECIFIC HAZARDS

- Fire departments should preplan suppression operations with detention facility supervisors.
- The greatest hazard at detention facilities comes from the potential behavior of detainees. A fire or other significant incident can create an opportunity for escape, hostage taking, or act of violence.
- Sentry dogs, crowd control gases, and weapons may be utilized by law officials during incidents at detention facilities.
- As entry and exit openings are likely to be in short supply and some buildings can be large, consider entry/exit and ventilation options based on openings that are available.
- Consider appropriate security/escorts before entering these facilities.

INST 41

RAPID STREET-READ GUIDE

BUILDING GROUP: Institutional Building—Attended Care Facility

BASIC CONSTRUCTION METHOD

- Wood frame, wood siding, vinyl siding, and stucco
- Brick and/or concrete block also common
- Concrete slab or perimeter foundations

ERA CONSIDERATIONS

- Industrial- and legacy-era buildings use a heavy grade of wood for structural members, some use poured-in-place concrete, concrete block, or brick masonry.
- Engineered lightweight-era buildings use smaller dimensional lumber and trusses for structural members in wood frame buildings, and concrete block is often used for larger buildings.

EXTERIOR FEATURES

- The sizes of these buildings can vary widely from a converted single-story, single-family dwelling (pictured) to a large multistory complex.
- Most have a central entryway and may include a covered drive-through in front of the primary building entrance. Windows are plentiful, and individual balconies can be encountered.
- Some include a separate ambulance transfer door remote from the primary entrance.
- One or more sides of the exterior may be inaccessible for apparatus due to gardens or other outside recreation areas. Likewise, an outdoor courtyard may be present that is not visible from the exterior perimeter of the building.
- Depending on the size and age of a particular building, different types of roof construction can be used but will normally be a flat, hip, or gable configuration.

INST 42

INTERIOR FEATURES

- Expect an open reception area inside the primary entrance door. Individual tenant rooms are typically served by central hallways. Common areas for socializing and recreation may be centralized or dispersed throughout the building.
- Expect a centralized dining and kitchen facility. It is rare to have cooking appliances in individual tenant rooms.
- Large buildings have multiple center hallways that access numerous rooms with one to four persons per room.
- Partitioned walls are found in abundance.
- Staff and security is on premises at all times.
- There are numerous windows and multiple doorways.
- Sprinklers are normally present, but older/smaller structures may not have sprinklers.
- Converted single-family dwellings can have multiple people in each room.

TACTICAL CONCERNS

Fire Spread: The spread of fire in these buildings is not a common problem as the interior is partitioned with numerous rooms, sprinklers are common, and constant supervision is common if not mandatory. However, if a fire were to start and extend into the hallways, it would rapidly spread to other areas served by the hallways. Kitchen areas are a prime location for fire.

Collapse: It will take a sizeable fire and time for collapse to occur in most of these buildings; however, collapse of a roof depends on the size and type of roof. Conventional wood roofs will resist fire for longer periods of time than metal roofs or wood roofs of smaller conventional and/or lightweight lumber. Smaller buildings with more conventional and/or lightweight wood roofs can fail depending on the size and duration of fire in a particular building.

Ventilation: Although these buildings have numerous doors and windows, ventilation operations will be assisted or hampered by the size of the building. Converted single-family dwellings are easily ventilated by either vertical, horizontal, and/or PPV, similar to operations used in average single-family dwellings. However, as the size increases up to and including large complexes, ventilation becomes more intensive. Contaminated center hallways in larger buildings must be ventilated early to allow access and egress to and from the individual rooms. This can be accomplished by a combination of horizontal ventilation and/or PPV.

Forcible Entry: Forcible entry is not a primary concern as there are multiple conventional doors and windows in these buildings. Additionally, the 24/7 staff or security personnel should be available to assist with entry/exit considerations.

Search: The need for search and rescue operations will be dependent on the size of a fire, the amount of contamination within a building, and the amount or lack of expertise of on-duty staffing in the building. Although the frequency of fires in these buildings is low, it will take minimal smoke/disruption to create a measurable amount of panic in the occupants. Rescue operations can be challenging and should be anticipated. Remember that the rooms within these buildings are filled with people in various stages of mobility and end of life considerations.

SPECIFIC HAZARDS

- Patients inside these buildings are the primary hazard as they can be nonambulatory and also numerous, depending on the size of a building.

- Single-family dwellings that have been converted into these facilities may not be easily identified from the exterior.

- Responsible staffing in converted dwellings are often ill-equipped to handle emergencies at night and have been directed to call 911 for any assistance.

- In some cases, expect minimal assistance from the staff in larger buildings for rescue considerations.

- Due to the type of patients, it takes little disruption (food burning on the stove, the presence of firefighters, etc.) to create undue worry. Anticipate the need for EMS resources for all types of fire department incidents at these facilities.

- The use of portable medical oxygen devices is likely to be abundant. Improper storage of spare oxygen cylinders is a strong concern.

RAPID STREET-READ GUIDE

BUILDING GROUP: Public Assembly—Restaurant

Note: This guide focuses on stand-alone purpose-built restaurants. See the Main Street Commercial section for restaurants that are part of another building.

BASIC CONSTRUCTION METHOD

- Wood/metal frame, wood or vinyl siding, stucco, or brick veneer
- Unreinforced masonry (URM) or conventional brick and/or concrete block
- Concrete slab or perimeter foundations

ERA CONSIDERATIONS

- Pre-1960 buildings use a heavy grade of wood for structural members, some use poured-in-place concrete, concrete block, and conventional brick masonry. Some older buildings will have unreinforced masonry construction.
- Smaller modern-era restaurants are likely to be wood/metal frame with lightweight trusses. Larger ones will be noncombustible with CMU or metal post and beam walls and metal trusses. Brick masonry found on modern era-buildings is likely a veneer on wood or metal framing.

EXTERIOR FEATURES

- The varying sizes of these occupancies can range from small "mom-and-pop" restaurants to those capable of seating hundreds of occupants.
- Many older buildings of unreinforced masonry construction have been converted into a restaurant-type occupancy.
- Different types of roof construction can be used and will vary between flat, pitched, and arched. Newer buildings typically use flat roofs. Conventional wood roofs and metal deck roofs are common on older buildings and lightweight wood and metal roofs are common on newer buildings.
- The rear of the building typically serves as a supply receiving area and can include auxiliary refrigeration equipment, grease/oil recycling storage, and waste dumpsters. Some of these areas are fenced or walled off for security.
- Purpose-built restaurants typically have grease hood exhausting to the roof. Older buildings that have been converted or multistory buildings may have exhaust vents on a side or rear wall.
- Facades of lightweight construction or rigid foam are common on many restaurant buildings.

INTERIOR FEATURES

- The interior normally consists of a waiting/dining area, kitchen, dishwashing station, and storeroom. Some include a separate lounge or bar area.
- The dining area will be filled with chairs, tables, and/or booths, often creating congestion for movement of personnel/hoselines, particularly if visibility is obscured.
- Elaborate decorations/architecture may conceal noteworthy voids.
- Based on the design of the restaurant and/or building, the number of windows and doorways varies widely.
- Food storage areas include a walk-in refrigerator—even in small buildings.
- Cooking areas are typically finished with noncombustible and durable materials.
- Fire sprinklers are typically required for purpose-built buildings that qualify as public assemblies. Older/smaller structures or those that have been converted from another use may not have sprinklers.

PUB 43

TACTICAL CONCERNS

Fire Spread: Although fires in restaurants are not a frequent occurrence, they usually start in the kitchen area, which is normally protected by sprinklers/suppression systems. The potential of vertical spread of fire in ducting that terminates on the exterior of a building must be checked from the interior to the exterior. Fire can also start in the electrical and gas appliances and can spread easily. Fires in the dining area are not common but can easily spread due to the openness of the room and lack of partition walls. Soffits and other decorative modifications must be checked for extension of fire in concealed voids. Remember to check for the spread of fire in voids above suspended ceilings.

Collapse: Roof collapse is the most common threat for restaurant fires due to long spans and the dead load of exhaust venting and HVAC equipment. Often, fires originate around grease hoods and smolder undetected while they degrade roof support elements. The collapse of a roof depends on the size and type of roof. Conventional wood roofs will absorb fire for longer periods of time than metal or wood roofs of smaller conventional and/or lightweight lumber. Poured-in-place concrete and concrete block best resist fire and collapse.

Ventilation: Although these occupancies may have numerous doors and windows, ventilation operations will be assisted or hampered by the size of the building. In most cases, conventional ventilation operations consisting of either vertical, horizontal, and/or PPV can be effectively utilized. However, as the size increases up to and including large complexes, ventilation becomes more intensive. Roof ventilation will be dependent on the type of roof encountered, with older wood roofs offering the most options, metal deck roofs offering less options, and lightweight wood roofs offering the fewest options.

Forcible Entry: Forcible entry can normally be accomplished by conventional methods, especially if there are multiple doors (metal frame with glass) and windows.

Search: The need for search operations will be dependent on the size of a fire, the amount of contamination within a building, and the size of the restaurant. Although the frequency of fires in these buildings is low and occupants are normally ambulatory, search in a dining area can be challenging due to obstacles such as tables, chairs, and/or booths. A potential search in these types of buildings should be practiced beforehand to develop the necessary expertise, as restaurants of all sizes are common.

SPECIFIC HAZARDS

- If a fire has involved the vent ducting in the kitchen area, be sure to check the (1) vent, (2) attic, and (3) roof for extension of fire.
- Three predominant hazards are high-voltage electrical, gas for cooking appliances, and the vents over the cooking appliances.
- Gas-fired cooking equipment is typically left on "warm" overnight to help minimize start-stop heating cycles.
- Deep fat fryers are open appliances that present a hot-liquid splash, spill, or immersion hazard.
- Lightweight facades are commonly found on these buildings.
- Look for the presence of a basement. Basements used for storage are not public spaces and are accessed by a stairway near the rear of the kitchen or back door.
- The presence of fire/smoke, or the activation of grease hood fire suppression system in eating establishments requires an inspection by the health department prior to reopening of the facility.

RAPID STREET-READ GUIDE

BUILDING GROUP: Public Assembly—Stadium/Arena

BASIC CONSTRUCTION METHOD

- Fire resistive (Type I) that combines reinforced concrete and coated steel
- Roofs made with steel trusses, domed concrete, or air-inflated membrane

ERA CONSIDERATIONS

- Legacy arenas are mostly reinforced concrete with a steel truss or domed, lightweight concrete roof. Legacy stadiums often lack a full roof and are mostly coated steel columns and beams on a reinforced concrete substructure.
- Newer stadiums and arenas have a reinforced concrete core (ground floor, concourses, and elevator towers) with coated steel seating supports, luxury boxes, and roof support columns.
- Where present, roofs are steel truss.

EXTERIOR FEATURES

- Large, expansive footprint with 360° apparatus access is usual.
- Windows (where present) only serve as daylighting for grand entryways or concourses.
- Dozens of entry and exit doors exist. Most are used by patrons and usually include some kind of turnstile or ticket counting system. Ramps may serve elevated doors.
- One portion of the exterior will include a service area with loading docks, utility access, and waste/recycle processing.
- Most HVAC systems are at ground level or are incorporated into the service area. Rooftop ventilators (fans only) may be present.

INTERIOR FEATURES

- Central seating areas are served by multiple access tunnels that connect to a perimeter concourse.
- Luxury suites, media boxes, and vendor spaces are accessed from the concourse.
- A system of service/supply hallways, tunnels, and elevators are typically hidden under the seating stands. Locker rooms, prop storage, vehicles, staff offices, and infrastructure support systems can also be found in these areas.

PUB 44

TACTICAL CONCERNS

Fire Spread: Covered arenas and stadiums rely on an automatic fire sprinkler system to suppress fires that originate in all areas except the seating stands. Open-air stadiums utilize sprinkler systems only for enclosed areas. The systems are designed to have multiple zones and they include multiple risers. The structures are built to compartmentalize fires to a given area using a combination of fire walls, fuse-link doors, and ventilation system dampers as well as the fire sprinkler system. Fire spread issues primarily exist with the utility conveyance system, waste/recycle collection areas, enclosed suites, and plastic seating.

Collapse: The risk of a general or significant collapse is almost nonexistent. Collapse threats exist for individual and isolated components within the structure such as scoreboards, luxury suite ceilings, light towers, and overhead equipment in service areas or tunnels.

Ventilation: The cavernous nature of arenas and enclosed stadiums renders most fire department ventilation techniques inadequate. For those occasions where an interior operation needs ventilation, some consideration can be given to engaging the building engineer to use facility ventilation systems to help move smoke. PPV fans can be used for fires in small spaces, such as under the general seating area or in vendor spaces on the concourse.

Search: The need to prioritize search is rare as customers/staff are likely to self evacuate during fires. Multiple emergency exit doors on each side of the stadium/arena should assist in occupant egress routes, although a panic-stricken stampede can produce a mass-casualty event. A difficult search challenge exists for the extensive tunnel and service areas beneath the seating stands. The use of preplans and large area search techniques is mandated to coordinate search efforts for potential victims in these areas.

Forcible Entry: Forcible entry can normally be accomplished by conventional methods for the main entrance doors. Emergency exit doors should include panic hardware that is easy to force. Stock receiving areas with overhead rolling doors require power saws and well-trained firefighters to defeat.

SPECIFIC HAZARDS

- Exiting patrons and parking lot congestion can hamper arriving apparatus.
- The circular nature of the concourse and the complexity of passageways can lead to directional disorientation, even with perfect visibility.
- Electrical distribution equipment is very complex and is typically high voltage and amperage.
- Fire suppressions and rescues in overhead equipment and scaffolding will require technical rope skills (scoreboards, light towers, lighting catwalk, etc.).
- Cooling equipment for ice arenas presents a hazmat potential.

RAPID STREET-READ GUIDE

BUILDING GROUP: Public Assembly—Auditorium/Theater

BASIC CONSTRUCTION METHOD

- Conventional brick, concrete block, and poured-in-place concrete for larger buildings
- May be of unreinforced masonry
- Wood/metal frame, wood siding, vinyl siding, or stucco for smaller buildings
- Concrete slab or perimeter foundations, some with storage basements

ERA CONSIDERATIONS

- Pre-1960 buildings use a heavy grade of wood for structural members, poured-in-place concrete, or concrete block or brick masonry.
- Modern engineered lightweight auditoriums and theatres are commonly CMU block with steel truss roofs. Steel or wood framing, or steel post and beam kit buildings may exist for smaller examples in rural areas.

EXTERIOR FEATURES

- The sizes of theaters can range from a single-story, single theater type building to a large multi-theater complex.
- Poured-in-place concrete, concrete block, and brick masonry are popular for larger buildings. Smaller buildings typically use wood or metal framing with stucco, metal or vinyl siding, or brick veneer.
- Depending on the size and age of a particular building, different types of roof construction can be used but will normally be a flat configuration. Conventional wood and metal deck roofs are common, with lightweight metal and wood roofs replacing the older heavier roofs in some cases.

INTERIOR FEATURES

- Common layout includes a reception/ticketing area and food concourse just inside the primary entrance doors. The concourse is used to access large open audience seating areas.
- Partitioned walls are used in smaller rooms outside the seating area(s).
- Theatres with a performance stage are likely to have a prop/costume storage area, individual dressing rooms, and a workshop behind or flanking the stage. Likewise, expect overhead lighting rigs, catwalks with straight ladder access, and curtain rigging in the stage area.
- Film storage and projection rooms are typically at the rear of the seating area and are typically accessed by a locked door and short stairwell.
- There are no windows; however, there are normally large exit doors.
- Sprinklers are typically required for larger structures.
- The projection area will have substantial electrical feeds and high-voltage projectors and substantial AC equipment considerations.

PUB 45

TACTICAL CONCERNS

Fire Spread: The spread of fire in these types of buildings is not a common problem as the interior is largely nonflammable with the exception of the audience seats and interior decorative elements. Historical theaters are a notable exception—they may not meet modern codes and are at risk for substantial fire spread and life loss. Likewise, kit-building theaters in rural areas may not have adequate code-compliant exits and fire code provisions. Sprinklers are common and can retard the spread of fire. Additionally, fire in masonry/concrete buildings does not accelerate as rapidly as fires in combustible buildings (wood frame, etc.). The primary locations for fire are the projectors and sound equipment above the viewing area and the HVAC equipment on the roof, which can easily spread fire/smoke via the ducting throughout a building.

Collapse: It will take a sizeable fire and time for collapse to occur in most of these buildings, and is not a common occurrence. However, collapse of a roof depends on the size and type of roof. Conventional wood roofs will resist fire for longer periods of time than metal roofs or wood roofs of smaller conventional and/or lightweight lumber. Smaller buildings with more conventional and/or lightweight construction and wood roofs can fail depending on the size and duration of fire in a particular building.

Ventilation: Ventilation operations will be assisted or hampered by the size of the building, and as the size increases up to and including large theater complexes, ventilation becomes more challenging. These buildings have minimal doors and a lack of windows, which will minimize horizontal ventilation operations. In most cases, horizontal ventilation and/or PPV will be the most beneficial operation. Vertical ventilation will likely only be a viable option on small to moderate size theaters. Remember that most of these types of buildings have substantial HVAC systems that can be used for ventilation operations.

Forcible Entry: As there is a lack of windows, forcible entry is limited to public assembly type exit doors, which are more difficult to force than single doors or doors with visible exterior locking devices. The advantage of exit doors is they normally provide a wide opening. The staff on duty during business hours should be available to assist with entry/exit considerations.

Search: The need for search and rescue operations will be dependent on the size of a fire, the amount of contamination within a building, and the amount or lack of expertise of on-duty staffing in the building. Although the frequency of fires in these buildings is low to nonexistent, it will take minimal smoke/disruption to potentially create a measurable amount of panic in the occupants. As an example, smoke from a faulty HVAC system can dramatically change this consideration. The success of a search operation will be based on restoring order to the incident and then verifying the status of all occupants. Rescue operations in the audience area(s) will be challenging due to the large open areas and a lack of visibility.

SPECIFIC HAZARDS

- Occupants are the primary hazard, although they are most often ambulatory and self evacuate. Expect occupant and vehicle congestion near the primary entrance if an incident requires evacuation. Rapidly escalating incidents may create panic and lead to a mass-casualty event.

- The primary origin for fire/smoke is in the rooms used for projection and sound equipment, the snack bar areas, and the HVAC equipment on the roof.

- A problem in the HVAC equipment would quickly spread to other areas of a building via the ducting, and in a relatively short period of time.

- Older theaters are often remodeled into bowling alleys or churches. The inverse is also common. In both cases, expect the hazards of remodels: poke-throughs, void spaces, dropped ceilings, and inadequate exits.

RAPID STREET-READ GUIDE

BUILDING GROUP: Public Assembly—Meeting Hall

BASIC CONSTRUCTION METHOD

- Wood /metal frame, wood/metal siding, vinyl siding, and stucco
- Conventional and unreinforced brick and/or concrete block, some poured-in-place concrete
- All-steel kit buildings in rural areas
- Concrete slab or perimeter foundations

ERA CONSIDERATIONS

- Pre-1960 buildings use a heavy grade of wood for structural members; some use poured-in-place concrete and concrete block or conventional brick masonry. Roofs are normally of heavy construction with flat, gable, or arched configurations.

- Post-1960 buildings use smaller dimensional lumber and lightweight trusses for structural members in wood frame buildings. Concrete block is often used, depending on the size of the building. Brick masonry tends to be a veneer on wood or metal framing.

EXTERIOR FEATURES

- There are thousands of these types of buildings of various styles and sizes across the country and can range from small buildings used for local club meetings to large multistory buildings. Some examples of these buildings are pool halls, nightclubs, granges, dance halls, armories, Masonic clubs, bingo parlors, and so on.
- Most include a primary entrance area, exit doors on each side and rear, and perhaps a shipping receiving/waste recycle area at the rear.
- Various types of roof construction can be used but will normally be a flat, gable, hip, or arch configuration. Conventional wood roofs and metal deck roofs are common on larger older buildings but have been replaced in many cases by lightweight flat or gable metal/wood trusses. HVAC units are typically roof-mounted.

INTERIOR FEATURES

- Expect an entryway foyer near the primary entrance that may include restrooms, coat storage, and/or partitioned rooms for vending.
- Kitchen facilities are common and can range from a typical residential style to a full commercial one capable of serving hundreds of meals.
- The defining characteristic of the building is a large open meeting space or multiple open meeting rooms. Some large areas may have movable separation walls used to break the large room into smaller ones. Some center or perimeter hallways are used to access partitioned smaller rooms.
- Some buildings have basements.
- Windows and multiple doorways are often present.
- Larger buildings are typically required to have fire sprinklers.
- The building fire load is typically limited to stackable tables and chairs.

PUB 46

TACTICAL CONCERNS

Fire Spread: The spread of fire in these buildings can be enhanced from two perspectives. One, the open areas used for meeting functions lack partition walls. Second, some of these occupancies still allow smoking in selected areas (such as lounges). If a fire were to start and extend into any of the hallways, it would rapidly spread to other areas served by the hallways. Additionally, fire in masonry or concrete buildings does not accelerate as rapidly as fires in combustible buildings (wood frame, etc.). A kitchen area, where present, is a prime location for the origin and spread of fire, particularly within a cooking ventilation ducting/system. Expect additional fire loads during holiday periods or special occasions (weddings, reunions, bar mitzvahs) that may not be code-compliant.

Collapse: The primary collapse threat is the large open span roof covering the meeting area(s). Depending on the size of fire, concrete and concrete block walls do not readily collapse. Conventional wood framing is not as fire resistive as concrete, and lightweight wood/metal framing is the least fire resistive. Older conventional wood roofs will resist fire longer than metal roofs; newer lightweight wood/metal roofs are the least fire resistive and can suddenly collapse when subjected to high heat levels and/or fire. Smaller buildings with lightweight wood/metal roofs are particularly prone to rapid collapse depending on the size and duration of fire.

Ventilation: Although these buildings often have numerous doors and windows, ventilation operations must be determined by the size of the building. The interior can be ventilated by either vertical, horizontal, and/or PPV, but will be dependent on the size/configuration of the assembly areas. As the size increases up to and including large buildings, ventilation becomes more challenging. Roof ventilation will be dependent on the size of building and type of roof, with conventional wood roofs offering more options than lightweight metal and wood roofs. Contaminated center hallways (if present) must be ventilated early to allow access to and from the individual rooms. This can be accomplished by a combination of horizontal ventilation and/or PPV.

Forcible Entry: Forcible entry is dependent on the availability of windows and doors. Because these buildings fall under the classification of a public assemblage, they must be equipped with enough doors to adequately handle a rapid evacuation of the interior occupant load. These same doors can also provide good forcible entry/exit avenues. In most cases, conventional forcible entry is sufficient for entry considerations. Due to the popularity of concrete block construction, doors are usually metal clad in metal frames and can be more difficult to open than conventional wood doors.

Search: Although the frequency of fires in these buildings is low, it will take minimal smoke/disruption to create a measurable amount of panic in a large group of occupants. Obviously some type of dilemma can dramatically and rapidly change this consideration. Rescue operations, if necessary, can be challenging and should be anticipated. It may be difficult to accurately determine the status of all occupants from those personnel responsible for a function and/or the building.

SPECIFIC HAZARDS

- Occupants are the primary hazard as they can be numerous depending on the size of a building and type of function. Fortunately, they are normally ambulatory. It is doubtful that personnel responsible for the building will know the status of all occupants.

- Facilities that serve semi-private or nonprofit social groups may have dubious additions or modifications that are not code-compliant. Likewise, storage of materials may be found in exit corridors.

- Decorations for holiday and other social events may introduce a significant fire load or create search obstacles.

- An entanglement hazard may present itself in open areas with various arrangements of movable tables and chairs.

- If a fire incident rapidly escalates in a meeting hall/nightclub, occupants are likely to rush the primary entrance door and clog the path (Cocoanut Grove, Road House, etc.). Fire departments should preplan for alternative entry points and mass-casualty incidents for non-fire-sprinklered meeting/nightclub facilities.

RAPID STREET-READ GUIDE

BUILDING GROUP: Public Assembly—Church

Note: This guide covers purpose-built churches. Buildings converted to a church are likely to fall into one of the Main Street Commercial guides.

BASIC CONSTRUCTION METHOD

- Either wood frame with wood, metal, or vinyl siding; metal structural members with metal or vinyl siding; or brick and/or concrete block
- Concrete slab or perimeter foundations

ERA CONSIDERATIONS

- Large historical-era examples are likely to be of stone and heavy timber construction.
- Industrial- and legacy-era buildings use a heavy grade of wood for structural members, poured-in-place concrete, and concrete block or brick masonry. Heavy timber trusses can be found with poured-in-place concrete and stone and masonry construction.
- Post-1960 buildings use smaller dimensional lumber and trusses for structural members in wood frame buildings, and concrete block is also used depending on the size of a building. Brick masonry tends to be a veneer on wood framing. Popular variants use all-metal structural members and some lightweight construction.

EXTERIOR FEATURES

- Places of worship are easily identified from the exterior either by design and/or designation.
- They are of varying ages and sizes, dating from the early 1800s and from small wood frame buildings to large complexes of various designs and materials used for their construction.
- Different types and styles of roof construction can be used but are often a gable configuration for an open and spacious interior. Conventional and lightweight wood roofs are common. Metal deck roofs are often used in flat roof designs. Many older church buildings use heavy wood timber trusses capped with a slate or tile roof.
- The presence of exterior masonry buttresses indicates an open-span heavy timber roof on unsupported masonry walls.
- Multiple exterior basement stairwells indicates that a meeting hall is below the worship area.
- Many churches have adjoining or attached meeting halls or classrooms.

INTERIOR FEATURES

- Most have an entry foyer that leads to a large worship area via multiple interior doors.
- Expect a large, open worship area with numerous fixed pews and/or chairs.
- The worship area is often surrounded or flanked with numerous smaller rooms for classes and other uses. Likewise, a choir or organ loft may be found in a mezzanine area located opposite or to the side of the worship stage.
- Partitioned walls are in abundance outside the worship area, but are not common in the worship area.
- Multiple windows and doorways are common.
- Multiple interior basement stairwells indicate a meeting hall below.
- Fire loads are relatively low. Fire sprinklers are likely to be found on large, modern churches or those that have been significantly updated.

PUB 47

TACTICAL CONCERNS

Fire Spread: The spread of fire is enhanced by the large open area of the worship area, but the spread of fire can be limited by partition walls that are common with numerous smaller rooms. Fires that originate in older large churches are often successful in destroying the entire building as a result of substandard wiring and numerous voids that allow a fire to grow and gain significant headway without detection. Fortunately, fires in places of worship are not a common occurrence. Where present, bell towers and steeples can enhance rapid fire spread due to limited access and their vertical nature.

Collapse: It will take a sizeable fire and time for collapse to occur in the older churches with heavy timber trusses. However, collapse of a roof depends on the size and type of roof. Conventional wood roofs will resist fire for longer periods of time than metal roofs or wood roofs of smaller conventional and/or lightweight lumber. Smaller buildings with more conventional and/or lightweight wood roofs can fail depending on the size and duration of fire in a particular building. Vertical features such as bell towers and steeples are a significant collapse hazard when attacked by fire or lightning strike. Unsupported masonry walls left freestanding after roof collapse or loss are extremely dangerous and prone to collapse.

Ventilation: Ventilation operations will be defined by the size of a particular building. Although these buildings often have numerous doors and windows that can be used for horizontal/PPV operations, expensive stained glass windows should be saved if possible. Horizontal/PPV operations can be successfully used in the individual rooms; however, these operations in an auditorium can be doubtful depending on size. If necessary, roof ventilation can be accomplished on flat and some pitched roofs of wood or metal construction with conventional coverings. Roof ventilation on older churches with steep pitched roofs, heavy sheathing, and tile/slate roofs will be doubtful at best.

Forcible Entry: Forcible entry is not difficult as there are multiple conventional doors and windows in these buildings that can be forced with conventional methods. The presence of on-duty staff can be anticipated on Wednesdays and Sundays in many churches. During the week, some churches have office staff available during the day who should be able to assist with entry/exit considerations.

Search: The need for search operations is minimal due to ambulatory occupants, but some type of collapse or explosion (or other rapid type of dilemma) can dramatically change this consideration. Although rare, some churches have 24/7 living accommodations for priests or other staff. Rescue operations in churches can be challenging due to large area worship/meeting spaces and smaller rooms.

SPECIFIC HAZARDS

- Older large masonry/timber churches are renowned for large fires that are difficult to extinguish and often result in a complete loss of a building.
- The presence of commercial-sized utility feeds may indicate a worship area with a large multimedia production stage or a meeting hall with industrial cooking equipment.
- Risk/benefit decisions might be influenced by pressure to save irreplaceable stained glass, historically significant features, or other high-value contents.

PUB 47

RAPID STREET-READ GUIDE

BUILDING GROUP: Miscellaneous Building/Structure—Pole Barn

BASIC CONSTRUCTION METHOD

- Wood post and beam with wood clapboarding or corrugated metal covering
- Steel post and beam with corrugated metal covering

ERA CONSIDERATIONS

- Historical- and industrial-era examples are usually heavy timber with mortise and tenon or dowel connections. Most have stone footers and foundations.
- Legacy-era examples are cut lumber with steel bolt/plate connections. Most have a perimeter concrete foundation.
- Engineered lightweight-era examples are typically pressure-treated round post columns with cut lumber beams connected with galvanized hangers and screws or nails. The roof is likely to be lightweight wood truss.

EXTERIOR FEATURES

- Simple square or rectangle shapes are most common.
- Roofs are gabled or lantern (or gambrel) styles with wood plank or plywood sheathing and shingles, corrugated tin, or aluminum weather coverings.
- Expect one or two large ground-level access doors plus a gable end access door (for lifting hay to a loft area).
- Wood clapboard siding is most common, although corrugated steel/aluminum can be found on newer or repaired barns.
- Many older barns have combustible roofs of shakes or shingles.

INTERIOR FEATURES

- Pole barns usually have a wide open space with little or no interior finish.
- An exposed ridge beam is used as an anchor for pulleys or other lifting equipment.
- Minimal utilities (usually just an electrical light and plug circuit) are typical.
- Most have a loft or multilevel storage bin.
- The floor can be earth, gravel, or concrete slab.

MISC 48

TACTICAL CONCERNS

Fire Spread: Any interior fire should also be considered a structural fire as all columns and beams are exposed. All-wood barns will become well involved in a short period of time. Barns used for stacked hay bale storage present a smoldering overhaul challenge. Older barns will have years of accumulated loose hay, sawdust, and other debris that can accelerate fire spread. Barns should be considered a wild card for predicting fire load—some are packed full of accumulated items. The potential for explosive and hazmat materials should always be expected.

Collapse: Older barns that have been well-maintained are quite stout and will burn for hours prior to collapse. Barns that are abandoned, age deteriorated, and those with previous collapses will fail quicker. Sagging is a common and usable warning for all barns except newer engineered lightweight ones. Collapses are usually general (complete failure). Storage in a loft will enhance early collapse of the affected area.

Ventilation: Most wood barns self-ventilate quickly because they are rarely insulted and the exterior siding and roof covering are immediately exposed to flame. If ventilation is desired for fire control, fire officers should weigh the risk against the benefit.

Forcible Entry: Forcible entry can normally be accomplished by conventional methods. Heavily secured doors may indicate the storage of high-value items.

Search: The wide open interior of barns makes for quick searches. Rarely is search a high tactical priority for barns and other outbuildings used for storage. Some barns are used for animal shelter, which may need to be considered.

SPECIFIC HAZARDS

- Alternative heating appliances (usually unvented) may be found.
- Fuel for farm equipment may be stored in the barn or nearby.
- Do-it-yourself additions and alterations that are sub-code or dubious are common
- Vermin infestation can be expected.
- Expect significant storage of combustible materials and fire loads that can accelerate fire spread or contribute to rapid collapse of interior above-grade bins or floors.
- Wood barns with wood combustible roofs are a recipe for a hot, fast-burning fire with noteworthy exposure problems.
- Many older barns are in a state of disrepair and lack structural integrity.

RAPID STREET-READ GUIDE

BUILDING GROUP: Miscellaneous Building/Structure—Kit Building

BASIC CONSTRUCTION METHOD

- Post and beam steel frame with light steel wall framing and purlins for the wall and roof coverings
- Concrete perimeter footers and concrete pads for inside columns

ERA CONSIDERATIONS

- Most are engineered lightweight buildings.
- Some examples have trussed roofs that replace the steel I-beams and purlins.
- The buildings are primarily used for storage and/or manufacturing processes.

EXTERIOR FEATURES

- Simple square or rectangle shapes are common.
- Roofs range from a very slight pitch to a gabled shape.
- Clear plastic or glass panels may be incorporated into the roof to add interior light.
- Expect minimal use of windows.
- Siding is normally sheet steel or aluminum. However, fiberglass can be used as a substitute for these materials.

INTERIOR FEATURES

- The interior will usually be a wide open space, with the outer walls finished with an insulation batting covered with a synthetic membrane.
- Tenants may build an interior building to form an office area, break room, or for area separation. These are usually wood frame with drywall.
- The floor can be earth, gravel, or concrete slab.
- Sprinklers may or may not be present.

MISC 49

TACTICAL CONCERNS

Fire Spread: Other than interior contents, the only combustible part of the building is the synthetic membrane that covers the wall insulation and plastic light diffusion panels in the roof. Fire intensity and spread concerns are mostly influenced by the occupancy fire load. Also consider the size of utilities.

Collapse: Exposed steel structural elements are subject to early failure, and a high contents fire load can speed steel failure. History shows that these building don't really fall down—they simply start to sag and eventually kneel down. This is due to the bolted-together construction and contiguous nature of the wall column and roof beam (see the pictures above). Sagging is a common and usable warning sign. Truss roof examples act more like other truss buildings and actually fall when they fail. A collapse issue may exist when extensive interior content such as overhead radiant heaters, lifts and hoists, drop-in ceilings, and other items have been hung from the steel roof beams.

Ventilation: Kit buildings are fairly simple to ventilate—the use of PPV fans can usually clear accumulated smoke. Rooftop ventilation is discouraged for kit buildings as the roof panels aren't really designed to support much weight. Roof and wall panels, however, are easy to cut through using a carbide-tipped or abrasive friction blade on a rotary saw.

Forcible Entry: Forcible entry can normally be accomplished by conventional methods through entry doors. Overhead rolling doors can be cut and disassembled, although some find that cutting a flap in a wall panel is quicker.

Search: The interior content and arrangement will dictate search challenges. Clues as to the search challenge may be found from the exterior signage on the building. Some kit buildings have an interior building that is used for an office, staff break room, or other use. These are typically wood frame and drywall and may add a search challenge.

SPECIFIC HAZARDS

- Most utilities can be secured from the exterior and can be an indicator of the interior content.
- Electrical shutoffs and feeds that appear disproportional to the building indicates high-voltage equipment within.
- Overhead hanging radiant-style space heaters (fired with propane or natural gas) can collapse.
- Poke-throughs for exhaust vents on the exterior wall or roof may indicate the use of industrialized manufacturing equipment.
- Do-it-yourself interior partitions and storage units may be sub-code, unlawful, or dubious.
- Buildings with no business name or signage warrant caution.
- The building may lack sprinklers.
- Mezzanines may be present in larger buildings.

RAPID STREET-READ GUIDE

BUILDING GROUP: Miscellaneous Building/Structure—Silo

BASIC CONSTRUCTION METHOD

- Concrete or steel materials common in modern-era models
- Wood staves used in historic-era silos
- Metal legs and/or slab foundations

ERA CONSIDERATIONS

- In this country, silos date back to 1843 and the use of wood staves was the most common construction material.
- In the early 1900s, concrete and steel replaced wood staves.

EXTERIOR FEATURES

- Silos come in a wide variety of sizes dictated by the needs of the owners.
- The round cylindrical shape is primarily designed for a combination of strength and the ability to provide bulk storage (silage).
- Grain elevators are used to transport silage to the top of silos, but are a part of the silo complex.
- Silos can either be stand-alone cylinders or part of a complex that combines storage and other related facilities (top picture).
- When silos are part of a complex, the construction of other buildings can typically vary from conventional wood/metal structural members to metal, and use a variety of materials for exterior sidings. Some larger complexes use concrete block for exterior walls, with conventional or lightweight roofs.
- Silos can be loaded from the top or bottom. This requires mechanized equipment to accomplish these operations. On many farms, spare tractors are used to supply the necessary power.

INTERIOR FEATURES

- Silos are commonly used for bulk storage of grain. They are also used for storage of carbon black, coal, cement, wood chips, sawdust, food products, and other similar items.
- Some silos are designed to keep their contents in a low-oxygen atmosphere to prevent mold and decay or reduce the risk of a dust explosion.
- Depending on the silage, dust is often a by-product of operating a silo and associated equipment.
- Some silos are equipped with water misting capabilities.

MISC 50

TACTICAL CONCERNS

Fire Spread: The spread of fire in silos is not an uncommon occurrence as it can be initiated from numerous sources such as spontaneous ignition, lightning, mechanical heat from friction/sparks (from loading/unloading equipment), and electrical shorts, to name a few causes. Typically, fires occur in the top 10 feet of silage, so access to a fire can be simplified by its location in a silo.

Collapse: Collapse of a silo can occur as a result of a dust explosion or a backdraft. Wood staves would be the most susceptible to an explosion and/or fire, with metal less susceptible and concrete potentially offering the most resistance to failure. However, if a silo is subjected to a fire and/or explosion, consider the structure to have been weakened and requiring due diligence for safety.

Ventilation: Interestingly, PPV has been effectively used to vertically ventilate a silo with the primary focus being to minimize the accumulations of smoke, heat, and steam during extinguishment operations. Natural ventilation can also be used if openable panels/hatches are in the proper locations.

Forcible Entry: These operations can normally be accomplished by using panels/hatches in various locations on a silo, particularly on/near the top and bottom portion.

Search: Instead of search operations, it is likely that rescue operations would be more likely, particularly entrapment considerations. However, search operations in associated structures that are a part of a silo complex would be similar to search operations in commercial buildings as both are likely to have large open areas.

SPECIFIC HAZARDS

- Departments that are responsible for responding to farm emergencies must be familiar with the methods required to resolve silo type emergencies and entrapment in farm equipment.

- In many cases, an interior fire can be suppressed by the implementation of a probing nozzle, but care must be exercised when directing water into a silo from the top that also introduces oxygen. If the interior atmosphere is oxygen deficient, this can cause an explosion.

- For additional information on silo fires, see the FEMA report "Hazards Associated with Agricultural Silo Fires," or use an internet search on "agricultural silo fires."

- Fires that occur in buildings in a silo complex will burn commensurate with their method of construction. Remember a fire in these surroundings will likely burn with more intensity due to a noteworthy presence of flammable dust.

RAPID STREET-READ GUIDE

BUILDING GROUP: Miscellaneous Building/Structure—Historical Building—Dwelling

BASIC CONSTRUCTION METHOD

- Wood frame construction with rough-sawn lumber for exterior walls and interior framing; exterior wood siding
- Masonry construction of exterior brick masonry walls with rough-sawn lumber for interior construction
- Roof construction of rough-sawn lumber in conventional configurations
- Perimeter foundation and/or basement

ERA CONSIDERATIONS

- Although these buildings/districts are typically older, ranging from the late 1700s to the early 1900s, it is not the era that sets these buildings/districts apart from other types of older buildings. Instead, it is the fact that they are classified as historic buildings and are listed on the National Register of Historic Places and/or have been identified by recognized preservation groups.
- Typically, these structures/districts maintain the original type of construction and materials of conventional and heavy weight construction materials and configurations.

EXTERIOR FEATURES

- The size of these buildings can range from a small, single-story dwelling to a very large, multistory dwelling (pictured).
- All-wood construction consisting of rough-sawn lumber for framing and wood siding is used. Some buildings have brick masonry walls of unreinforced masonry (URM) with interior and roof of conventional wood construction.
- Conventional wood roofs of pitched gable or hip configurations are used. Some roofs are covered with slate or tile materials. Combustible wood roofs of shakes or shingles were also used.
- Double-hung windows were typically used in these buildings.

INTERIOR FEATURES

- Multiple floors are common, and a basement and/or cellar can be expected.
- The size of these residential buildings can vary from 1,000 sq ft to over 5,000 sq ft.
- Lath and plaster is the material of choice for walls and ceilings.
- The interior normally consists of a dining area, family room, kitchen, and bedrooms.
- Expect solid core doors and substantial locking hardware.
- Windows may be comprised of plate glass, and original double-hung windows are typical.
- Numerous types of moldings (crown, base, etc.) are used.

TACTICAL CONCERNS

Fire Spread: The spread of fire in these buildings is very similar to typical residential type buildings. Lath and plaster will better retard the spread of fire as compared to the drywall used in newer buildings. Balloon frame construction will enhance the vertical spread of fire in the walls into an attic area and between floors in multistory buildings. Basement/cellar fires will readily extend upward. Additionally, combustible wood roofs can create exposure problems, particularly in windy conditions.

Collapse: These buildings will last longer than newer lightweight buildings as the construction is older conventional. It will take a sizeable fire and time for collapse to occur in most of these buildings. Although collapse of a roof depends on the size and type of roof, conventional wood roofs are the most common and collapse can be enhanced by multiple layers of roofing materials. Sagging is a dependable collapse warning sign.

Ventilation: In most cases, conventional ventilation operations consisting of either vertical, horizontal, and/or PPV can be effectively utilized, depending on the size of a building. However, as the size increases up to and including large areas and/or multiple floors, ventilation will become more intensive. If necessary, roof ventilation will be enhanced by the fact that the roof will be of conventional construction—lightweight metal or wood trusses are rare on these buildings. Some departments will open a ventilation opening on the roof and over exterior walls to keep upward travel of fire in balloon frame construction from spreading into an attic area.

Forcible Entry: These operations can normally be accomplished by conventional methods on common doors and windows. However, remember that the doors and locks in these buildings can be more substantial than modern applications, and it is not uncommon to encounter plate glass in the windows.

Search: Search operations will be dependent on the size and location of a fire, the amount of contamination within a building, and the anticipated occupant load. Search operations will be very similar to a typical residential dwelling as the floor plans are similar. The presence of a basement and/or cellar will add to the difficulty of a search in minimal or no visibility conditions.

SPECIFIC HAZARDS

- Due to age, anticipate balloon frame construction on wood buildings.
- Unless the building has been modernized, expect knob and tube wiring to be present.
- In buildings that are much older, post and beam and/or mortise and tenon construction can be encountered.
- Unreinforced masonry brick construction may include soft bricks and substandard mortar. Arched-top window frames and doorways may have keystones that can dislodge and cause collapse of load-bearing brick.
- Assume multiple layers of roofing materials that increase roof dead load.
- Historical dwellings are often used for bed and breakfast businesses which may present interior configuration and occupant search challenges.
- Moldings such as base, crown, and those used around windows must be removed to search for void space extension if exposed to fire.
- Basement/cellar fires will readily extend upward and can also be difficult to extinguish and ventilate.

RAPID STREET-READ GUIDE

BUILDING GROUP
Miscellaneous Building/Structure—Historical Building—Commercial

BASIC CONSTRUCTION METHOD

- Exterior of brick masonry construction with rough-sawn lumber for interior construction for larger buildings
- Wood frame construction with rough-sawn lumber for exterior walls and interior framing, and exterior wood siding for smaller buildings
- Roof construction of rough-sawn lumber in conventional configurations
- Perimeter foundation and/or basement

ERA CONSIDERATIONS

- Similar to historic residential buildings, it is not the era that sets these buildings (or districts with multiple buildings) apart from other types of older buildings, it is the fact that they are classified as historic buildings/districts. They are listed on the National Register of Historic Places (NRHP) and/or have been identified by recognized preservation groups.
- Typically, these structures maintain the original type of conventional construction and heavy weight materials and typically do not use modern light-weight construction materials and configurations.

EXTERIOR FEATURES

- The varying sizes of these occupancies can range from a moderate size building to large buildings, such as the pictured railroad station in Indiana and the Elks Lodge building in Oregon.
- Brick masonry is popular for larger buildings. Small to moderate size buildings use either brick masonry or wood framing with stucco or wood siding.
- Different types of conventional roof construction can be used and will vary between flat, pitched, and arched. Roof coverings are usually slate, tile, or multiple layers of composition.
- Ground floor businesses with showroom windows are likely to include cast iron columns, lintels, and door frames.

INTERIOR FEATURES

- The interior floor plan is dependent on the intended use of the structure, although most will have narrow hallways and doorways between spaces.
- Expect a heavy grade of wood construction that can use milled timber materials.
- Lath and plaster wall finishes are common as are stamped-tin ceilings.
- Some of these buildings have been converted to house multiple smaller businesses.
- Expect interior doorway transoms for light and air distribution.
- Elaborate decorations/architecture that can conceal noteworthy voids are often used.
- Many of these buildings have added poke-through HVAC systems to improve the interior environment.

MISC 52

TACTICAL CONCERNS

Fire Spread: The spread of fire in these buildings will be dependent on the size of a particular building. In many cases, conventional construction of heavy wood structural members and lath/plaster wall/ceiling finishes will resist the spread of fire as compared to modern materials. These older buildings normally have open attics so rapid fire spread in an attic can be expected. Transoms over interior doors can speed fire spread. Stairways and hallways are typically open and can help accelerate horizontal and vertical fire spread. Lastly, wood frame construction is likely a balloon frame configuration that will rapidly spread fire vertically into an attic via an involved exterior wall.

Collapse: It will take a sizeable fire and time for collapse to occur in most of these buildings due to the type and size of construction. However, once a fire develops momentum, it can rapidly begin to weaken even the older types of heavy conventional construction that can burn with a significant intensity. Floors can collapse when exposed to fire but the size of construction is superior to modern, more flammable floor construction and can equate to longer time frames as compared to lightweight construction. Exterior walls of unreinforced masonry construction can collapse outward (more than twice their height!) when floor/roof structural members begin to fail. Collapse of a roof depends on the size and type of roof. Although conventional wood roofs will resist fire for longer periods of time than newer roofs, many of these roofs are supporting an extremely heavy dead load of slate, tile, or numerous layers of other roofing materials. Smaller buildings with more conventional wood roofs can rapidly fail depending on the size and duration of fire in a particular building.

Ventilation: Although these occupancies can have multiple doors and windows, ventilation operations will be assisted or hampered by the size of the building. In most cases, conventional ventilation operations consisting of either vertical, horizontal, and/or PPV can be effectively utilized. However, as the size increases up to and including large complexes, ventilation will become more intensive. If necessary, roof ventilation will be dependent on the type of roof encountered, with older wood roofs offering the most options. It is doubtful that many metal deck roofs will be encountered and lightweight wood roofs were not used on these types of buildings. Fires that have captured vertical combustible void spaces will likely destroy the building if vertical ventilation can't be accomplished.

Forcible Entry: Forcible entry can normally be accomplished by conventional methods. However, expect the presence of hardwood for interior doors and frames and substantial locking devices.

Search: The need for search operations will be dependent on the size of a fire and the type of building as it can be a common commercial type building or a commercial residential building. Remember that a potential search in larger commercial types of buildings should be practiced beforehand to develop the necessary expertise for large open areas.

SPECIFIC HAZARDS

- A primary hazard will be the size of some of these buildings.
- Expect a significant roof dead load.
- Consider the presence of balloon frame construction in wood frame buildings.
- Masonry buildings will likely be comprised of unreinforced masonry construction. Additionally, look for the presence of spreaders, joist anchors, and rafter tie plates.
- Buildings that are not equipped with operative sprinklers pose an additional risk although there should be a suppression system in a kitchen (if present).
- Anticipate numerous voids depending on the interior decorations/architecture, particularly when the interior has been remodeled for multiple commercial businesses.
- Some NRHP buildings may be exempted from modern fire and life-safety requirements even if significant restoration has taken place.

GLOSSARY

abandoned. A building status for those that have outlived their usefulness, fallen into disrepair, and show signs that the owner has basically given up on the building.

advanced framing methods (AFM). Refers to a variety of wood framing techniques that reduce the amount of lumber (and waste) used to construct a wood frame building and increase its energy efficiency. AFM is also known as optimal value engineering (OVE).

alternative building methods. Building construction materials, assemblies, and systems that are nontraditional, unusually innovative, or don't readily fit into the five classic types. Also called hybrid construction.

aluminum. A natural element that exists in many minerals and ores. In fact, aluminum is the most abundant metal that exists on earth.

arched truss. A truss in which the top chord is arched and the bottom chord is straight (horizontal). Arched trusses can be bowstring (tied) or rigid.

attic. A large space that is created by a steep pitched roof (arched, gable, etc.) for drainage and/or appearance.

autoclaved aerated concrete (AAC). Building blocks made from a mixture of sand, Portland cement, gypsum, water, expansion agents, and air that forms a solid block that is one fifth the weight of a similar size concrete block.

axial load. A load that is imposed through the center of the material.

balloon frame. A wood framing method where exterior wall studs are continuous from the sill plate to the roof plate. Floors are attached to ribbon board, with no fire-stopping structure within the wall.

bar truss (or bar joist). A steel parallel chord truss assembled with angle iron for the chords and cold-drawn round billet for the web.

base of operations. Concept of ensuring the platform you are working on (roof or floor) will safely support you for the duration of your operations.

basement. A habitable space that is either completely or partially below the ground floor. For this text, basement and cellar are interchangeable terms.

beams. Structural elements that deliver loads perpendicularly to their imposed load and in doing so, create opposing forces within the element.

bowstring truss. A tied truss with an arched upper chord and a horizontal tension bottom chord that connects the ends of the arched cord, creating compression in the top chord. Diagonal web members are added to help transfer loads.

bridge truss. A roof style characterized by sides that are sloped from the exterior walls to a flat roof portion. The sloped sides are derived from the trapezoidal shape of the truss (unequal parallel chords, with the bottom chord longer than the top).

brittle. A material that will fracture or fail as it is deformed or stressed.

building block systems. Any of various alternative materials used to form a stacked wall. For example, mortar-less concrete blocks engineered with unique internal shapes filled with expanded polystyrene (EPS).

building triage. The process of evaluating current and changing conditions and making judgments about the risks and integrity of various portions of a building.

bungalow construction. An older wood frame construction style that uses rough-sawn 2 × 3 in. or 2 × 4 in. rafters spaced up to 36 in. on center. Each is butted together at the ridge without a ridge board, and typically use 1 × 4 in. spaced sheathing nailed to the rafters.

buttress. An exterior wall bracing feature used to assist with lateral forces created where roof beams or trusses rest on a wall. Also known as a counterfort. Buttress are structural in nature and can take on numerous shapes (a diagonally ascending stack of stone or brick is most common).

cantilever beam. A beam supported at only one end. (Or a beam that extends well past a support in such a way that the unsupported overhang places the top of the beam in tension and the bottom in compression.)

carbon-fiber reinforced polymer (CFRP). Composite materials that include a reinforcing material (the carbon fibers) that is bound together with a polymer (like epoxy).

cast iron. A material usually formed from molten pig iron, which has a high carbon content and is thus brittle.

ceiling. An interior surface (lining) that covers the top of a room and is not considered a structural element such as walls, floors, and foundations.

cellar. (see basement)

chimney. A structural component used for the venting of hot flue gases or smoke from a stove, boiler, furnace, fireplace, or other appliance.

cockloft. A small space that is created when a roof is raised above the level of ceiling joists and rafters to provide a pitch for drainage.

column. Any structural element that is loaded axially, along its length, in compression.

combustible. Will burn, flammable.

component connections. Also known as framing junctions, where two or more structural members are joined and how they are joined.

compression. A stress that causes a material to flatten or crush.

concentrated load. A load that is applied within a small area or at one point.

concrete. A mixture of Portland cement, sand and aggregate (gravel), and water that cures into a solid mass.

concrete infill. A type of masonry wall construction that consists of gaps between parallel courses of masonry units that are filled with concrete and pieces of brick or concrete with vertical and/or horizontal runs of rebar.

continuous beam. A beam supported by three or more columns.

conventional construction. Solid lumber of 2 × 6 in. or larger used in a standard framing configuration.

GLOSSARY

cool roof. One designed to reflect more of the sun's rays than a common roof, such as one using composition or shingles.

crawl space. The unfinished space below a ground floor that allows access to under-floor utilities (pipes, ducts, etc.). Crawl spaces are of limited height and typically have a soil surface.

cricket. The junction of a vertical member (such as a skylight riser, parapet wall, etc.) and a horizontal member (such as a roof) where the intersection junction is covered by a roofing material.

cross-laminated timber (CLT). An engineered wood product using several layers (three to seven or more) of boards that are layered crosswise (typically rotated 90°) and glued.

curtain wall. An exterior wall used to enclose multiple stories.

daylight basement (or walkout basement). A basement arrangement found on buildings built on slopes and are under the grade floor (or main entrance), which allows occupants to walk out of the basement on the lower grade level through a doorway to the outside.

dead loads. The weight of the building itself and anything *permanently* attached to the building.

decking. The horizontal or pitched platform for floors or roofs. Decking is applied directly to beams to provide a surface to accept loads (building contents and people) or a durable cover (roofing).

decorative sheathing. Thin wood paneling used to finish interior walls or the outside of cabinets.

diagonal sheathing. A series of 1 × 6 in. boards that run at a 45° angle from the exterior walls to the primary structural members and provide increased structural stability (compared to straight sheathing) as they cross more roof structural members.

distributed load. A load spread over a large surface area or over multiple points.

division wall. An occupancy division wall is used to provide a major subdivision within a building for tenant needs. A fire division wall is used to subdivide a building and/or attic to restrict the spread of fire.

door. A moving panel or other moveable cover used to close an opening in a wall.

double-stud wall construction. A building technique that uses two parallel walls, spaced about 3½ in. apart, that are built with dimensional lumber and configured with either opposing (aligned) or offset (staggered) studs. The gap between the walls can be filled with insulation and provide a high R-value.

ductile. A material that will bend, deflect, or stretch as a load is applied—yet retain some strength.

eccentric load. A load that is imposed off-center, causing a material to want to bend.

electrochromic smart glass windows. (Also known as suspended particle display windows.) Windows that are primarily designed to allow an occupant to change the amount of light a window reflects. This is accomplished by using tiny transparent electrodes sandwiched between two panes of glass.

engineered wood. A term used by the fire service to describe a host of wood products that use modern methods to transform wood chips/slivers, veneers, shavings, and even recycled wood products into components that replace sawn lumber, sheathing, and other composite structural materials.

engineered wood product (EWP). Derivative wood product primarily manufactured by binding fibers, strands, particles, or veneers of wood together with adhesives. Also referred to as manufactured board, man-made wood, and composite wood.

era. The historic time period during which a building was built. Predominate eras include the pre-WWI (historical), pre-WWII (industrial), post WWII (legacy), and the new engineered lightweight.

facade. An exterior construction feature that is used on the walls of a building to alter its visual appearance. For this text, facade encompasses

fascias, false mansards, cantilevers, eyebrows, and overhangs.

fiber-reinforced product (FiRP). A FiRP (pronounced "furp") is a wood beam that has layers of high-strength synthetic fiber material or carbon graphite strands sandwiched and bonded to layers of cut timber or laminated strand lumber (LSL). FiRP beams can carry twice the load of a solid wood beam.

field stones. Easily accessed stones that are common to the area of construction. Granite, quartz, limestone, and various forms of river rock are examples of field stones.

finger-jointed lumber (FJL). FLJ has become a common method to produce long lengths of wood members from multiple short pieces of native wood lumber. When joining these short pieces, the joining ends are mitered in an interlocking fingers configuration and pressed together with an adhesive as a bonding agent.

fire load. The potential amount of heat energy (measured in British thermal units—BTUs) that may be released when a material is burning. The term fire load is not a building engineering term—it's purely a fire behavior term.

fire resistance rating (FRR). The length of time to burn *through* a given material—rated in minutes or hours.

fireplace. An architectural structure or appliance designed to contain a fire for heating and/or cooking.

flame spread rating (FSR). The length of time it takes to burn across the surface of a given material—rated in minutes or hours.

floor. The platform and substructure that serves as a base for accommodating people movement, furnishings, and fixtures within a building.

floor covering. The covering that serves as a durable (and attractive) surface to protect the subfloor.

floor supports. Structural elements responsible for carrying the load of a floor. Supports may be a foundation wall, structural beam, or a stud load-bearing wall.

footers (or footings). Weight-distributing pads that serve as the bottom of foundations.

foundation. A building's anchor to earth and base for all elements built above that anchor.

foundation walls. These are walls installed below grade to serve as structural support for other structural elements and also to hold back soil and other materials.

frame. The structural case or border into which a window is set.

framed. A building built on site one piece at a time—also known as stick-built. The building is enclosed by simple siding attached right to the framing.

girder. A beam that carries other beams.

glazing. The process of setting glass and/or thermoplastic into a window frame. The glass (transparent material) of a window assembly is known as the glazing.

glued laminated timber (GLT or glulam). Glued laminated timber is comprised of multiple layers of dimensional timber bonded together with moisture-resistant adhesives.

gravity connection. The connection of two or more materials that relies on the gravitational weight of the upper element to hold it to the other.

green insulation. Insulation that uses recycled materials such as cotton and denim in place of fiberglass.

ground gradient. A term used to describe electricity that is returning to zero potential through nonconductive surfaces like soil, concrete, and masonry. Downed wires can create a ground gradient in concentric waves from their contact on those materials and travel for several yards. Firefighters who feel tingling through their boots should shuffle-step away from the gradient and notify others.

header. See lintel.

hybrid building. An unofficial term that refers to a building that combines various NFPA 220 types in one structure or a building that is constructed

in a manner that, from a firefighter's view, doesn't really fit into any of the NFPA 220 types. It is also described as alternative construction.

impact load. A moving or sudden load applied to a building in a focused or short time interval. For example, wind, large crowds, and fire stream water are all impact loads.

insulated concrete form (ICF). Forms made of permanent expanded polystyrene (EPS) that are used for poured concrete and come as blocks, panels, or planks.

interstitial space. A space created between building materials that can hide utilities or other building components. Also, the space created when a suspended ceiling is hung from overhead beams.

jack rafter. Roof rafter used in hips or valleys to span between ridge boards or wall plates.

jamb. The structural case, border, or track into which a door is set. A jamb supports and may contain the stop for a door.

joist. A wood or steel beam used to create a floor or roof assembly that supports sheathing or decking. Joists span between primary supporting members such as foundations, load-bearing walls, or structural beams.

knob and tube wiring. An older electrical wiring style identified by a two-wire lead into a structure from a pole. In addition, this wiring usually leads to a fuse box (with removable glass fuses), often on the porch of residential structures. Within a building, a knob and tube includes a single, minimally-insulated wire supported by ceramic knobs used as spacers/supports between the wire runs. Porcelain or cloth tubes were used wherever a wire passed through a wall stud, floor, or box. A cloth tube or loom was used where two wires crossed or where a wire entered a junction or outlet box. This is an ungrounded system.

lamella roof (also known as a Summerbell roof). An arched roof that uses a weave of eggcrate, geometric, or diamond-patterned roof supports that is higher (or steeper) than the common bowstring roof.

laminated strand lumber (LSL). LSL is an engineered structural composite lumber manufactured from flaked and chipped strands of native wood blended with an adhesive. Mostly, LSL uses strands oriented in a parallel fashion (also known as parallel strand lumber—PSL).

laminated veneer lumber (LVL). An engineered wood product consisting of thin sheet veneers of native wood that are stacked with grains aligned and then glued with a phenolic resin.

ledger. A beam attached to a wall column that serves as a shelf (ledge) for other beams or building features.

lightweight construction. Solid or engineered products used to form assembly-built structural elements that are lower in mass than previous construction methods.

lightweight trusses. Trusses that are comprised of members of 2 × 4 in. (or smaller) and are often made from engineered lumber (or metal).

limited-combustible. Materials that have about one-half the heat potential of wood, or not over 3,500 Btu/lb.

lintel. A beam that spans an opening in a load-bearing wall, such as over a garage door opening (often called a "header"). Lintels can also be commonly found over windows and doors in unreinforced masonry construction and in newer CMU construction.

live green roof. A roofing system that employs a layer of soil and planted vegetation to insulate and protect the structure from heat and cooling loss.

live load. Any load applied to a building other than dead loads. Live loads are typically transient, moving, impacting, or static (like furniture).

loads. Static and dynamic weights that come from the building itself and anything that is placed within, or acts upon a building.

lock. Various types of locking devices used to provide security.

lookout basement. A basement arranged such that the walls extend above the grade level so that some of the windows are above grade.

masonry. A common term that refers to brittle materials like brick, tile, concrete block, and stone.

modular panel systems. Factory-built panels that are assembled on a job site to form load-bearing walls.

monolithic. A poured-in-place concrete and steel building that forms a "single stone."

mortar. A workable paste made from a mixture of sand, cement or lime, and water. Once cured, mortar serves as a binding agent for masonry blocks.

noncombustible. Materials that will not ignite, burn, support combustion, or release flammable vapors when heated.

occupancy. The intended use or purpose of a building. Common occupancy classifications are:

  assembly. Buildings used for the assembly of people for civic, social, religious, recreational, food or drink consumption, or awaiting transportation.

  business. Buildings used for offices, professional or service transactions, storage of records, or ambulatory health care.

  day care. Buildings used for the supervised care of children with no overnight care. (NFPA recognizes day care facilities separate from educational.)

  educational. Buildings used for educational purposes of six or more people (at any one time) through the 12th grade and some day care facilities. (ICB puts day cares in an E classification.)

  factory. Buildings used for assembling, disassembling, fabricating, finishing, manufacturing, packaging, or repairing operations that are not classified as Group H (hazardous) or Group S (storage).

  hazardous. Buildings used for the manufacturing, processing, generation, or storage of materials that constitute a physical or health hazard in quantities in excess of allowed control areas.

  institutional. Buildings where people are cared for or live in a supervised environment because of health, age, medical treatment, or those detained for penal or correctional purposes.

  mercantile. Buildings used for the display and sale of merchandise and involving the stocking of goods that are accessible to the public.

  residential. Buildings used for sleeping purposes when not classified as Group I.

  storage. Buildings used for storage that is not classified as a hazardous occupancy.

  utility and miscellaneous. Buildings, and buildings of an accessory character, that are not classified by any other occupancy use. This group includes structures such as agriculture buildings, carports, sheds, tanks, and towers.

occupied. A building that is occupied or has a *high probability* of being occupied during an incident.

oriented strand board (OSB). Known mostly by its acronym, OSB is sheathing that is formed with wood shavings and a urea-formaldehyde adhesive.

optimum value engineering (OVE). See advanced framing methods (AFM).

panel wall. A single-story exterior wall used to enclose a space.

parallel chord truss. A truss in which the top and bottom chords run in the same plane.

parapet wall. A continuation of a wall above a roof line.

particle board (PB). Wood sheathing made from a coarse sawdust and glue.

partition wall. A wall used to divide areas or rooms into smaller areas or to separate one portion of an area from another and usually not load-bearing.

party wall. A wall shared by two buildings or two occupancies within the same building. If the

party wall carries beams or structural assemblies, it is a structural element.

pilaster. A decorative column that protrudes in relief from a wall to give the appearance of a separate post column. Over time, the fire service began using the term pilaster to describe any interior or exterior thickening of wall used to add lateral support for roof beams and trusses.

pilings. Vertical posts that are driven down into the earth to serve as the foundation or foundation anchor of buildings.

pillar. A freestanding vertical post, monument, or architectural feature.

pinned connection. Those that use a screw, nail, nut and bolt, rivet, or similar device to pass through the elements being connected. Pinned connections concentrate transferred loads to a single point.

plastic. A synthetic or semi-synthetic material that is made of moldable polymers (a molecule with many connected atoms). Most plastics are derived from petroleum.

plenum space. An interstitial space used as an air return for HVAC systems.

plywood. A wood product made from layering sheet veneers of wood such that grain directions alternate 90° with each layer.

post and beam. A building built with a series of post columns and beams (no load-bearing walls). The building is enclosed by panel or curtain exterior walls. The same as a skeletal frame building.

pre- and post-tensioned concrete. Concrete that has steel cables placed through the plane of the material and then tensioned, compressing the concrete to give it the required strength. Cables can be pre-tensioned (at a factory) or post-tensioned (at the job site).

precast concrete. Slabs of reinforced concrete that are poured at a factory and then shipped to a job site.

pressure-equalized rain screen insulated structure technique (PERSIST). A building technique developed by the National Research Council in Canada. The method consists of 2 × 4 in. framing, OSB sheathing covered with a peel-and-stick membrane (rubberized asphalt adhesive backed by a layer of high-density cross-laminated polyethylene), and single or multiple layers of rigid extruded polystyrene (XPS) foam insulation. Finally, the building is finished with any type of preferred siding.

protected. Having a fire resistance rating of at least one hour based on its structural elements or protective envelope for the structural elements.

purlin. A beam placed horizontally and perpendicularly to trusses or beams to help support roof sheathing or to hang ceilings.

radiant barrier sheathing (RBS). Plywood or OSB sheathing with an aluminum type foil affixed to one side that is designed to reflect radiant heat away from the foil.

rafter. A sloped wood joist that supports roof coverings between a ridge beam and wall plate on peaked and hipped roofs.

raker. A diagonally-oriented column (loaded in compression).

rebar. (Short for reinforcing bar.) A steel bar that is used as a tensioning material in reinforced concrete and masonry to increase stability and strength. The rough surfaces on the rebar aid in bonding the rebar to the concrete.

reinforced concrete. Concrete that is poured over steel rebar, which becomes part of the cured concrete mass.

reinforced masonry construction. A wall construction method using stacked brick or block and mortar with steel rebar reinforcement placed in open cells and then filled with concrete, or steel embedded in the mortar joints.

repeated load. Loads that are transient or intermittently applied (like people on an escalator).

ridge beam. The uppermost beam of a pitched roof. Rafters attach to the ridge beam.

rigid or rib arch truss. A truss with a curved, self-supporting top chord (not tied by the bottom chord) and horizontal bottom chord along with web members that are all rigidly connected.

rigid connection. A connection formed by bonding two materials together. Examples include bead welds in steel, glues, poured concrete over steel, and the like. Rigid connections tend to spread transferred loads over a greater area.

roof. The top portion of a structure that is responsible for providing an active role in sheltering interior spaces and includes structural supports and coverings.

roof slope-pitch. This refers to the degree of slope or pitch for a roof and is expressed as a ratio. For example, a 4:12 pitch means that the roof rises 4 in. vertically for every 12 in. of horizontal distance.

sash. The metal, wood, or plastic framework that surrounds and supports glazing (window glass).

sheathing. All manner of materials used to cover or encase walls, ceilings, and roofs of framed structures. It is the first layer of covering for studs, joists, trusses, or rafters.

shear. A stress that causes a material to tear or slide part.

shear wall. A reinforcement wall that adds building stiffness to help resist the impact load of wind.

simple beam. A beam supported by columns at the two points near its ends.

skeletal frame. A building built with a series of post columns and beams (no load-bearing walls). The building is enclosed by panel or curtain exterior walls.

slabs (when used as a foundation). Flat horizontal elements that simply rest on the ground.

slab-on-grade foundation. A concrete slab that is poured over a suitable rock base on the ground, and then the walls, floor(s), and roof are erected on top of the slab foundation.

spalling. Refers to a pocket of concrete that has crumbled into fine particles through the exposure to heat.

static load. A constant load that rarely moves.

straight sheathing. A series of 1 × 6 in. boards that run at a 90° angle to supporting structural members.

strike. The receptacle that receives a dead bolt or latch from a locking mechanism. A protrusion that stops a door and keeps it from swinging past the jamb.

steel. A metallic material made from iron ore, carbon, and an alloy agent (metallic solid solution).

structural assembly. An engineered collection of interconnected building components that form a cohesive structural unit such as a roof or floor.

structural elements. The essential underpinnings of a building that allow it to stand erect and resist imposed loads and gravity. Foundations, columns, beams, and connections are the primary structural elements of any building.

structural hierarchy. A concept that defines the progressive order in which building loads are delivered to earth.

structural insulated panels (SIPs). Specially engineered panels used to form load-bearing walls and the roof. Each panel consists of two outer skins of OSB with an insulating core made from expanded or extruded polystyrene (EPS or XPS, EPS is most common).

strut. A horizontal column (loaded in compression).

subfloor. The horizontal platform material that is attached to the top of floor joists and can be made from tongue-and-groove (T&G) planking, plywood, OSB, or even lightweight concrete.

suspended beam. A beam that has one or both ends supported from above by a cable or rod (sometimes called a hung beam).

suspended load. A load that is hanging from something above.

GLOSSARY

tension. A stress that causes a material to pull apart or stretch.

tied arch roof. A roof construction method that uses arch-shaped beams that are held in compression with tensioned horizontal tie-rods.

tilt-up. A structure built using prefabricated, load-bearing wall sections (typically reinforced concrete) that are tilted upright, then pinned together.

timber truss. Large dimension lumber used to form a truss. Commonly found in older roofs, this type of roof construction is normally made from full-dimensional lumber, and is often comprised of multiple members bolted together to form one structural member.

titanium. An abundant metal found in many minerals. It is lightweight, low density, noncorrosive, and nonmagnetic. Titanium alloys are known for a high strength-to-weight ratio and tremendous resistance to heat.

torsion load. A load that is imposed in such a way that causes a material to twist.

traditional wood products. Refers to the century-old development and improvement of manufactured wood products for a specific application that cut lumber cannot fill.

transom. A beam above a door used to support glass (to help spread light) or a louver (to help improve ventilation) as opposed to a mullion, which refers to a vertical structural member between panes of glass. (A transom is the traditional American term for a transom light, which is the window over this horizontal member.)

triangular truss. The most common type of truss used to form a peaked roof.

truss. An engineered structural element that uses groups of rigid triangles to distribute and transfer loads. The triangles create an open web space. Trusses are used in lieu of solid beams in many buildings.

truss loft. An attic space created by the open web nature of trusses.

unoccupied. A building that is normally occupied but the occupants are *likely to not be in the structure currently.*

unprotected. A material that when exposed (or can be exposed) in its natural state to the effects of heat and/or fire will cause a degradation of its structural integrity.

unreinforced masonry. A wall construction method using stacked brick or block and mortar without Portland cement, steel rebar, or strapping. Also, a modern masonry wall that is not designed for load-bearing structural applications.

vacant. A building that is likely to still be in an acceptable condition in terms of structural integrity and marketability. A vacant building is likely to be secured (doors/windows).

veneer wall. A decorative-only wall added to help improve the building's appearance.

walk-up basement. A basement that is accessed by an exterior stairway entrance. The exterior entrance may be unprotected, partially covered, or fully enclosed.

wall. A vertical or upright surface designed to enclose or divide a compartment. Walls can be load-bearing (a wall column structural element that supports floor or roof beams) or non-load-bearing (supports its own weight plus anything attached to it).

wall-bearing. A building where beams or roof/floor assemblies rest on the load-bearing walls (as opposed to posts).

wind/snow load. Atmospheric loads that stress a building.

wythe. A continuous vertical section of masonry, one unit in thickness. A single wythe can be separate from or interconnected with an adjoining wall. In a multiple-wythe wall, the wythes are interconnected for additional strength and stability, and are often used for a structural load-bearing wall.

ACRONYMS

AC alternating current

ACC autoclaved aerated concrete

AFM advanced framing methods

ARRA American Recovery and Reinvestment Act

Btu British thermal unit

CFRP carbon-fiber reinforced polymer

CLT cross laminated timber

CMU concrete masonry unit (cinder block)

COM commercial

DC direct current

EPS expanded polystyrene

ELB engineered lightweight building

EWIB engineered wooden I-beam

EWP engineered wood products

FiRP fiber reinforced plastic (usually pronounced "furp")

FJL finger-jointed lumber

FPS feet per second

FRR fire resistance rating

FSR fire spread rating

GLT glue laminated timber

glulam glue laminated

HP horsepower

HUD Housing and Urban Development (government agency)

HVAC heating, ventilation, and air conditioning

IAP incident action plan

IBC International Building Code

IC incident commander

ICF insulated concrete forms

INST institutional buildings

ISO incident safety officer

LODD line-of-duty death

LSL laminated strand lumber

LVL laminated veneer lumber

LW lightweight

MANF manufacturing/warehouse building

MFD multifamily dwelling

MGP metal gusset plate

MISC miscellaneous building

NFPA National Fire Protection Association

NIOSH National Institute for Occupational Safety and Health

NIST National Institute of Standards and Technology

NRHP National Registry of Historical Places

OFF office and hotel buildings

OSB oriented strand board

OVE optimal value engineering

OWC open web construction

OWS open web steel

PB particle board

PPE personal protective equipment

PPV positive pressure ventilation

PSL parallel strand lumber

PUB public assembly building

PV photovoltaic (solar electrical panel)

PVB polyvinyl butyral

PVC polyvinyl chloride

RBS radiant barrier sheathing

RF radio frequency (as in electrical waves)

SBB straw bale building

SCBA self-contained breathing apparatus

SFD single-family dwelling

SIP structural insulated panel

SOG standard operating guideline

SOP standard operating procedure

T&G tongue and groove

TIC thermal imaging camera

UL Underwriter's Laboratory

UPS uninterrupted power supply

URM unreinforced masonry

USFA United States Fire Administration

UTL utility

UV ultraviolet

WMD weapon of mass destruction

WWI World War I

WWII World War II

XPS extruded polystyrene

INDEX

A

abandoned buildings, 265–266
AC/DC currents, 219
adhesives and glues
    hydrated, 176
    melting temperature of, 168, 177
    softening and failure of, 64, 80, 81, 120, 122
    toxicity and flammability of, 17, 18, 20
    trusses connected with, 177
adjoining walls, 152, 153, 154
advanced framing methods (AFM), 130, 132–133, 155
advertising signs, 199, 248
aerial devices
    cornice dangers involving, 65, 228–229
    facades needing, 233
    lamella roofs needing, 182
    for ventilation, 171, 182
aging, and collapse, 91–92, 98
air shafts, 197, 240–241
air-blast resistant windows, 206
alterations. *See also* remodels; renovations
    clues to, 260
    collapse and, 74, 91, 97
    division walls and, 64, 195
    timber trusses with, 188
    Type III buildings with, 64
    Type IV buildings with, 68
    Type V buildings with, 70
    void space due to, 64, 68, 238
alternative building construction methods
    aspects of, 87–88, 119–120, 129
    firefighting safety and, 2–3
    flashover caused by, 130, 131, 272
    insulated concrete forms as, 122–124
    lightweight steel as, 120–122
    Saebi alternative building systems as, 135
    straw bale buildings as, 126–127
    straw clay buildings as, 120, 127–128
    structural insulated panels as, 124–126
alternative energy systems
    fuel cell, 226
    solar, 224–226
aluminum
    aspects of, 22
    corrugated roofing of, 190–191
    sidings of, 156
American Four Square homes, 299–300
American Institute of Timber Construction, 67–68
American Recovery and Reinvestment Act (ARRA), 129
Americans with Disabilities Act of 1990 (ADA), 97
anchor bars, 39, 92, 159
annealed glass, 205
antennas, 246
arched roofs
    bowstring, 36–38, 182, 183, 184–185
    lamella, 182
    tied, 182–183
arched trusses, 36–38, 182–184
arena buildings, 381–382
asbestos
    hazmat aspects of, 155
    shingle sidings, 155, 235
asphalt-felt sidings, 156
assembly buildings. *See* public assembly buildings
assembly-built construction, 40, 99, 118
atriums, 242
attended care facility buildings, 377–378
attics
    aspects of, 166
    balloon framed vulnerability of, 73–74, 154
    common (shared), 64, 107, 238, 272
    division walls in, 153
    facade hazards and, 231–232, 272
    fire in, 190, 267

rain roofs as, 191
renovations in, 235–236, 238, 273
skylight issues regarding, 198, 241–242
truss lofts as, 167
in Type III buildings, 63–64
vents on, 197, 198, 233, 267
auditorium buildings, 383–384
autoclaved aerated concrete (AAC), 133
axial loads, 13, 31–32, 36, 37

B

ballistic-resistant glass, 206, 209
balloon frames
    collapse in, 74–75
    fire spread in, 74, 154–155
    identification of, 74, 154
    stud spacing and length in, 73, 154
    Type V buildings with, 73–75, 154–155
bar trusses (joists), 35–36
base of operations, 140, 141, 142, 272
base station antennas, 246
basements
    common (shared), 108, 146
    considerations of, 145–146
    egress from, 145
    escape windows in, 145
    historical perspectives of, 144
    sub-basements and, 145
    types of, 143
beams. *See also* I-beams and I-joists; post and beam construction
    aspects and types of, 33–34
    fire cut, 38–39, 91, 92, 158, 159
big-box commercial buildings, 106, 343–344
block/masonry buildings, 345–346
boiling liquid expanding vapor explosion (BLEVE), 221
bond beam caps, 158, 160
bowstring trusses, 36–38, 182, 183, 184–185
Brannigan, Francis, 1, 27n1, 270
brick. *See also* unreinforced brick walls
    clay foundations, 140
    king row, 158, 160
    noggin, 153
    reinforced walls of, 160–161
brick and joist construction. *See* Type III (221 or 200) ordinary construction
bridge truss roof, 180–181, 189
British thermal units (BTUs), 12
brittle materials, 14–15
    cast iron as, 15, 21–22, 27n1
    concrete as, 23
    masonry as, 24
brownstone homes, 311–312
Brunacini, Alan, 258
bucks, window and door, 126, 211
building block systems, 133
building code occupancy uses, 101–102

building materials. *See also* evolving building methods/materials
    aspects of, 14–15
building triage. *See* triaging building fires
buildings. *See also* size-up
    classifying of, 49–51, 88, 101–103, 259
    eras of, 88–99
    NFPA types of, 52–82
    size considerations of, 103–106
    uses of (occupancy types), 99–102
bulkhead doors, 198, 215
bungalow construction, 74, 93, 168
burns, radio frequency, 247
business buildings, 101
buttresses
    roofs supported by, 162, 182, 183, 184
    walls with, 32, 36, 162

C

cantilever beams, 34
Cape Cod homes, 297–298
carbon-fiber reinforced polymers (CFRP), 25
cast iron
    brittle nature of, 15, 21–22, 27n1
    columns, collapse of, 27n1, 91
    as wrought iron, 37–38
Castro, Joe, 276–277
C-channel steel studs, 21, 120, 121
ceilings
    concrete, 151
    defined, 140, 149
    directly fastened, 149–151
    drywall, 150–151
    lath and plaster, 149–150
    metal wire mesh and plaster, 150
    pulling of, 235, 266
    suspended, 55, 151–152, 236
    tin and decorative wood, 150
cellars. *See* basements
cellular antennas and communication equipment
    antennas of, 246
    base stations of, 246
    coaxial cables of, 247
    equipment huts/rooms of, 246–247
    hazards of, 223, 247
    radio frequency waves, thermal effect and, 247
cement. *See also* concrete; masonry
    fiber siding, 156
    Portland, 23, 91, 157, 158, 160, 161
center core floor plans, 55
center hallway floor plans, 55
center hallway structure homes, 319–320
Chicago construction. *See* balloon frame
chimneys/fireplaces, 196–197
chords
    in I-beams or joists, 175–176
    in trusses, 35–36, 186–187

INDEX

church buildings, 387–388
cinder block. *See* masonry
circuit hallway floor plans, 55
classifying buildings, 49–51, 88, 101–103, 259
clay brick foundations, 140
clay tile roof coverings, 193
coaxial cables, 247
cocklofts, 63–64, 108, 166
Coleman, John, 271
collapse
    aging and, 91–92, 98
    alterations and, 74, 91, 97
    in balloon frame buildings, 74–75
    cast iron columns and, 27n1, 91
    communicating potential, 279
    construction classification predicting, 278
    of engineered wood platform (lightweight) construction, 82, 98–99, 148
    of exterior elements, 91
    gravity connections and, 91
    of insulated concrete form buildings, 124
    in lightweight steel buildings, 121–122
    of masonry, modern or unreinforced, 59
    mortar, water-soluble and, 91, 158
    occupancy shifts, alterations and, 97
    open spans and, 97
    overloading and, 97–98
    of post and beam construction, 73
    in post-WWII buildings, 97–98
    predicting, 270, 277–279
    of pre-WWI buildings, 91–92
    in pre-WWII era buildings, 94
    safe areas in, 160
    in stacked log buildings, 72
    in straw bale buildings, 127
    in straw clay buildings, 120, 127–128
    of structural insulated panel buildings, 126
    structural involvement predicting, 278
    of tilt-up concrete slab construction, 59, 61
    time elapsed in predicting, 278–279
    of Type II building walls, 59
    in Type IV, areas of, 69
    visualizing loads in predicting, 278
Collapse of Burning Buildings: A Guide to Fireground Safety (Dunn), 277
Colonial homes, 295–296
columns
    cast iron, 21–22, 27n1, 91
    cross-laminated timber as, 19
    glued laminated timber as, 20
    H-column, 21, 31, 121
    heavy timbers as, 67
    monolithic buildings, use of, 24
    in post and beam construction, 72
    rakers as, 32
    steel, 21, 53–54
    struts as, 32
    types of, 31–32
commercial buildings, 106–108, 260–261, 331–336, 341–344

communication/data systems. *See* cellular antennas and communication equipment
component connections, 166, 187
composite sheathings, 41
composite wood. *See* engineered wood products
composites, 25–26, 41
composition shingle roof coverings, 193
compression forces, 13
concealed spaces. *See* void spaces
concentrated loads, 12
concrete. *See also* insulated concrete form; tilt-up concrete slab construction
    autoclaved aerated, 133
    ceilings, 151
    formed walls, 162
    in foundations, 141
    heat retention of, 54
    infill masonry walls, 157
    nonstructural lightweight flat roofs, 180
    overhaul operations of, 24, 54
    spalling of, 24, 54, 56, 57
    tile roof coverings, 194
    tilt-up buildings, 349–350
    types of, 23–24
concrete masonry units (CMU), 24
condo homes, 327–328
connections
    metal hangar joist, 147
    tilt-up slab, 162–163
    trusses with glued, 177
    types of, 38–39
contents of buildings
    overhaul operations affected by, 60
    as primary hazard in Type II buildings, 59–60
    synthetics, fire load, spread and, 82, 96
continuous beams, 34
conventional construction, 79, 166, 167
conventional platform construction
    exterior walls of, 154–155
    Type V buildings, 76–77
converted mill buildings, 353–354
cool roofs, 131
cornices
    newer synthetic, 65, 228–229
    older stone or wood, 227–228
    on Type III buildings, 65
corrugated
    roofs, 190–191
    sidings, 156
counterforts. *See* buttresses
Craftsman homes, 299–300
craftsman-built trusses, 79
crawl space vents, 142, 143, 145–146
crickets, 197
cross-laminated timbers (CLT), 19–20, 134
curtain walls
    about, 41
    in Type I construction, 55, 56, 60

D

day care buildings, 101
daylight basements, 143
dead loads
    defined, 12, 166
    of heating, ventilation, and air conditioning systems, 222
    live green roofs as, 127, 131
    roofs as, 166
    solar panels as, 120
decking. *See* sheathings
decontamination. *See* hazmat decontamination triggers
decorative sheathings, 17–18
deep-pocket fire, 127
detention (jail) facility buildings, 375–376
diagonal sheathings, 166
dimensional lumber, 16
distributed loads, 12
division walls, 64, 152, 153, 195
doors
    assembly parts of, 214
    bucks in, 126, 211
    bulkhead, 198, 215
    construction materials of, 214–215
    glass, 215
    impact-resistant glass in, 209–213
    inward swing or outward swing, 215–216
    locks on, 214, 216
    metal, 214–215
    overhead, 214–215
    penthouse, 198
    security bars on, 215
    in Type IV buildings, 68–69
    wooden, 214
double-pane glass, 96, 206
double-stud wall construction, 130, 152, 154
drug labs, 248
drywall
    ceilings, 150–151
    in engineered lightweight construction, 81, 82
    as firewall, 153
    in lightweight steel buildings, 121, 122
ductile materials, 14–15
    aluminum as, 22
    concrete with steel as, 23
    plastics as, 25
    steel as, 21
ducting, 199
Dunn, Vincent, 277

E

eccentric loads, 13
eco-friendly construction. *See* green construction
educational buildings, 101
electrical utilities
    distribution wires of, 218
    ground gradients, 219–220
    knob and tube wiring, 218, 237
    main shutoff of, 217–218
    outlets/switches/junction boxes, 218–219
    power distribution circuit protection of, 218
    renovations to, 237
    shock hazards of, 219
    supply feed of, 217
    transformers of, 218
electrochromic smart glass, 131–132
electrocution, 219
elevator houses, 198
energy-efficient glass, 206
engineered lightweight building (ELB) era, 98–99, 148
engineered wood platform (lightweight) construction
    aspects of, 79–81
    collapse of, 82, 98–99, 148
    drywall issues in, 81, 82
    exterior walls of, 154–155
    fire spread in, 81–82
    I-joist flooring in, 147, 148
    trusses in, 80, 81–82, 98
    Type V (111 or 000) wood frame construction with, 79–82
engineered wood products (EWP)
    cross-laminated timber, 19–20, 134
    defined, 18, 166
    fiber-reinforced products, 134
    finger-jointed lumber, 20
    glued laminated timber, 20
    laminated strand lumber, 19
    laminated veneer lumber, 19
    oriented strand board, 18
engineered wood systems, 133–134
engineered wooden I-beam (EWIB). *See* I-beams and I-joists
equipment huts/rooms, 246–247
era considerations
    attended care facilities, 377
    auditorium/theater buildings, 383
    big-box buildings, 343
    block/masonry buildings, 345
    brownstone homes, 311
    center hallway structure homes, 319
    church buildings, 387
    Colonial and Georgian homes, 295
    concrete tilt-up buildings, 349
    converted mill buildings, 353
    Craftsman and American Four Square homes, 299
    detention (jail) facilities, 375
    fast food buildings, 339
    garden apartment homes, 321
    high rise-1st generation buildings, 365
    high rise-2nd generation buildings, 367
    high rise-3rd generation buildings, 369
    historical buildings-commercial, 397
    historical buildings-dwelling, 395
    hospital buildings, 373
    industrial/legacy strip style buildings, 335
    kit buildings, 391
    legacy townhome/condo/apartment homes, 327

INDEX

lightweight townhome/condo/apartment homes, 329
manufactured (mobile) homes, 309
McMansion homes, 307
meeting hall buildings, 385
mega-box buildings, 341
modern lightweight homes, 305
modern strip-style stores, 337
pole barns, 389
post-WWII low rise buildings, 361
prairie style homes, 301
pre-WWI ordinary commercial buildings, 331
pre-WWII low rise buildings, 359
pre-WWII ordinary (taxpayer) buildings, 333
project housing-high density homes, 325
project housing-low density homes, 323
public storage-multistory buildings, 357
public storage-single story buildings, 355
railroad flat homes, 317
restaurant buildings, 379
row frame homes, 315
school buildings, 371
silo buildings, 393
split level homes, 303
stadium/arena buildings, 381
steel manufacturing/warehouse buildings, 347
tenement homes, 313
21st century buildings, 363
Victorian/Queen Anne and Cape Cod homes, 297
wooden manufacturing/warehouse buildings, 351
eras of construction
    engineered lightweight building, 98–99, 148
    post-WWII (after 1949): the legacy era, 94–98
    pre-WWI (pre-1914): the historic era, 89–92
    pre-WWII (pre-1939): the industrial era, 93–94
era/use/type/size classification methods, 88, 259, 262
escape windows, 145
evolving building methods/materials. *See also* green construction
    advanced framing methods, 130, 132–133
    building block systems, 133
    engineered wood systems, 133–134
    modular panel systems, 133
expanded polystyrene forms (EPS), 122–124, 125–126, 133
exterior tie plates, 39, 92, 159, 233
extruded polystyrene (XPS), 124, 130

F

facades
    aerial devices for, 233
    aspects of, 230–231
    attachment concerns of, 231–232
    attic considerations of, 231–232, 272
    fire spread investigation in, 233–234
    height and shape concerns of, 232
    overhangs on, 231
    roofline concealment of, 233
    size-up of, 272
    supports for, 232
    Type III buildings with, 66
factory buildings, 101
false floors, 55
fan shafts, 198, 199
fast food buildings, 339–340
feet per second (FPS), 210
fiber cement sidings, 156
fiberglass
    corrugated roofing, 190–191
    sidings, 156
    skylights, 241
fiber-reinforced product (FiRP), 134
field stone foundations, 140
finger-jointed lumbers (FJL), 20
fire, deep-pocket, 127
fire blocking in walls, 76, 154
fire cut beams, 38–39, 91, 92, 158, 159
fire division walls, 152, 153, 195
Fire Engineering, 7
fire escapes
    age of, 244
    aspects of, 243
    ladders of, 244
    location of, 244–245
    party (balcony) type, 243–244
    screened stairway type, 244
    special concerns regarding, 245
    standard type, 244
    wooden, 243
fire loads
    as heat energy released, 12
    occupancy groupings and, 60, 102
    synthetic contents and, 82, 96
    in Type IV buildings, 69
fire resistance rating (FRR), 52
fire resistive construction. *See* Type I (442 or 332) fire resistive construction
fire spread
    in balloon frame buildings, 74, 154–155
    contents, modern and, 96
    in engineered wood platform (lightweight) construction, 81–82
    in facade interiors, 233–234
    factors dictating, 269
    glass and, 96
    in hallways, central, 94
    in insulated concrete form buildings, 124
    in lightweight steel buildings, 121–122
    in post-WWII buildings, 96–97
    in pre-WWI buildings, 90–91
    in pre-WWII buildings, 93–94
    rate of change in, 268
    rule of six sides in, 269
    in stacked kitchens and baths, 64, 317, 318
    in stacked log buildings, 72
    in stairwells, open, 94
    in straw bale buildings, 127
    in straw clay buildings, 127–128

in structural insulated panel buildings, 125–126
technology updates causing, 96
in utility chases, 94
in void spaces, 94
fire spread rating (FSR), 52
fireground clock, 3, 267–268
fireground operations, 3
fireplaces/chimneys, 196–197
fires previous, in buildings, 65–66
first-due decision maker
    180 and 360 size-ups by, 258–259, 285n1
    burning materials briefing by, 266
    indexing system, use by, 290
    occupancy use briefing by, 101
    response area knowledge by, 6
    triage by, 275
flame spread rating (FSR), 52
flashover
    alternative building methods causing, 130, 131, 272
    compressed time to, 1
    glass, double-pane and tempered causing, 96, 206
    injuries and death due to, 3
    particle board and, 17
flat panel walls, 123, 124
flat roofs. *See* roofs, flat
floodlighting size-ups, 262
floors
    coverings of, 147, 148
    defined, 140, 146
    joists of, 146, 147, 148
    multiple, in buildings, 147–148
    on sloping ground, 147–148
    subflooring of, 40–41, 146, 147
    supports of, 146
foam homes, 135
footers, 30, 31
forces resisting loads, 13
forcible entry
    in impact-resistant glass, 212–213
    of Saebi alternative building systems homes, 135
    security bars and, 207
    in size-up, 270, 272
    tempered glass and, 205
    Type I buildings, types of, 56–57
foundations
    basement/cellar types, 143–146
    era of, 140–141
    as relevant size-up, 143
    types of, 30–31, 141–146
    walls, 30–31
framed (stick-built) construction, 42
framing
    junctions, 166
    steel, 21, 53–54, 58–59
    stick-built as, 42
freshwater systems, 223
front-loading size-up. *See also* size-up
    era/use/type/size method for, 259, 262
    predicting collapse through, 278
    pre-incident planning as, 258
fuel cell energy systems, 226
full dimension lumber, 16

G

gable roofs, 159, 167–169, 186
gambrel roofs, 167, 170–171
garden apartment homes, 321–322
gas utility systems
    boiling liquid expanding vapor explosion, 221
    liquefied natural gas, 220–221
    liquefied petroleum gas (propane), 220–221
Georgian homes, 295–296
girders, 34, 42, 58, 67
girts, 72
glass
    annealed (plate), 205
    ballistic-resistant, 206, 209
    doors, 215
    double-pane, 96, 206
    electrochromic smart, 131–132
    energy-efficient, 206
    fire spread and, 96
    hurricane-resistant, 206
    impact-resistant, 209–213
    laminated safety, 205
    panel skylights, 241
    tempered, 96, 205
    windows, types of, 204–206
    wired, 205
glazing. *See* glass
glued laminated heavy timber, 16–17
glued-laminated timber (GLT), 20
glues. *See* adhesives and glues
glulam (glued laminated heavy timber), 16–17
granite foundations, 140–141
gravity connections
    aspects of, 38–39
    collapse of, 91
    fire cut beams in, 38–39, 91, 92, 158, 159
green construction
    aspects of, 129–130
    types of, 130–132
green insulation, 131
grid block walls, 123, 124
ground fault interruption (GFI), 218–219
ground gradients, 219–220
grouping buildings
    in initial size-up, 259–260
    as institutional, 261
    as Main Street commercial, 260–261
    as manufacturing/warehouse, 261
    as miscellaneous, 261
    as multifamily dwellings, 260
    as office/hotel, 261
    as public assembly, 261
    as single-family dwellings, 260

INDEX

subsets of, 292–293
zoning as, 260
gypsum board. *See* drywall

H

hallways, 55, 94, 319–320
hangers, metal, 55, 147, 175, 178
hazards. *See also* overhead hazards
    attended care facility, 378
    auditorium/theater, 384
    big-box building, 344
    block/masonry building, 346
    brownstone home, 312
    buildings with chemical/material, 101
    cellular antenna, 223, 247
    center hallway structure home, 320
    church building, 388
    Colonial and Georgian home, 296
    concrete tilt-up building, 350
    converted mill building, 354
    Craftsman and American Four Square home, 230
    detention (jail) facility, 376
    electrocution, 219
    fast food building, 340
    garden apartment, 322
    high rise-1st generation building, 366
    high rise-2nd generation building, 368
    high rise-3rd generation building, 370
    historical buildings-commercial, 398
    historical buildings-dwelling, 396
    hospital building, 374
    industrial/legacy strip style building, 336
    kit building, 392
    legacy townhome/condo/apartment, 328
    lightweight townhome/condo/apartment, 330
    manufactured (mobile) home, 310
    McMansion home, 308
    meeting hall building, 386
    mega-box building, 342
    modern lightweight home, 306
    modern strip-style store, 338
    pole barn, 390
    post-WWII low rise building, 362
    prairie style home, 302
    pre-WWI ordinary commercial building, 332
    pre-WWII low rise building, 360
    pre-WWII ordinary (taxpayer) building, 334
    project housing-high density home, 326
    project housing-low density home, 324
    public storage-multistory building, 358
    public storage-single story building, 356
    railroad flat home, 318
    remodel, 238–239
    restaurant building, 380
    row frame home, 316
    school building, 372
    silo building, 394
    split level home, 304
    stadium/arena building, 382
    steel manufacturing/warehouse building, 348
    tenement home, 314
    21st century building, 364
    Victorian/Queen Anne and Cape Cod home, 298
    wooden manufacturing/warehouse building, 352
hazmat decontamination triggers
    abandoned buildings as potential, 265
    asbestos shingle siding, 155, 235
    barn storage as potential, 390
    drug labs as, 248
    ice arena cooling equipment as, 382
    manufacturing warehouses as, 348
    PCB transformers, 218
    public storage units as potential, 355
H-columns, 21, 31, 121
headers, 34, 158, 160
heating, ventilation, and air conditioning systems (HVAC)
    appliances and systems of, 221–222
    dead load hazards of, 222
    plenum spaces with, 151–152
    Type I buildings with, 54–55
heat-strengthened glass, 205
heavy timber/mill construction. *See* Type IV (2HH) heavy timber/mill construction
hierarchy, structural, 42, 258
high rise-1st generation buildings, 365–366
high rise-2nd generation buildings, 367–368
high rise-3rd generation buildings, 369–370
high-rise buildings, 106
hip roofs, 167, 170
the historic era. *See* pre-WWI (pre-1914): the historic era
historical buildings-commercial, 397–398
historical buildings-dwelling, 395–396
Hoevelmann, Jason, 2
horizontal ventilation, 57
hospital buildings, 373–374
hot mop roof coverings, 193
Housing and Urban Development (HUD) windows, 207
hurricane-resistant glass, 206
hybrid buildings
    alternative construction methods of, 87–88
    combining NFPA 220 building types, 85–87
    size-up challenges of, 263
hybrid panelized roof system, 178–179

I

I-beams and I-joists
    chords in, 175–176
    as engineered wood, 5–7
    flat roofs with, 175–176
    floor failure of, 147, 148
    oriented strand board in, 18–19
    poke-throughs in, 82
    steel, 21, 121, 176
    types of, 34

impact loads, 12
impact-resistant doors and windows
    aspects and testing of, 209–212
    firefighter considerations regarding, 212
    tactical tips regarding, 212–213
impact-resistant windows, ventilation issues of, 212
incident action plan (IAP), 279
incident commander (IC), 7–8, 29, 190, 209, 242, 275
Incident Management for the Street-Smart Fire Officer (Coleman), 271
incident safety officer (ISO), 275
industrial commercial buildings, 335–336
the industrial era. *See* pre-WWII (pre-1939): the industrial era
institutional buildings, 101
    attended care facility, 377–378
    detention (jail) facility, 375–376
    hospital, 373–374
    school, 371–372
    size-up of, 261
insulated concrete form (ICF)
    as alternative building method, 122–124
    collapse of, 124
    fire spread in, 124
    pyrolysis in, 124
    wall types of, 123–124
insulation
    double-stud walls and, 130
    flashover due to, 3, 4, 120
    green, 131
    insulated concrete form walls as, 122–124
    pressure-equalized rain screen insulated structure technique as, 130
International Building Code (IBC), 86, 101–102
interstitial areas. *See also* void spaces
    false floors, 55
    suspended ceilings, 55, 151
    in Type I buildings, 55
inverted roofs, 191–192
iron. *See* cast iron

J

jack rafter, 166, 185
jails, 375–376
jambs, door, 214
Joe's Building Fire Triage Program, 276–277
joists. *See also* I-beams and I-joists
    bar, 35–36
    defined, 34
    flooring, 146, 147, 148
    as I-joists and floor failure, 147, 148
    metal hanger connections, 147
junction boxes, electrical, 218–219

K

king row, brick, 158, 160
kit buildings, 391–392

knee walls, 64, 80, 94, 235
knob and tube wiring (K&T), 218, 237
kraft paper, 178–179

L

lamella roofs, 182
laminated safety glass, 205
laminated strand lumber (LSL), 19
laminated veneer lumber (LVL), 19
ledgers or ribbon boards, 34, 72, 73, 154
legacy apartment homes, 327–328
legacy condo homes, 327–328
the legacy era. *See* post-WWII (after 1945): the legacy era
legacy platform construction
    exterior walls of, 154–155
    Type V buildings, 77–79
legacy strip style commercial buildings, 335–336
legacy townhomes, 327–328
lightweight apartment homes, 329–330
lightweight concrete roofs, 180
lightweight condo homes, 329–330
lightweight construction, 166
lightweight steel buildings
    C-channel studs in, 120, 121
    collapse of, 121–122
    drywall issues in, 121, 122
    fire spread issues of, 121–122
    post and beam construction in, 95, 120
lightweight steel sidings, 156
lightweight townhomes, 329–330
lightweight trusses, 166, 268
lightweight wood platform buildings. *See* engineered wood platform (lightweight) construction
lightwells, 197, 240–241
limited combustible materials, 52
line-of-duty death (LODD), 1, 81
lintels, 34, 158, 160
liquefied natural gas (LNG), 220–221
liquefied petroleum gas (propane) (LPG), 220–221
live green roofs, 127, 131
live loads, 12, 129, 166
load-bearing wall. *See* wall columns
loads
    forces resisting of, 13
    imposition of, 12–13
    as predicting collapse, 278
    types of, 12–13
locks, door, 214, 216
log buildings. *See* stacked log buildings
look-out basements, 143
lumber
    cut, sawn or native wood, 15–16
    nominal and full dimensional, 16, 78
    old-growth and new-growth, 15–16
    rough-sawn, 77

M

Main Street commercial buildings
    big-box, 106, 343–344
    fast food, 339–340
    industrial/legacy strip style, 335–336
    mega-box, 106, 341–342
    modern strip-style store, 337–338
    pre-WWI ordinary, 331–332
    pre-WWII ordinary (taxpayer), 333–334
    size-up of, 260–261
man-made wood. *See* engineered wood products
mansard roof, 171, 181
manufactured board. *See* engineered wood products
manufactured (mobile) homes, 309–310
manufacturing/warehouse buildings
    block/masonry, 345–346
    concrete tilt-up, 349–350
    converted mill, 353–354
    public storage-multistory, 357–358
    public storage-single story, 355–356
    size-up of, 261
    steel, 347–348
    wood, 351–352
masonry
    aspects of, 24–25
    cinder block, 24
    concrete infill, 157
    fireplace, 196–197
    parapet walls of modern, 230
    Portland cement in, 23, 91, 157, 158, 160, 161
    rebar in, 157
    reinforced construction, 157
    unreinforced, 59, 157
    veneers, 24, 41, 156, 157–158, 161
    wall collapse of, 59
    wythe walls of, 157
masonry/block buildings, 345–346
materials
    aluminum and titanium, 22–23
    aspects of, 14–15
    cast iron, 15, 21–22, 27n1
    composites, 25–26, 41
    concrete, 23–24
    engineered wood, 18–20
    masonry, 24–25
    noncombustible, 52
    protected, 52
    steel, 21
    unprotected, 52, 58–59
    wood, 15–18
McMansion homes, 307–308
meeting hall buildings, 385–386
mega-box commercial buildings, 106, 341–342
mercantile buildings, 101
metal
    deck fire in, 179
    doors, 214–215
    hangers, 55, 147, 175, 178
    roofing, 194
    sheathings, 41
    straps on roofs, 159
    temperature for failure of, 21, 179
    types of, 21–22
metal gusset plates (MGP), 177, 178
mill buildings. *See* Type IV (2HH) heavy timber/mill construction
mill (converted) buildings, 353–354
mini-malls. *See* Type III (221 or 200) ordinary construction
miscellaneous buildings/structures
    historical buildings-commercial, 397–398
    historical buildings-dwelling, 395–396
    kit buildings, 391–392
    occupancy grouping of, 101–102
    pole barns, 389–390
    silos, 393–394
    size-up of, 261
mobile homes. *See* manufactured (mobile) homes
modern lightweight homes, 305–306
modern strip-style store buildings, 337–338
modular panel systems, 133
monitor roofs, 167, 171–172, 195
monolithic construction, 23–24, 42
mortar, 24–25
    Portland cement as, 23, 91, 157, 158, 160, 161
    water-soluble, collapse of, 91, 158
mortise and tenons, 72, 73
multifamily dwellings (MFD)
    brownstone homes, 311–312
    center hallway structure homes, 319–320
    garden apartment homes, 321–322
    legacy townhome/condo/apartment homes, 327–328
    lightweight townhome/condo/apartment homes, 329–330
    project housing-high density homes, 325–326
    project housing-low density homes, 323–324
    railroad flat homes, 317–318
    row frame homes, 315–316
    size-up of, 260
    tenement homes, 313–314
multiple-wythe walls, 161

N

National Fire Protection Association 220 (NFPA)
    building classifications, 49–51, 101–103
    definitions of, 51
    numerical designations of, 52–53
    Type I (442 or 332) fire resistive construction, 53–57, 103
    Type II (222, 111 or 000) noncombustible construction, 58–61, 103
    Type III (221 or 200) ordinary construction, 62–66, 103
    Type IV (2HH) heavy timber/mill construction, 67–70, 103
    Type V (111 or 000) wood frame construction, 70–82, 103
National Institute for Occupational Safety and Health (NIOSH), 285n1

National Institute of Standards and Technology (NIST), 2, 271
National Registry of Historic Places, 91
native woods, 15–16
NFPA 220: *Standard on Types of Building Construction*, 50, 51–52, 71
NFPA 5000: *Building Construction and Safety Code*, 86, 101–102
nominal dimension lumber, 16, 78
noncombustible construction. *See* Type II (222, 111, or 000) noncombustible construction
noncombustible materials, 52
nonstructural lightweight concrete flat roofs, 180
numerical designations, NFPA 220, 52–53

O

occupancy. *See also* uses of buildings (occupancy types)
    division walls, 153
    separation, 86–87
    size-up of, 265
occupied buildings, 264
office/hotel buildings
    high rise-1st generation, 365–366
    high rise-2nd generation, 367–368
    high rise-3rd generation, 369–370
    post-WWII low rise, 361–362
    pre-WWII low rise, 359–360
    size-up of, 261
    21st century, 363–364
offset walls or studs. *See* double-stud wall construction
old-growth and new-growth lumber, 15–16
open span collapse, 97
open web
    bar joist roofs, 179–180
    flat roofs constructed with, 176
    steel joists, 176
    trusses, 34–35
open web construction (OWC), 176
open web steel (OWS), 176
openings. *See* void spaces
optimum value engineering (OVE), 130
ordinary construction. *See* Type III (221 or 200) ordinary construction
oriented strand boards (OSB), 18–19, 40, 120–121, 124–126
outlets, 218–219
overhaul operations
    concrete heat retention and, 24, 54
    contents impacting, 60
    strawbale, straw clay houses and, 120, 127, 128
overhead doors, 214–215
overhead hazards
    cornices, 65, 227–229
    facades, 230–234
    parapet walls, 229–230
overloaded building collapse, 97–98

P

pack rat conditions, 248–249
panel walls, 41
panelized flat roofs, 178–179
parallel chord trusses, 35–36, 186–187
parallel strand lumber (PSL), 19
parapet walls
    bond beam on, 158, 160
    of concrete tilt-up slabs, 64, 229
    gable roofs hidden by, 159
    masonry, modern construction of, 230
    Type III buildings with, 64–65
    of unreinforced brick or masonry, 158, 229
particle boards (PB), 17
partition walls, 41, 152, 153
party walls, 41, 153–154
penthouse doors, 198
performance-designed construction. *See also* assembly-built construction
    firefighting time not addressed in, 118
    prescriptive building compared to, 117–118
perimeter (deep) foundations, 141–143
personal protective equipment (PPE), 248, 267
photovoltaic panels (PV), 225–226
pilasters, 32, 36, 37, 163. *See also* buttresses
pilings, 31
pillars, 31
pinned connections, 38–39
pipe chases, 64, 198
placarding abandoned buildings, 265–266
plaster sidings, 156
plasterboard. *See* drywall
plastic skylights, 241
plastics, 25
platform construction. *See* conventional platform construction; legacy platform construction
plenum space, 151, 152
plywoods, 17
    T-11 sidings, 156
poke-throughs, 56, 79, 82
pole barn construction. *See* post and beam construction
pole barns, 389–390
polyvinyl butyral (PVB), 210
polyvinyl chloride (PVC), 156, 195
Portland cement, 23, 91, 157, 158, 160, 161. *See also* mortar
post and beam construction
    aspects of, 42, 72
    collapse in, 73
    exterior walls, 72, 73, 123–124
    fire spread in, 73
    insulated concrete form walls using, 123, 124
    lightweight steel buildings with, 95, 120
    mortise and tenons in, 72, 73
    pole barn construction as, 72
    Type V buildings, 72–73
post-WWII low rise buildings, 361–362
post-WWII (after 1945): the legacy era

aspects of, 94–95
collapse issues in, 97–98
fire spread issues in, 96–97
foundations in, 141
legacy platform construction, 77–79
trusses in, 97
Type I legacy buildings, 95
Type II legacy buildings, 95
Type III legacy buildings, 95–96
Type IV legacy buildings, 96
power distribution circuit protection, 218
prairie style homes, 301–302
pre- and post-tensioned concrete, 23
precast concrete, 23
pre-incident study, 50, 258, 269
prescriptive building, 117–118
pressure-equalized rain screen insulated structure technique (PERSIST), 130
pre-WWI ordinary commercial buildings, 331–332
pre-WWI (pre-1914): the historic era
    aspects of, 89–90
    collapse issues of, 91–92
    fire spread issues of, 90–91
    foundations in, 140–141
    transoms in, 91
pre-WWII low rise buildings, 359–360
pre-WWII ordinary (taxpayer) buildings, 333–334
pre-WWII (pre-1939): the industrial era
    collapse in, 94
    fire spread in, 93–94
    foundations in, 141
project housing-high density homes, 325–326
project housing-low density homes, 323–325
protected materials, 52
public assembly buildings
    auditorium/theater, 383–384
    church, 387–388
    defined, 101
    meeting hall, 385–386
    restaurant, 379–380
    size-up of, 261
    stadium/arena, 381–382
public storage-multistory buildings, 357–358
public storage-single story buildings, 355–356
pulling ceilings, 235, 266
purlins, 34
pyrolysis, 124

Q

Queen Anne homes, 297–298

R

radiant barrier sheathing (RBS), 130–131
radio frequency waves (RF), 247
rafters
    defined, 34
    jack, 166
    tails of, 71, 74, 154, 169, 170, 263
    tie plates on, 39, 92, 159, 233
railroad flat homes, 317–318
rain roofs, 191
raised roofs, 191–192
rakers, 32, 163
Rapid Street-Read Guide
    52 buildings of, 289–290
    design of, 290
    index to, 292–293
    using of, 290–291
razor wire, 245–246
reading a building. *See* size-up
rebar, 23, 157, 158
reinforced brick walls, 160–161
reinforced concrete, 23
reinforced masonry construction, 157
relevant size-up. *See* size-up
remodels. *See also* renovations
    benefits of, 238–239
    hazards of, 238–239
renovations
    attic rooms with, 235–236, 238, 273
    commercial, 238
    covered windows as, 236
    to electrical utilities, 237
    residential, 237–238
    to roofs, 235
    to siding, 235
    soffits in, 236
    suspended ceilings in, 236
    tie plates and rods in, 236–237
    void spaces created by, 64, 68, 238
    to windows, 234–235, 236
repeated loads, 12
residential buildings, 101, 106–107
restaurant buildings, 379–380
RF burns, 247
rib arch trusses, 183–184
ribbon boards or ledgers, 34, 72, 73, 154
ridge beam, 34
rigid (rib) arch trusses, 36–37, 183–184
rigid connections, 38–39
roofs. *See also* roofs, flat; roofs, styles of; roofs, unique construction
    advertising signs on, 199, 248
    air shafts/lightwells on, 197
    appendages of, 195–199
    attic vents on, 197, 198, 233, 267
    buttresses supporting, 162, 182, 183, 184
    chimneys/fireplaces, 196–197
    cool, 131
    coverings of, types, 193–195
    cricket junction in, 197
    dead loads of, 166
    definitions of, 140, 166–167
    ducting on, 199

elevator houses on, 198
fan shafts on, 198, 199
heating, ventilation, and air conditioning systems on, 221–222
live green, 127, 131
metal, 194
metal straps on, 159
penthouse doors, 198
reading of, 165–166
renovations to, 235
scuttle, 198
sheathings of, 166
skylights, 197–198
slope/pitch of, 166
terms, important to, 166–167
vent pipes on, 198
ventilators on, 198, 199
roofs, flat
conventional, 174–175
general, 167, 173–174
with metal gusset plates or glue, 177, 178
nonstructural lightweight concrete, 180
with open web bar joist, 179–180
with open web construction, 176
panelized, 178–179
shed, 174
with wooden I-joists, 175–176
roofs, styles of
arched, 182–186
bridge truss, 180–181, 189
gable, 159, 167–169, 186
gambrel, 167, 170–171
hip, 167, 170
mansard, 171, 181
monitor, 167, 171–172, 195
sawtooth, 167, 172–173
roofs, unique construction
corrugated, 190–191
gable truss (older), 186
inverted (raised), 191–192
parallel chord truss (older), 186
rain, 191
single-ply, 194–195
timber trusses (older), 186–190
rough-sawn lumber, 77
row frame homes, 315–316
rule of six sides, 269

S

Saebi alternative building systems (SABS), 135
safe areas, in collapse, 160
safety glass, 205
sash, 204
sawn lumber (native wood), 15–16
sawtooth roofs, 167, 172–173
school buildings, 371–372
scuppers, 233

scuttle, 198
sealed buildings, 57
search considerations, 271
Searching Smarter, (Coleman), 271
Sears Catalog mail order homes, 93
security bars
on doors, 215
on windows, 206–207
self-contained breathing apparatus (SCBA), 155, 248, 271
sewer systems, 223
shafts, 56, 68–69, 197, 198, 199, 240–241
shear forces, 13
shear walls, 41
sheathings
composites as, 41
decorative, 17–18
diagonal, 166
metal as, 41
radiant barrier, 130–131
of roofs, 166
straight, 166
as structural assemblies, 40–41
types of, 17–18
wood as, 40–41
shed roofs, 174
shingles
asbestos, 155, 235
composition, 193
wood or shiplap sidings, 155–156
shock hazards, 219
sidings, wood frame
asbestos shingle, 155, 235
asphalt-felt, 156
corrugated (fiberglass, aluminum, lightweight steel), 156
fiber cement, 156
masonry veneer on, 156
plaster or stucco, 156
plywood (T-11), 156
renovations to, 235
vinyl, 156
wood shiplap or shingles, 155–156
signs, advertising, 199, 248
silos, 393–394
simple beam, 34
single-family dwellings (SFD)
Colonial and Georgian homes, 295–296
Craftsman and American Four Square homes, 299–300
manufactured (mobile) homes, 309–310
McMansion homes, 307–308
modern lightweight homes, 305–306
prairie style homes, 301–302
size-up of, 260
split level homes, 303–304
Victorian/Queen Anne and Cape Cod, 297–298
single-ply roofs, 194–195
single-wythe walls, 161
size considerations
big-box buildings, 106
commercial buildings, 107–108

high-rise buildings, 106
mega-box buildings, 106
residential buildings, 106–107
two-four-six method of, 104–105
vagaries of, 103–104
size-up. *See also* first-due decision maker; size-up, 360 degrees
    access/egress in, 272–273
    base of operations in, 272
    burning materials determination in, 266
    conventional or lightweight construction in, 266
    crawl space vents used for, 142, 143, 145–146
    facades in, 272
    fire rate of change in, 268
    fire spread in, 269
    fireground clock and, 267–268
    first-due decision maker and, 258–259, 285n1
    floodlights for, 262
    forcible entry in, 270, 272
    foundations in, 143
    front-loading of, 258, 259, 262, 278
    grouping buildings in, 259–260
    hazards affecting tactics, 271–274
    initial 180 (use and size), 101, 259–262
    name/utilities in, 273–274
    reading a building, 257–258
    search considerations, 271
    soft, 260
    standard operating guidelines of, 258
    survivable spaces determination in, 264, 268, 271
    unconscious competence in, 274
    uses of buildings (occupancy types) in, 101–102, 260–262
    ventilation, 270–271
size-up, 360 degrees
    building status in, 264–266
    construction method as, 262–263
    doors and windows in, 263
    era considerations as, 263
    exterior features as, 263
    interior features as, 263–264
    mailboxes, multiple in, 263
    tactical judgments of, 264
    utility feeds in, 237, 263–264
skeletal construction. *See* post and beam construction
skylights, 197–198, 241–242
slab foundations, 31
slab-on-grade foundations, 141
slate roof coverings, 193
snow/wind loads, 12
soffits, 78, 236
solar energy systems, 224–226
solar panels, 120, 224, 226
spalling, 24, 25, 54, 56, 57
spans, open, 97
split level homes, 78, 303–304
spreaders, 92
stacked kitchens and baths, 64, 317, 318
stacked log buildings, 71–72

stadium buildings, 381–382
stairwells, 94
standard operating guidelines (SOG), 145, 258
standard operating procedures (SOP), 145
static loads, 12
steel
    aspects of, 21
    C-channel studs, 21, 120, 121
    columns, 21, 53–54
    corrugated roofing of, 190–191
    decorative enhancements, void spaces of, 122
    framing, 21, 53–54, 58–59
    hot- and cold-rolled, 21
    I-beams, 21, 121, 176
    open web joists of, 176
    protection materials of, 53–54
    temperature for failure of, 21, 179
    as unprotected in Type II buildings, 58–59
steel, lightweight. *See* lightweight steel buildings
steel framing
    Type I buildings, protected, 53–54
    Type II buildings, as unprotected in, 58–59
steel manufacturing/warehouse buildings, 347–348
stick-built (framed) construction, 42, 167
storage buildings, 101
straight sheathings, 166
straw bale buildings (SBB)
    aspects of, 126–127
    collapse in, 127
    fire spread in, 127
    overhaul operations in, 120, 127
straw clay buildings (SCB)
    aspects of, 119–120, 127–128
    collapse, 120, 127–128
    fire spread in, 127–128
    overhaul operations in, 120, 128
strikes, door, 214
strip malls. *See* Type III (221 or 200) ordinary construction
structural assemblies, 40–41
structural elements
    beams, 33–34
    columns, 31–32
    connections, 38–39
    fire cut beams, 38–39, 91, 92, 158, 159
    foundations, 30–31
    trusses, 34–38
structural hierarchy, 42, 258
structural insulated panel (SIP) buildings
    aspects of, 124–125
    collapse of, 126
    fire spread in, 125–126
struts, 32
stucco sidings, 156
studs
    advanced framing methods, use of, 132, 155
    balloon framing, spacing of, 73, 154
    C-channel steel, 21, 120, 121
    double-stud wall construction, 130, 152, 154
    laminated strand lumber as, 19

laminated veneer lumber as, 19
in legacy platform buildings, 77
offset, 152, 154
spacing of, 132
subfloor, 40–41, 146, 147
Summerbell roofs, 182
supply feed, electrical, 217
surface-to-mass ratio, 14
survivable spaces, 264, 268, 271
suspended beams, 34
suspended ceilings, 55, 151–152, 236
suspended loads, 12
suspended particle display windows, 131–132
switches, electrical, 218–219
synthetics, 82, 96

T

T-11 plywood sidings, 156
tactical concerns
attended care facilities, 378
auditorium/theater buildings, 384
big-box buildings, 344
block/masonry buildings, 346
brownstone homes, 312
center hallway structure homes, 320
church buildings, 388
Colonial and Georgian homes, 296
concrete tilt-up buildings, 350
converted mill buildings, 354
Craftsman and American Four Square homes, 230
detention (jail) facilities, 376
fast food buildings, 340
garden apartment homes, 322
high rise-1st generation buildings, 366
high rise-2nd generation buildings, 368
high rise-3rd generation buildings, 370
historical buildings-commercial, 398
historical buildings-dwelling, 396
hospital buildings, 374
industrial/legacy strip style buildings, 336
kit buildings, 392
legacy townhome/condo/apartment homes, 328
lightweight townhome/condo/apartment homes, 330
manufactured (mobile) homes, 310
McMansion homes, 308
meeting hall buildings, 386
mega-box buildings, 342
modern lightweight homes, 306
modern strip-style stores, 338
pole barns, 390
post-WWII low rise buildings, 362
prairie style homes, 302
pre-WWI ordinary commercial buildings, 332
pre-WWII low rise buildings, 360
pre-WWII ordinary (taxpayer) buildings, 334
project housing-high density homes, 326
project housing-low density homes, 324
public storage-multistory buildings, 358
public storage-single story buildings, 356
railroad flat homes, 318
restaurant buildings, 380
row frame homes, 316
school buildings, 372
silo buildings, 394
split level homes, 304
stadium/arena buildings, 382
steel manufacturing/warehouse buildings, 348
tenement homes, 314
21st century buildings, 364
Victorian/Queen Anne and Cape Cod homes, 298
wooden manufacturing/warehouse buildings, 352
taxpayer buildings, 62, 333–334
temperature
for failure of steel, 21, 179
for human life threshold, 271, 285n1
for melting of glues, 168, 177
for plate or annealed glass failure, 205
tempered glass, 96, 205
tenement homes, 313–314
tension forces, 13
theater buildings, 383–384
thermal bridging, 132–133
thermal effect of radio frequency waves, 247
thermal imaging cameras (TIC), 6, 7, 131
thermoplastic compounds, 205
tie plates and rods, 39, 92, 159, 183, 233, 236–237
tied arched roofs
bowstring compared to, 182–183
identification tips for, 183
trusses in, 36, 162
tilt-up concrete buildings, 349–350
tilt-up concrete slab construction
basics of, 23, 42
parapet walls of, 64, 229
rakers (support jacks) and, 163
Type II buildings with, 58, 59
wall collapse of, 59, 61
wall connections of, 162–163
timber trusses, 166–167
timber trusses (older)
alteration issues of, 188
dimensional size and component connection issues, 187
failure potential of, 187–188
location issues of, 188
mass issues of, 188
offensive operations in, 189–190
parallel chord and gable, 186–187
size issues of, 188
visual identifiers of, 187
time. *See* fireground clock
titanium, 22
toe-up, 126
tongue and groove (T&G), 40–41, 67, 77
torsion loads, 13
toxicity, of adhesives and glues, 17, 18, 20
traditional wood products, 16–18

transformers, electrical, 218
transoms, 91
triaging building fires
    incident safety officer, role in, 275
    Joe's Building Fire Triage Program for, 276–277
trusses
    arched, 36–38, 182–184
    bar, 35–36
    bowstring, 36–38, 182, 183, 184–185
    bridge roof, 180–181, 189
    chords in, 35–36, 186–187
    craftsman-built, 79
    in engineered wood platform (lightweight) construction, 80, 81–82, 98
    glued connections of, 177
    legacy platform type, 77
    lightweight, 166, 177
    loft, 63–64, 167
    open web, 34–35
    in post-WWII legacy era construction, 97
    in pre-WWII era buildings, 94
    rigid (rib) arch, 36–37, 183–184
    in stacked log buildings, 71
    tied, 36, 162
    timber, 166–167
    timber (older) roofs, 186–190
    triangular, 35, 167
    in Type III building lofts, 63–64
tubular daylight devices (TDD), 241
21st century office/hotel buildings, 363–364
two-four-six method, 104–105
type (of building) considerations, 102–103
Type I (442 or 332) fire resistive construction
    aspects of, 103
    case studies of, 56–57
    center core/center hallway floor plans of, 55
    concrete spalling and heat retention of, 54
    curtain wall construction of, 55, 56, 60
    forcible entry of, 56–57
    heating, ventilation, and air conditioning systems of, 54–55
    identification of, 57
    interstitial areas of, 55
    legacy buildings, 95
    open areas of, 54
    primary hazards of, 54
    as sealed buildings, 57
    shafts in, 56
    steel framing, protected in, 53–54
    ventilation of, 57
Type II (222, 111, or 000) noncombustible construction
    aspects of, 103
    case studies of, 60–61
    contents of, as primary hazard in, 59–60
    fire extension and by-products of, 60
    girders in, 58
    legacy era buildings, 95
    steel framing, as unprotected in, 58–59
    tilt-up concrete slab construction in, 58, 59
    wall collapse in, 59
Type III (221 or 200) ordinary construction
    alterations and concealed spaces/voids in, 64
    aspects of, 62, 103
    attics, cocklofts, and truss lofts in, 63–64
    case studies of, 66
    cornices on, 65
    facades of, 66
    legacy era buildings, 95–96
    mini- and strip malls as, 62
    parapet walls in, 64–65
    post-WWII era buildings, 95–96
    taxpayer buildings as, 62
    vertical extensions of, 62–63
Type IV (2HH) heavy timber/mill construction
    American Institute of Timber Construction regarding, 67–68
    aspects of, 67, 103
    case studies, 69
    collapse, areas of, 69
    doors in, 68–69
    fire loads in, 69
    fire size in, 69
    girders in, 67
    incident scope, 70
    legacy era buildings, 96
    openings, shafts, and voids in, 68–69
    renovations and concealed spaces/voids in, 68
    tongue and groove planks in, 67
Type V (111 or 000) wood frame construction
    aspects of, 70–71, 103
    balloon frame in, 73–75, 154–155
    case studies of, 80–81
    conventional platform construction, 76–77
    engineered wood platform (lightweight) in, 79–82
    foundations in, 141
    legacy platform construction, 77–79
    post and beam, 72–73
    renovations in, 70
    stacked log, 71–72
    western platform construction, 75–76
    wood framing in, 70–82

U

ultraviolet rays (UV), 18, 206
Underwriters Laboratory (UL), 2, 271
uninterruptible power supply (UPS), 219, 246–247
unoccupied buildings, 265–266
unprotected materials, 52, 58–59
unreinforced brick walls
    bond beam caps on, 158, 160
    collapse safe areas in, 160
    king row in, 158, 160
    metal straps on roofs of, 159
    post-1933, 158
    post-1959, 159
    post-1971, 159–160

pre-1933, 158
water-soluble mortar in, 91, 158
weight bearing in, 161
window frames in, 160, 207
wythes in, 157–158, 160, 161
unreinforced masonry (URM)
construction of, 157
wall collapse of, 59
uses of buildings (occupancy types)
changes in and to, 99–100
International Building Code regarding, 101–102
NFPA 5000 codes regarding, 101–102
size-up reports regarding, 101
types of, 101–102
utility and miscellaneous buildings, 101–102
utility chases, 94
utility systems. *See also* electrical utilities; gas utility systems
communications/data systems, 223
freshwater systems, 223
wastewater (sewer) systems, 223

V

vacant buildings, 265
veneer, masonry, 24, 41, 156, 157–158, 161
vent pipes, 198
ventilation
aerial devices for, 171, 182
horizontal and vertical, 57
hot mop roof issues of, 193
impact-resistant windows and, 212
in lightweight concrete roofs, 180
of metal roofs, 194
of monitor roofs, 171, 172
rain roofs and, 191
roof ducting and, 199
in sealed buildings, 57
of single-ply roofs, 195
size-up considerations of, 270–271
skylights as, 241
solar panels hindering, 224, 226
of Type I buildings, 57
ventilators, 198, 199
vents
attic, 197, 198, 233, 267
crawl space, 142, 143, 145–146
vertical ventilation, 57, 74
Victorian homes, 297–298
vinyl sidings, 156
void spaces
alterations and renovations resulting in, 64, 68, 238
in balloon frames, 74, 154–155
fire spread in, 94
interstitial and plenum, 151, 152
pipe chases as, 64, 198
steel decorative enhancements and, 122
in Type III buildings, 64
in Type IV buildings, 68

W

walkout basements, 143
walk-up basements, 143
wall columns, 31, 140, 154
wall-bearing construction, 42
wallboard. *See* drywall
walls. *See also* collapse; tilt-up concrete slab construction
adjoining, 152, 153, 154
balloon frame exterior, 73–75, 154–155
brick noggin in, 153
buttresses on, 32, 36, 162
concrete, 162–163
defined, 140
division, 64, 152, 153, 195
drywall in, 153
exterior, masonry, 157–161
exterior, siding of, 155–156
exterior, wood framing of, 154–155
fire blocking in, 76, 154
fire division, 152, 153, 195
flat panel, 123, 124
foundation, 30–31
grid block, 123, 124
insulated concrete forms in, 123–124
interior, 152–154
knee, 64, 80, 94, 235
of lightwells, 240–241
masonry, types of, 157–158
masonry veneers, 24, 41, 156, 157–158, 161
occupancy division walls, 153
offset, 152, 154
parapet, 64–65, 158, 159, 160, 229–230
partition, 41, 152, 153
party, 41, 153–154
post and beam, exterior, 72, 73, 123–124
reinforced brick, 160–161
shear, 41
spreaders and exterior plates on, 92
straw bale, 126–127
tie plates on, 39, 92, 159, 233
types of, 41
unreinforced brick, 157, 158–160
weight bearing of masonry, 161
wythes in, 157–158, 160, 161
wastewater (sewer) systems, 223
water-soluble mortar, 91, 158
western platform construction, 75–76
windows. *See also* glass
air conditioners in, 208
air-blast resistant, 206
barricaded, 207
basement escape, 145
bucks in, 126
clues based on, 204, 235
covered, as renovated, 236
electrochromic smart glass, 131–132
frames, 204
glazing, 204

    Housing and Urban Development, 207
    impact-resistant, 209–213
    recessed frames of, 160
    renovations to, 234–235
    rooms identified by, 208
    sashes, 204
    security bars on, 206–207
    suspended particle display, 131–132
    types of, 204
    in unreinforced masonry construction, 207
wind/snow loads, 12
wired glass, 205
wood
    panelized roof system, 178–179
    shake shingle roof coverings, 193
    sheathings, 40–41
    shingles or shiplap siding, 155–156
    traditional products of, 16–18
wood frame construction. *See* Type V (111 or 000) wood frame construction
wood manufacturing/warehouse buildings, 351–352
wrought iron, 37–38
wythes, 157–158, 160, 161

Z

zero-clearance fireplace, 196

ABOUT THE AUTHORS

DAVE DODSON, BATTALION CHIEF (RET.)

Dave Dodson is a 34-year fire service veteran—25 years on the street—starting his fire service career with the United States Air Force. After the USAF, Dave spent almost seven years as a fire officer and training/safety officer for the Parker Fire District in Parker, Colorado. He became the first career training officer for Loveland Fire & Rescue (CO) and spent time as an engine officer, hazmat technician, duty safety officer, and emergency manager for the city. He accepted a shift battalion chief position for the Eagle River Fire District in Colorado before starting his current company, Response Solutions, which is dedicated to teaching safe and practical incident handling.

Dave is the author of Jones & Bartlett's *Fire Department Incident Safety Officer* and Fire Engineering's Training DVD series *The Art of Reading Smoke*. He continues to develop and deliver classes on firefighter safety and survival issues and first-due officer procedures.

Chief Dodson has served as the chairman of the NFPA 1521 Task Group (Fire Department Safety Officer) and served on the Fire Service Occupational Safety and Health Technical Committee for NFPA. Dave is also a past president of the Fire Department Safety Officer's Association. In 1997, Dave was chosen as the ISFSI George D. Post Fire Instructor of the Year.

JOHN MITTENDORF, BATTALION CHIEF (RET.)

John Mittendorf is a 30-year veteran of the Los Angeles City Fire Department and held the rank of battalion chief until his retirement in 1993. He has been a member of the National Fire Protection Research Foundation on Engineered Lightweight Construction Technical Advisory Committee and has provided training programs for the National Fire Academy, UCLA, and the British Fire Academy in England. He has also acted in an advisory capacity for five college fire science advisory boards and is the author of numerous fireground articles for magazines in the United States and Europe.

Chief Mittendorf is the author of the books *Ventilation Methods and Techniques*, *Truck Company Operations*, editions 1 and 2, *Facing the Promotional Interview*, and the recently released the DVD program, *Ten Commandments of Truck Company Operations*, from Fire Engineering/PennWell. He currently lectures in the United States and the United Kingdom on strategy and tactics, truck company operations, fireground operations, ventilation operations, and the Complete Fire Officer. John is also a member of the editorial advisory board of *Fire Engineering* magazine. In 2008, Chief Mittendorf received the Fire Department Instructors Conference (FDIC) Lifetime Achievement Award.